ABOUT ISLAND PRESS

Island Press is the only nonprofit organization in the United States whose principal purpose is the publication of books on environmental issues and natural resource management. We provide solutions-oriented information to professionals, public officials, business and community leaders, and concerned citizens who are shaping responses to environmental problems.

In 1999, Island Press celebrates its fifteenth anniversary as the leading provider of timely and practical books that take a multidisciplinary approach to critical environmental concerns. Our growing list of titles reflects our commitment to bringing the best of an expanding body of literature to the environmental community throughout North America and the world.

Support for Island Press is provided by The Jenifer Altman Foundation, The Bullitt Foundation, The Mary Flagler Cary Charitable Trust, The Nathan Cummings Foundation, The Geraldine R. Dodge Foundation, The Charles Engelhard Foundation, The Ford Foundation, The Vira I. Heinz Endowment, The W. Alton Jones Foundation, The John D. and Catherine T. MacArthur Foundation, The Andrew W. Mellon Foundation, The Charles Stewart Mott Foundation, The Curtis and Edith Munson Foundation, The National Fish and Wildlife Foundation, The National Science Foundation, The New-Land Foundation, The David and Lucile Packard Foundation, The Pew Charitable Trusts, The Surdna Foundation, The Winslow Foundation, and individual donors.

PROTECTING PUBLIC HEALTH
& THE ENVIRONMENT

*For all those who work on behalf of the
environment, public health, and sustainable agriculture.
And to all those beings that suffer from environmental damage,
may the Precautionary Principle bring a better world.*

Protecting Public Health & the Environment

Implementing the Precautionary Principle

EDITED BY

Carolyn Raffensperger and Joel A. Tickner

FOREWORD BY

Wes Jackson

ISLAND PRESS

Washington, D.C. / Covelo, California

Library of Congress Cataloging-in-Publication Data
Protecting public health and the environment : implementing the precautionary principle / edited by Carolyn Raffensperger, Joel Tickner.
    p.   cm.
   Includes bibliographical references and index.
   ISBN 1–55963–688–2
   1. Environmental sciences—Decision making. 2. Environmental policy—Government policy. 3. Risk assessment. 4. Health risk assessment.   I. Raffensperger, Carolyn.   II. Tickner, Joel.
   GE105.P76   1999   99–19514
   363.7'056—dc21   CIP

Contents

~

Acknowledgments

~~~

Here we can throw precaution to the winds and say a million thank you's to all of those who have given ideas, love, support, time, energy, good poetry, and music toward the development of this book.

A special thank you to the founders of the Science and Environmental Health Network (SEHN). Your vision, stick-to-itiveness, and generosity have made this book possible. The SEHN Steering Committee is chaired by Paul Locke at the Environmental Law Institute and Peter deFur at Virginia Commonwealth University. Wingspread participants who are also on the Steering Committee include Mary O'Brien, Steve Lester, and Peter Montague. We should note that it was a question posed by Peter Montague at a SEHN meeting that spawned our interest in the Precautionary Principle. He asked, "What environmental decision-making tools do we have other than risk assessment?"

OMB Watch, under the direction of Gary Bass, serves as the parent organization to SEHN. We are grateful for its ongoing relationship and support.

What would we have done without the foundations who believed in the Precautionary Principle Project and SEHN's capability to carry it out? Peter Myers at the W. Alton Jones Foundation, Roxanne Turnage at the C.S. Fund, and Brian Reilly at the Johnson Foundation all worked with us to craft the conference and the plan of work that followed. Their strategic and sub-stantive questions were as important as the money they so generously gave us. All three foundations cosponsored the conference with SEHN.

Lois DeBacker at the C.S. Mott Foundation has been the best of col-leagues. The Mott Foundation has contributed general support to SEHN for several years. Since we didn't have to worry about paying the rent, we could double our efforts on behalf of the environment and public health.

The U.S. Environmental Protection Agency has supported Joel Tickner's work on the Precautionary Principle through its STAR Fellowship Program, which freed up resources and allowed us to be far more thorough in our research and outreach on the principle.

The Lowell Center for Sustainable Production at the University of Mass-achusetts, Lowell, cosponsored the Wingspread Conference. But that does-n't describe all of their contributions, which included mundane but essential services such as copying and mailing.

Nancy Myers is a friend, coconspirator, and writer extraordinaire. She joined us at Wingspread and helped translate precautionary arcana into something that would make sense.

Bette Hileman also attended Wingspread as a reporter. Her insights and journalistic integrity actually give reporters a good name. The world is a bet-ter place because Bette writes.

Sue Maret and Katherine Barrett, besides participating at the Wingspread Conference, documented the conference and posted the discussions to a group of people who participated virtually (by e-mail). This was a grueling task requiring them to think quickly, synthesize, and write for long, long hours. Their work is reflected in the Afterword.

Ken Geiser deserves a special tribute as an advisor to many graduate stu-dents (including Joel) and to the environmental movement. Ken facilitated the most difficult sessions of the Wingspread Conference and helped us think through the agenda and goals.

We would like to thank all of the participants of the Wingspread Confer-ence on Implementing the Precautionary Principle, including the virtual participants, who gave selflessly of their creativity, vast experience, and problem-solving abilities. We also thank all those working on the principle around the world who have provided insight, comments, and documents during the preparation of this book.

CAROLYN WRITES
I offer my thanks and heart to Fred Kirshenmann, husband and wise com-panion. He never wavered in his support of me or the ideas reflected in this

work. His contribution to the book and to my life are immeasurable and probably proof that there is a loving God.

My year and a half partnership with Joel has been an unexpected gift. From the day he called to ask if I would participate in his dissertation, to the amazing joys and frustrations of the Wingspread Conference, through the editing of this book, Joel has been easy-going and the hardest worker on behalf of the environment and public health one can imagine. It's hard to describe how closely we worked together and how well we got along. A pivotal moment in our relationship came the day that I helped edit something Joel had written. When I saw the printed manuscript, Joel had put my name first. I did not deserve it. Joel, thank you for your open heart, your generous mind, your willingness to participate in North Dakota farm life, and your need for little sleep. Most of all, thank you for taking the big risk of sharing your ideas and giving others the credit. May it be returned to you tenfold.

JOEL WRITES
My wife Judit provided both moral and spiritual support. Her companionship and piano music were a welcome relief from the long hours spent on the Precautionary Principle. My thanks to her for her ever-present sweetness and love.

Of course, two editors living half a continent apart and in almost opposite living situations cannot keep from taking pride in each other's unfailing dedication to this work. This book and the Wingspread Conference would never have happened had it not been for Carolyn's vision, determination, and ability to think big. Since Carolyn's magnificent idea to apply to the Johnson Foundation to host a Wingspread Conference, there has been no stopping her creative process and the successes in distributing and applying the Precautionary Principle. Tough, thorough, and demanding, Carolyn ensured that the book was tightly edited, complete, and to Island Press on time. Our multiple daily calls and philosophical discussions coupled with Carolyn's enthusiasm, encouragement, thankfulness, great wisdom, and guidance have made this an experience that few coeditors have the privilege to live.

Finally, we both give our heartfelt thanks to Heather Boyer and others at Island Press who have demonstrated their commitment to innovative environmental solutions and have provided constant, gracious assistance throughout the writing process.

We can only hope that this book will encourage readers to expand the boundaries of their thinking about environmental and public hazards and take precautionary actions for the ecological security of our planet.

Foreword

~

As we close out the millennium and prepare for another century, many of us are asking ourselves what it will take to contain the excesses and not just those fueled by oil or uranium but also fueled by our discoveries, insights, and inventions. It seems doubtful that we can contain any excesses so long as we operate as though our knowledge is adequate to run the world. We continue to pluck one apple after another from the tree of knowledge and ignore the fruit of the other tree of the garden, the tree of life. Some apples from the knowledge tree rot faster than others.

The dating of the scientific revolution is often, if not usually, set at 1500 to 1700. The industrial revolution is often said to begin one hundred years later in 1800. (These are very general dates but let's allow them for the moment.) During the scientific revolution era, most scientists were simply trying to understand how the world *is*, delving deeper into the nature of God's laws. (As we will see, there were important notable exceptions from within that two-hundred-year period.) It is easy to understand that the transition from "how the world *is*" to "how the world *works*" is a subtle but profound shift for the human mind, for the mind is now made ready for the era of the inventor who becomes increasingly present in that century from 1700 to 1800. Instrumentalism or utilitarianism becomes such a force to deal with, that it forces its way into science. The distinction between basic and applied science had to be made.

Dr. Roald Sagdeev is a plasma physicist who once led the U.S.–Soviet Apollo Soyuz mission, a former member of the U.S.S.R.'s Academy of Sciences and a man who played a major political role during the first five years of *perestroika*. He is now a professor of physics at the University of Maryland. In his book, *The Making of a Soviet Scientist*, he has this to say:

> The development of a revolution in science is controlled by its own internal logic. To build a new revolutionary concept, to make a breakthrough, requires a certain hidden incubation period, during which time there is the accumulation of experimental data, the painful assessment of difficulties, careful invention, and then the injection of new scenarios and explanations. Science and physics have always progressed in this way. Somehow we, the nuclear physicists of the twentieth century, were spoiled by quick successes like the Manhattan project and, a few years later, a parallel breakthrough with a nuclear bomb on the Soviet side. Many of us, even wise and experienced leaders, . . . thought that if an appropriate budget were given it would almost guarantee immediate technical progress in resolving any problems nature presented us. If we physicists had subconsciously become somewhat arrogant, our punishment was not long in coming. Controlled thermonuclear fusion, unlike the uncontrolled one which was an apocalyptic hydrogen bomb explosion was not an easy nut to crack.

As it turned out, the very nature of plasma—the hottest state of matter and at the same time the least controllable substance—destroyed the legend of Almighty Science versus Nature. What was originally perceived as a quick victory of the new, inexhaustible source of energy became a long, protracted war against plasma instabilities. It created almost a deadlock and proved to be a setback and a warning of even bigger future failures of humanism, which saw itself as the master of nature.

Here is an apple that rotted early. But now the metaphor has the potential to fail us because the valuable lesson learned over a period of a few years by a few physicists in the Soviet Union could be dismissed as an insight from those doing science at the edge of the envelope. But what about away from the envelope's edge where the legend of Almighty Science versus Nature still dominates the scientific culture? Here the assumption remains that knowledge is adequate to run the world. Those of us who embrace or look to nature as a standard or measure rather than seek to bring more science to subdue nature are in a distinct minority. Our arguments for alternatives to the dominant paradigm are often something like "what if the dominant paradigm does not work, should we not hedge our bets?" These are last retreat arguments. The seeds for embracing the Precautionary Principle have been planted. The seeds are in the ground; some may even be sprouting.

The legend of Almighty Science versus Nature with the weight on science as the superior factor is unlikely to die very soon for most of us. It was short for those Soviet fusion scientists. It will be longer for fission because if we expand nuclear power after the portable liquid fossil fuel epoch, eventually Murphy's Law will go into effect. The legend as it pertains to agriculture will die too, if we don't learn the lessons as to how the world *is* or *was* or even how the world *has worked* over the millions of years the world has existed. To apply those lessons, it will likely require that the evolutionary, ecological worldview along with the modern molecular synthesis be embraced.

I have often thought how much easier life would be if the implementation of a technology or practice either allowed us to live or drop dead immediately. Smoke one cigarette and you die. Use 2-4D on a crop once and you die. Instead, we're faced with a world of uncertainty, and we have to resort to fields like epidemiology or other sciences that depend on statistics. Because we don't drop dead, we allow ourselves to draw our boundaries of consideration much narrower than they should be. Boundaries over space and time are nearly always much narrower than the boundaries that include the cause. When the boundaries are made appropriately larger, they embrace more of our ignorance and more ambiguity. Now the Precautionary Principle kicks in.

Even biologists, including many evolutionary biologists, are not what I would call "Deep Darwinians." Such Darwinians would not stand by when dead sheep are fed to cattle (resulting in Mad Cow disease). Nor would they stand still as huge cattle, hog, and chicken confinement operations are put in place, operations that require heavy doses of antibiotics, some of which are already useless in hospitals. Deep Darwinians would count any human-made chemical with which humans have not evolved to be regarded as guilty until proven innocent. Yes, humans are highly adaptable. (Our balance, vision, upright stance, much of which evolved east of the rift in Africa, is such that we can ride bicycles.) Yes, there are many technological activities we engage in with impunity to ourselves and the ecosphere. This reality and the need to wait into the long run for many effects to take hold mask, in the short run, whether we have gotten away with something or not.

In mentioning Darwin, evolution, ecology, and the molecular oneness of life at a basic level, I am mentioning the basis for an operating philosophy that deserves enough standing to more than match the position Francis Bacon (1561–1626) and Rene Descartes (1596–1650) hold in our modern minds, especially our scientific minds. Bacon gave us the outline for the institutional requirements of practical knowledge right in the midst of the period when scientists were simply wanting to know how the world *is*. As biographer Perez Zagorin says of Francis Bacon, he was the first to see science as "a public, collaborative, and progressive enterprise!" Public here easily translates into public *money*. Zagorin also sees Bacon "as a thinker about science: the conditions favorable to its growth; the changes and procedures required to insure its progress; its contribution to the inauguration of a new regime of knowledge." As early as 1592, at age 31, he wrote in a letter that he was taking "all knowledge to be my province." Such a challenge helped lay the groundwork for the "knowledge as adequate" worldview that nearly all of us hold today.

Utilitarian science was also in the mind of Descartes. In the *Discourse on Method* (1637), this 41-year-old mathematician who revolutionized algebra insisted that a combination of math and experimentation would provide us with "knowledge which is very useful in life." Had Descartes stopped with knowledge as merely "useful," we might be able to forgive him for the excesses many of us have promoted, excesses derived by following his ideas. In fact, Descartes thought that we could "render ourselves the masters and possessors of nature."

Few of us in industrial society really want to do without science. I certainly don't. And so the question comes, "What would an ignorance-based worldview be like when it comes to doing science" or stated in the language of this volume, "What would it look like were we to truly employ the Precautionary Principle as a dominant way of operating in our science? In our daily lives?"

The Precautionary Principle is not new. One contributor to this volume noted that it goes back at least to the Hippocratic Oath, whose injunction is "First, Do No Harm." It is also widespread in our daily lives, men get checked for the prevalence of prostate specific antigen in their blood. Scopes check our colons and intestines. There are countless other examples. The question becomes then why is it not a sufficiently dominant way of being, that countless problems build up before we catch them, too often when it is too late to reverse the consequences?

My younger daughter and her husband, both high school teachers, recently moved to a small Kansas town with their three-year-old son. In a few months this son, my grandson, will welcome his new baby brother. My older daughter, her husband, and their three-year-old daughter recently visited them. Four young adults (one of them pregnant) and two three-year-olds did what Sunday visits often occasion—go to the park, get hot and sweaty, and become thirsty. But in this public park town officials have warned that small children and pregnant women should not drink the water. When the younger daughter and her family moved in, they received a letter warning them that the nitrate level was unsafe for children, including those in utero. The letter also said that for one year my pregnant daughter is entitled to free bottled water from the grocery store. I know this sort of thing has been going on for hundreds of years, but where is the outrage that would prevent, stop, and reverse the problem? Where is the money required to remove the nitrate?

The Des Moines River runs through Des Moines, Iowa. Nitrate pollution is a problem for that river. The city of Des Moines, however, is sufficiently rich that millions of dollars can adequately remove the toxic nitrates. Small town, Kansas, cannot. Who is to blame for the nitrate problem in the first place? The farmers who spread chemicals on the fields? Area feedlots? Eighty percent of the nitrogen becomes airborne and will fall who knows from where. The reality of nonpoint-source pollution means that no one is to blame even though all are to blame.

Where does the Precautionary Principle fit here? It should be an automatic derivative of the Darwinian evolutionary ecological worldview since high nitrate concentration runs beyond what *Homo sapiens*, a product of evolution, have experienced. With the Baconian-Cartesian worldview, precaution is more a derivative of experience, something *empirically* learned. Precaution, informed by a deep sense of our evolutionary history and evolutionary mechanisms, can eliminate some of the guess work.

Wes Jackson
The Land Institute

Preface

ESTABLISHING A GENERAL DUTY OF PRECAUTION IN ENVIRONMENTAL PROTECTION POLICIES IN THE UNITED STATES

A Proposal
Ken Geiser

The United States Congress should amend its environmental laws to establish a general duty to act precautiously toward the environment. A general duty to act with caution would clarify the responsibilities of all parties to assure environmental protection, even where there are no statutory regulations with which to comply and where there is no scientific certainty on which to rely.

The chapters in this book offer a wide range of stimulating perspectives on the Precautionary Principle, which calls for protective actions toward the environment, even when the evidence of harm remains uncertain. Proponents of precaution appear well aware that adopting such a posture would significantly alter our current approach to environmental policy making. Just how this proposed approach might be introduced into national policy remains more sketchy. One strategy would be to think big: to acknowledge that this new approach requires more than a simple shift in procedures and to assert that this is, indeed, a change in the very principles of environmental protection. Consider how this might occur.

Between 1969 and 1979, the U.S. Congress enacted some twenty-three major laws for regulating pollutants and developments that would otherwise threaten human health and ecological systems. It was a period of environmental activism during that the Environmental Protection Agency (EPA) and the Occupational Safety and Health Administration (OSHA) were established and filled with legal authorities.

While OSHA was established through the passage of one omnibus law, the Occupational Safety and Health Act of 1970 (OSHAct), the EPA was created by executive order and empowered by a wide range of statutes that had very different objectives and philosophies. Many of these laws were media specific (e.g., air, surface water, groundwater, land) statutes, whereas others focused on specific hazards (pesticides, nuclear power, toxic chemicals). Only one of the laws, the National Environmental Protection Act of 1969 (NEPA), addressed the entire domain of environmental protection. Indeed, NEPA attempted to set out a comprehensive national mission for environmental protection.

Section 2 of NEPA declares the purpose of the act:

> To declare a national policy which will encourage productive and enjoyable harmony between man and his environment; to promote measures that will prevent or eliminate damage to the environment and the biosphere and stimulate the health and welfare of man; to enrich the understanding of the ecological systems and natural resources important to the Nation.

Such grand ambitions are followed in Title I by:

> Section 101(a) The Congress . . . declares that it is the continuing policy of the Federal Government . . . to use all practical means and measures . . . to create and maintain conditions under which man and nature can exist in productive harmony, and fulfill the social, economic, and other requirements of present and future generations of Americans.

Finally in Section 101(c), NEPA states:

> The Congress recognizes that each person should enjoy a healthful environment and that each person has a responsibility to contribute to the preservation and enhancement of the environment.

The law goes on to state that it is a responsibility of federal agencies to assure that all projects that involve federal resources receive a full environmental review prior to approval. While this idea was quite controversial at the time NEPA was enacted, today, it is routinely accepted that projects initiated with federal resources undergo an environmental impact assessment that evaluates the environmental effects of the proposed action against alternative strategies.

For large projects, these environmental impact assessments can be quite extensive taking into consideration a wide array of ecological, social, and cultural conditions and the process must be open to public review and comment. It is clear that the intention of this procedure is to slow down proposed actions and require that proponents engage in a thoughtful review of the environmental impacts with the assumption that the least damaging strategy will be selected. This is not so different from the spirit that lies under the now popular precautionary approach.

The precautionary approach is a concept currently written into several international agreements and foreign environmental statutes as the "Precautionary Principle." The Precautionary Principle asserts that parties should take measures to protect public health and the environment, even in the absence of clear, scientific evidence of harm. It provides for two conditions. First, in the face of scientific uncertainties, parties should refrain from actions that might harm the environment, and, second, that the burden of proof for assuring the safety of an action falls on those who propose it. This is sympathetic to the philosophy put forward in NEPA: to assure that those who promote an action assess the environmental and health impacts of that action before proceeding.

But NEPA is limited here to federal, state, or private projects engaging federal resources. It does not apply to fully private actions. The government is assigned a responsibility, but no other parties are provided with rights or responsibilities. In this respect NEPA differs dramatically from the OSHAct.

The OSHAct also states a clear purpose in its opening paragraphs. Section 2 (b) states that:

> The Congress declares it to be its purpose and policy . . . to assure
> as far as possible every working man and woman in the Nation
> safe and healthful working conditions and to preserve our human
> resources—

There are then listed a set of objectives. The first two are quite direct:

1. by encouraging employers and employees in their efforts to reduce the number of occupational safety and health hazards at their places of employment, and to stimulate employers and employees to institute new and to perfect existing programs for providing safe and healthful working conditions;

2. by providing that employers and employees have separate but dependent responsibilities and rights with respect to achieving safe and
   healthful working conditions.

There are federal responsibilities as well, and these begin with the third
objective:

3. by authorizing the Secretary of Labor to set mandatory occupational
   safety and health standards. . . .

But it is in the fifth Section of the OSHAct that it differs so markedly
from NEPA, and from all other environmental statutes for that matter. In
section five the act creates duties:

Section 5(a) Each employer—

1. shall furnish to each of his employees employment and a place of
   employment which are free from recognized hazards that are causing or
   are likely to cause death or serious physical harm to his employees;
2. shall comply with occupational safety and health standards promulgated under this act.

Section 5(b) Each employee—

shall comply with occupational safety and health standards and
all rules, regulations, and orders issued in pursuant to this act
which are applicable to his own actions and conduct.

Section five has come to be known as the "General Duty Clause" because
it creates a general obligation on both employers and employees to follow
the law and to strive toward safe and healthy workplaces. This was not a new
concept. The Senate committee report of the OSHAct identified thirty-six
states that had such general duty clauses written into their state health and
safety statutes. The concept also followed a common law doctrine that held
that there was a general duty to refrain from actions that would cause harm
to others.

On occasion, OSHA inspectors cite employers under the general duty
clause when they recognize a workplace hazard, but can find no applicable
standard. The federal courts have likewise used the general duty clause to
uphold citations against employers where there is no clear standard. Thus,
the general duty clause acts as a kind of default authority that supports the
wide array of mandatory standards that guide health and safety considerations at work.

The NEPA language offers no parallel default authority to those who promote environmental protection. There have been efforts to use Section 101(c) as a general statement of rights comparable to common law nuisances, but these have been conventionally rejected by the courts. Indeed, none of the other federal statutes offer anything like the OSHAct general duty clause.

From this perspective, it is not difficult to see that there would be value in formally establishing a general duty to protect the environment. If such a general duty to protect the environment were created, it would provide for some of the benefits that the OSHAct general duty clause now offers. First, it would clarify the rights and responsibilities of parties other than the government in the protection of the environment. Second, it would establish an obligation or duty to act in a protective manner toward the environment. Third, it would set a default authority for restraining environmentally damaging behavior, even where there are not applicable standards or regulations to condition that behavior.

Admittedly, a general duty to protect the environment is a fairly broad and ambiguous concept. There is virtue in such breadth, but the ambiguity may work against its application. Herein lies the value of the precautionary approach. The Precautionary Principle offers the practicality that a general principle of environmental protection lacks.

NEPA lays out some general principles of environmental protection and "recognizes" that "each person" should enjoy a healthful environment and that "each person" has a responsibility to preserve and enhance the environment. But NEPA does not create a "duty" to protect the environment. Instead, NEPA is content to create a responsibility for project proponents to conduct environmental impact assessments. The precautionary approach goes beyond the impact analysis written into NEPA that only requires that the impact assessment of options be completed; the precautionary approach creates an obligation to consider competing options and to act cautiously whenever possible. This is the responsibility that is currently missing in NEPA and throughout U.S. environmental policy.

The obligation to act cautiously is implied in many environmental statutes. The Clean Air Act permits ambient air standards to be set without actual proof of health hazards, so long as the standard can be defended as a scientifically supportable "adequate margin of safety." The Superfund legislation requires that "potentially responsible" contributors to dumpsites must step forward and engage in remediation. The Endangered Species Act requires that projects not proceed if they would threaten listed species. The concept of "unreasonable risk" is invoked thirty-eight times in the Toxics

Substances Control Act, which states as national policy that "adequate data should be developed with respect to the effect of chemical substances and mixtures on health and the environment and that the development of such data should be the responsibility of those who manufacture and those who possess such chemical substances and mixtures." Yet, nowhere in these voluminous laws is there stated a general duty on all persons to act in a prudent and cautious manner where there is a reasonable expectation of harm to the environment. Therefore, the U.S. Congress should amend the National Environmental Policy Act to establish a general duty of precaution toward the environment for all parties. The language needs to be thoughtfully crafted, but it could be similar to the following statement:

> The right to enjoy a safe, clean, and sustaining environment requires that persons accept a general duty to take reasonable precautions to protect the environment even in the absence of clear evidence of harm and even where specific regulations do not require such actions.

Such language should be added to the NEPA statement of national policy. It would link the right to a healthy environment with the obligation to protect the environment. It would clarify that there exists a responsibility for precautious action on all persons recognizing that corporations, trusts, associations, states, and municipalities are conventionally covered under this term. It would indicate that the lack of full scientific certainty or knowledge of a specific law could not be used as reason to postpone measures to prevent environmental degradation. Finally, it would create a default provision for restraining environmentally damaging behavior, even where there are no specific regulations.

We may debate for decades the absolute or relative harm any product or action may have on our health and the environment that supports us. Uncertainty will long plague us, and, for many, the issues of safety and risk will rely more on personal beliefs than the results of a few scientific studies. That cannot, and should not, deter us as a nation from acting protectively. Where there are options, we should feel a general duty to act with precaution.

Introduction

◆

TO FORESEE AND TO FORESTALL

When Rachel Carson completed her book, *Silent Spring*, she dedicated it to Albert Schweitzer who said, "Man has lost the capacity to foresee and forestall. . . . He will end up destroying the earth." To foresee and forestall is the basis of the Precautionary Principle. It is the central theme for environmental and public health rooted in the elemental concepts of "first do no harm" and "an ounce of prevention is worth a pound of cure." In its simplest formulation, the Precautionary Principle has a dual trigger: If there is a potential for harm from an activity and if there is uncertainty about the magnitude of impacts or causality, then anticipatory action should be taken to avoid harm. Scientific uncertainty about harm is the fulcrum for this principle. Modern-day problems that cover vast expanses of time and space are difficult to assess with existing scientific tools. Accordingly, we can never know with certainty whether a particular activity will cause harm. But we can rely on observation and good sense to foresee and forestall damage.

We have failed to heed Carson and Schweitzer's warning. Industrial development increased rapidly following World War II, with little regard for human health or the environment. Growth was considered akin to prosperity, and some small environmental damage was a small price to pay for the benefits of industrialization. Research and legislation developed during the

1

late 1960s and early 1970s acknowledged that there were substantial adverse impacts associated with unlimited growth. With increasing knowledge about the complexities of ecosystems, the human body, and the impacts of various stressors, we have realized that we actually understand much less than we thought we did about these systems.

During the 1970s and 1980s, tools such as risk assessment and cost-benefit analysis were developed to assist decision makers in making more rational decisions about industrial activities and their impacts. However, their incorporation into decision-making structures was based on the misguided belief that humans could fully understand the impacts of their activities on the environment and establish levels of insult at which the environment or humans could rebound from harm. Too much emphasis was placed on the role of science to model and predict harm in extremely complex ecological and human systems. Risk assessment, which was originally developed for mechanical problems such as bridge construction where the technical process and parameters are well defined and can be analyzed, took on the role of predictor of extremely uncertain and highly variable events. The risk-based approach, now central to environmental and public health decision making in the United States, has in part led to a regulatory structure based on pollution control and remediation, rather than fundamental prevention.

The quantitative, risk-based approach to environmental and public health regulation has taken on an importance in government agency operations. It allows agencies to justify and defend their decisions to the courts, businesses, and the public in the guise of objective, unbiased numbers, avoiding mention of the values implicit in decisions affecting public and environmental health. This approach is viewed as the "sound science" approach to decision making, where decisions are made on the basis of what we can quantify, without considering what we do not know or cannot measure. That is lumped under the category of "uncertainty," which can be addressed in a neutral way through additional information and modeling. The risk assessment process, however, is as much policy and politics as it is science. A typical risk assessment relies on at least 50 different assumptions about exposure, dose-response, and relationships between animals and humans. The modeling of uncertainty also depends on assumptions. Two risk assessments conducted on the same problem can vary widely in results.

Current environmental and public health decision-making processes, based primarily on the level of risk, suffer from several limitations, which constrain their ability to identify, anticipate, and prevent potential harm to human health and the environment. Decisions to take action to restrict potentially dangerous activities are often taken after science has established a causal association between a substance or activity and a well-defined, singular adverse impact. Proving causality takes both extensive time and resources. During this research period, action to prevent potentially irreversible human and environmental harm is often delayed in the name of uncertainty and the harmful activity continues. For a variety of reasons, it may not even be possible to demonstrate a causal association in complex human/ecological systems.

For example, even basic knowledge about the impacts of the most widely used toxic chemicals is unavailable. Analysis of the impacts of human activities on health and the environment is wrought with uncertainty. This ignorance leads to an important question for decision makers, "How can science establish an 'assimilative capacity' (a predictable level of harm from which an ecosystem can recover) or a 'safe' level when the exact effect, its magnitude, and interconnectedness are unknown?"

Further, regulatory programs often demand the achievement of statistical significance in experimental and observational research. Even though an effect is not significant to a statistician, it still may be significant to the person or community. This "laboratory" model of science places an emphasis on minimizing Type I errors (incorrectly concluding that there is an effect when one does not exist) and thus unnecessary regulation at the expense of increasing the potential for Type II errors (incorrectly concluding that there is not an effect when there is one), placing humans and the environment in jeopardy. Achieving adequate statistical power (the predictive potential of an experiment) for a study to be considered acceptable is difficult if the number of subjects or effect is small.

Even low level exposure to stressors may cause adverse impacts. These impacts may be impossible to monitor or control. For example, there is growing evidence that some synthetic chemicals may disrupt the hormone system at very low levels of exposure and not at high doses (an inverted U-shaped dose response), with effects happening when exposure takes place during sensitive periods in the development of a fetus. It is virtually impossible to know what level of exposure will affect a fetus or what impacts that exposure will cause.

Science has not begun to address the wide range of physical and chemical stressors to which humans and ecosystems are exposed since it focuses on single chemicals/stressors in single media. If we are ignorant about the impacts of only single human activities on health and the environment, we are even more ignorant about the cumulative effects of many potentially harmful activities.

Finally, risk assessments and other analyses are very time consuming, contentious, and costly. For example, a single risk assessment on a single chemical might take up to five years and cost upwards of $5 million. This excludes the cost of the harm that may be caused by the activity under study. Focusing on opportunities to prevent harm (e.g., using the Precautionary Principle) is a much more cost-effective use of limited resources.

There is a need for decision makers to bridge the gap between uncertain science (and the need for more information) and the political need to take action to prevent harm. As trustees of ecosystem and public health, government agencies have an obligation to prevent harm despite the existence of uncertain impacts. They must consider that there could be large political and economic consequences if they are wrong. The question of what society should do in the face of uncertainty regarding cause and effect relationships is a question of public policy, not science. A decision not to act in the face of uncertainty, to await further scientific evidence, is as much a policy decision as taking preventive action.

HISTORY OF THE PRECAUTIONARY PRINCIPLE

The term "Precautionary Principle" is relatively new to the national and international environmental policy arena, though the concept has its roots in hundreds of years of public health practice. Even early environmental legislation encompassed a precautionary approach to environmental protection. For example, in the legislative history to Sweden's first environment act, the Minister of Justice noted that environmental policy should lead to actions in the face of uncertainty and shift the burden of proof of safety to those who create risks.

The principle emerged as an explicit basis of policy during the early 1970s in West Germany as "Vorsorgeprinzip" or the "foresight" principle of German water protection law. At the core of early conceptions of this principle in Germany was the belief that society should seek to avoid environmental damage by careful "forward-looking" planning, blocking the flow of potentially harmful activities. The Vorsorgeprinzip has been invoked to justify the implementation of vigorous policies to tackle river contamination, acid rain, global warming, and North Sea pollution. Implementation of the foresight principle has given rise to a globally competitive industry in environmental technology and pollution prevention in Germany.

The Precautionary Principle was first introduced internationally in 1984 at the First International Convention on Protection of the North Sea, designed to protect the fragile North Sea ecosystem from further degradation due to the input of persistent toxic substances. At the Second North Sea Conference, ministers noted that "in order to protect the North Sea from possibly damaging effects of the most dangerous substances . . . a precautionary approach is addressed which may require action to control inputs of such substances even before a causal link has been established by absolutely clear scientific evidence." Following this conference, the principle was integrated into numerous international conventions and agreements including the Maastricht Treaty, the Barcelona Convention, and the Global Climate Change Convention, among others.[1] The principle guides sustainable development in documents like the 1990 Bergen Ministerial Conference on Sustainable Development and the 1992 United Nations Conference on Environment and Development. It has become a central theme of environmental law and policy in the European Union and many of its member states.

The Precautionary Principle itself is a relatively new concept to environmental protection in the United States. However, as in many other countries, the general notion of precaution underlies much of the early U.S. environmental and public health legislation. For example, the former Delaney Clause of the Food, Drug, and Cosmetics Act prohibited the incorporation into processed food of any level of a substance that had been found carcinogenic in laboratory animals. The National Environmental Policy Act (NEPA) requires that any project receiving federal funding and that may pose serious harm to the environment undergo an environmental impact statement, demonstrating that there were no safer alternatives. The Clean Water Act (CWA) established strict goals in order to "restore and maintain the chemical, physical, and biological integrity of the Nation's waters." The Endangered Species Act requires the protection of threatened species beyond economic interests. The Occupational Safety and Health Act (OSHA) was designed to "assure so far as possible every working man and woman in the Nation safe and healthful working conditions." The OSHA draft Carcinogen Standard (which was never put into practice) required precautionary actions any time a chemical used in the workplace was suspected of being a carcinogen in animals.

Early court decisions also gave substantial deference to the Environmental Protection Agency (EPA) to take action to prevent harm even before considerable evidence of cause and effect was gathered. For example, in the Reserve Mining Case, the court ruled that "the public's exposure to asbestos created a sufficient health risk to justify taking precautionary and preventive measures to protect the public health." In a case over EPA regulations requiring reductions in lead additives in gasoline, the court noted, "Where a statute is precautionary in nature, the evidence difficult to come by, uncertain or conflicting because it is on the frontiers of scientific knowledge, . . . we will not demand rigorous step-by-step proof of cause and effect."

Much of the early precautionary nature of U.S. environmental and occupational safety and health policy was lost during the 1980s, when the Reagan administration disarmed these protections. In addition, a U.S. Supreme court case involving occupational health standards for benzene, and the rise in supremacy of quantitative risk assessment and cost-benefit analysis in environmental and occupational health, eroded the early precautionary nature of environmental and public health protections. The protection of health and environment has not fully recovered from these actions.

A strong public backlash to losses in environmental and health protections, coupled with the industrial disasters in Chernobyl and Bhopal, led to a rejuvenation of the grassroots activism in the United States and new calls for the public's right to know and expanded environmental protections. Creating a public right to know led to an understanding that companies were emitting enormous amounts of pollutants into the environment. This, coupled with a realization that pollution-control strategies were not eliminating but rather shifting pollution, led to the passage of the Pollution Prevention Act of 1990, which sets prevention as the highest priority in environmental programs.

Responding to the public's strong pro-environment sentiment, the U.S. government signed the Rio Declaration at the United Nations Conference on Environment and Development in 1992. Section 15 of the declaration calls for states to adopt the Precautionary Principle. The U.S. Environmental Protection Agency has admitted that it is bound by the Rio Declaration and must identify ways to implement the Precautionary Principle. In 1996, the U.S. President's Council on Sustainable Development, a multi-stakeholder presidential board, issued a statement of principles for sustainable development, among them is an implicit definition of the Precautionary Principle: "There are certain beliefs that we as Council members share that underlie all of our agreements. We believe: (number 12) even in the face of scientific uncertainty, society should take reasonable actions to avert risks where the potential harm to human health or the environment is thought to be serious or irreparable."

Perhaps the most noteworthy work on the Precautionary Principle in the United States has occurred in the Great Lakes Region and on the state level. In the Great Lakes, the International Joint Commission (IJC), a 100-year-old bi-national body established to protect waters along the Canadian–U.S. border, determined that attempts to manage persistent and bioaccumulative pollution in the Great Lakes had failed and these could not be managed safely. As a result, the commission issued a call to sunset all persistent toxic substances, noting that action is needed to protect health and environment "whether or not unassailable scientific proof of acute or chronic damage is universally accepted. Gordon Durnil, who was appointed by President Bush to head the U.S. delegation to the commission, relates how the IJC reached its conclusions (see chapter 17). First he asked the various scientists serving on committees within the IJC to describe what they knew. He received a myriad provisos on lack of evidence and absence of significant proof linking chemicals to harm. Next he asked these scientists, what they believed. They believed that there was harm linked to these substances, even if they could not prove it.

On the state level, at least 25 states have established some type of pollution prevention legislation. California passed Proposition 65, which requires companies and other establishments to label any products that contain substances that could cause cancer or developmental harm. The Commonwealth of Massachusetts has passed several laws that are precautionary in nature. For example, its wetlands statute requires those building near wetland areas must demonstrate that no harm to wetland integrity will occur. The Commonwealth's Rivers Act requires that anyone building within a river buffer zone demonstrate that there is no other option for building. And the Toxics Use Reduction Act requires firms using certain toxic chemicals to identify alternatives to reduce or eliminate their use. Most recently, a bill was introduced in Massachusetts establishing the Precautionary Principle as a general duty for government agencies and businesses.

Business organizations have also begun to recognize the importance of implementing the precautionary approach as a corporate responsibility and its benefits business. Both the International Chamber of Commerce and the World Business Council for Sustainable Development have endorsed the Precautionary Principle as a management tool necessary to achieve sustainable development. There are numerous examples of individual companies and industries (e.g., British Petroleum with regards to global warming) taking precautionary action to avoid environmental and health harm. Author Stephen Schmidheiny explains that business leaders are "used to examining uncertain negative trends, making decisions, and then taking action, adjusting, and incurring costs to prevent damage." They support the Precautionary Principle not only because it can help them avoid liabilities, but also because of opportunities for innovation, improved corporate image, and product development.

History of the Wingspread Conference and This Volume

In 1993, the New York Times published a series of articles by Keith Schneider that stated that many environmental problems of modern times were exaggerated. This series of articles led to the publication of more articles, books, and the establishment of so-called "sound-science" organizations. The attack on environmental and occupational health regulation during the 104th Congress incubated a need for the development of pro-active measures to fight this attack. Emerging issues, such as global climate change and endocrine disruption, reinforced the demand for new approaches to decision making in the face of uncertainty. Environmental groups, as well as the scientists and lawyers working with them, felt that the Precautionary Principle represented an important paradigm that addresses the limits of science while promoting action to prevent harm.

Environmentalists recognized that the Precautionary Principle had achieved some prominence in Europe and in international treaties but not in the United States among other countries. While they understood the underlying basis of the principle, it was unclear what precaution actually meant in practice. There was also no clear structure to integrate the principle into decision making in the way that risk assessment had been integrated over the past 15 years. To bring the Precautionary Principle to the forefront of environmental and public health decision making in the United States, advocates felt that a meeting was needed to develop a structure and methods for operationalizing the principle. From this need the Wingspread Conference on Implementing the Precautionary Principle was born.

During the weekend of January 23–25, 1998, the Science and Environmental Health Network convened 35 academic scientists, grassroots environmentalists, government researchers, and labor representatives from the United States, Canada, and Europe to discuss ways to formalize the Precautionary Principle. The workshop focused on understanding the history and scientific and political contexts under which the principle developed, its basis, and how it could be implemented in toxic chemicals policy, agriculture, and biodiversity. The Wingspread participants issued a consensus statement calling for and defining the Precautionary Principle. It states: "When an activity raises threats of harm to human health or the environment, precautionary measures should be taken even if some cause and effect relationships are not fully established scientifically."

The Wingspread Statement on the Precautionary Principle represents an important definition for the principle because it amplifies and clarifies both the Rio Declaration and the President's Council's statement. The Wingspread statement starts off with a call to action, because we have already reached our capacity for environmental insults. As defined at Wingspread, the Precautionary Principle has four components: (1) Preventive action should be taken in advance of scientific proof of causality; (2) the proponent of an activity, rather than the public, should bear the burden of proof of safety; (3) a reasonable range of alternatives, including a no-action alternative (for new activities) should be considered when there may be evidence of harm caused by an activity; and (4) for decision making to be precautionary it must be open, informed, and democratic and must include potentially affected parties.

Since the Wingspread Conference, the Precautionary Principle has been invoked in places the convenors could not have predicted before January 1998. While toxics have traditionally been the domain for the Precautionary Principle internationally, in the United States it is gaining its greatest support among sustainable agriculture advocates. In Washington state, people protesting the use of hazardous waste in the manufacture of fertilizers called for decision making to be based on the principle. It has also been identified by advocates as the single most important issue in the enormous grassroots response to the U.S. Department of Agriculture's draft organic agriculture rule. Opponents of genetic engineering have also used the Wingspread Statement in international meetings.

In the United States it is likely that the principle will first be solidified at the local and state levels, given the more conservative nature of federal government policies and the entrenchment of risk assessment. Once successes are made at these levels, pressure can be brought on the federal government to institutionalize the principle. This differs from international experience with the principle, where it starts as a global concept and then works its way down to the local level. Nonetheless, several federal agencies are considering how to incorporate the principle into children's environmental health and other environmental concerns, and the Precautionary Principle was included in the "description" of endocrine disruption developed under the U.S. EPA's Endocrine Disruptor Screening and Testing Committee. In the summer of 1998, the Indiana Republican Committee adopted the Precautionary Principle as the basis for its environmental platform. The concept of precaution is beginning to take hold in the United States; however, it will be some years before it reaches the level of prominence held in Europe and other regions of the world. We hope that this volume will help to demystify and provide structure and credence to the notion of precaution.

STRUCTURE OF THIS BOOK

This book is an outgrowth of the Wingspread Conference and the need to operationalize the precautionary approach in environmental and public health decision making. The majority of the chapters contained in this volume were written by Wingspread participants to guide discussions at the conference. Others were solicited after the conference to address other important areas. This volume differs from previous discussions on the principle in that it struggles with the difficult questions of implementation and fundamental change required to support a more precautionary approach to environmental and public health hazards.

The book consists of four parts that provide a series of steps toward understanding and developing precautionary decision making. Taken together, the parts provide a compass to guide the reader toward a new way of thinking about ecosystem and public health protection, that is scientifically rigorous and grounded in ethics.

Part I, An Overview of the Precautionary Principle, provides an analysis of lessons learned from implementing the Precautionary Principle in Europe as a whole, and in Sweden, the country whose approach to regulating toxic substances has most clearly approximated precaution. These examples provide information on how precaution has been implemented to date.

Part II, Law and Theory, provides an overview of the factors that have led to the current way science and law are incorporated into decision making. The authors of these chapters argue that fundamental changes in law and science will be needed to support precautionary decision making. These include general legal duties for initiators of potentially harmful activities to act in a precautionary manner, shifting burdens of proof to initiators of activities, and expanding and modifying the science used in decision making so that it favors protection of health and the environment.

Part III, Integrating Precaution into Policy, describes practical approaches toward implementing precaution. These chapters provide examples of structures for integrating the principle in decision making, as well as the tools to implement precaution. These chapters describe a fundamentally new way of making decisions in the face of uncertainty—a way that is consistent with respect, common sense, and prudence. In addition, lessons learned from the U.S.–Canada International Joint Commission's approach to precaution are described.

Finally, Part IV, The Precautionary Principle in Action, provides examples of opportunities to take precaution as well as examples of occasions when it was not taken. These chapters provide an overview of the political, economic, and scientific barriers to Precautionary Principle implementation from the perspective of a farmer, policy analysts, a doctor, and a scientist.

When Rachel Carson used the words "to foresee and forestall" at the beginning of Silent Spring, she described humankind's hubris in thinking it could spread synthetic chemicals to serve human needs without disrupting fragile, interconnected ecosystems. While she was attacked by petrochemical interests and some in government for her observations, we now know that she was right. Synthetic pesticides, such as DDT, have caused untold ecological and toxicological harm. In essence, the result of her systematic, careful observations was to foresee the potential harm caused by these synthetic chemicals. *Silent Spring* was a call to forestall this rapid development and deployment of pesticides and to return to an ethic of working with nature, not against it. That is the ultimate goal of precaution.

NOTE

1. The language contained in some of these international treaties discussed in this chapter is contained in Appendix B.

Part I

AN OVERVIEW OF THE PRECAUTIONARY PRINCIPLE

Chapter 1

❧

THE PRECAUTIONARY PRINCIPLE IN CONTEMPORARY ENVIRONMENTAL POLICY AND POLITICS

Andrew Jordan and Timothy O'Riordan

This chapter outlines the history of the Precautionary Principle and discusses its current status in national and international environmental policies. We need to know about the genesis of the principle before considering how to improve its application in the future. To this end, we open this chapter by discussing the broad meaning of the Precautionary Principle as it has emerged in the last twenty years. We discuss the origins of the principle in the environmental policy of the former West Germany. We go on to outline seven "core" elements of precautionary thinking and the extent to which they find expression in contemporary environmental policy. Finally, we begin the process of considering how to improve existing tools of decision making to incorporate precautionary thinking. Paradoxically, we conclude that the application of precaution will remain politically potent so long as it continues to be tantalizingly ill-defined and imperfectly translatable into codes of conduct, while capturing the emotions of misgiving and guilt. Indeed, as we move into an era of greater scientific engagement with various political interests over the application of precautionary measures in particular decision-making situations, and when futures have to be selected rather than decreed, so the Precautionary Principle will become part of the mainstream and the more participatory politics of the transition to sustainability.

DEFINING THE PRECAUTIONARY PRINCIPLE

The modern environmental movement proceeds by capturing ideas and transforming them into principles, guidelines, and points of leverage. Sustainability is one such idea, now being reinterpreted and implemented in the aftermath of the 1992 Rio Conference (for reviews see O'Riordan, Jordan and Voisey, 1998; O'Riordan and Voisey, 1998). So too is the Precautionary Principle being reinterpreted and implemented. Like sustainability, it is neither a well-defined nor a stable concept. Rather, it has become the repository for a jumble of adventurous beliefs that challenge the status quo of political power, ideology, and environmental rights. Neither concept has much coherence other than it is captured by the spirit that is challenging the authority of science, the hegemony of cost-benefit analysis, the powerlessness of victims of environmental abuse, and the unimplemented ethics of intrinsic natural rights and intergenerational equity. It is because the mood of the times needs an organizing idea that the Precautionary Principle is getting attention.

To stop the sustainability concept from becoming completely meaningless, Norton (1992, 98) calls for the following:

> a set of principles, derivable from a core idea of sustainability, but sufficiently specific to provide significant guidance in day to day decisions and in policy choices affecting the environment.

Precaution is one such principle, for it provides an intuitively simple approach to ensuring that human intervention in environmental systems is made less damaging. Admittedly, precaution lacks a specific definition. As yet, it cannot prescribe specific actions or solve the kind of moral, ethical, and economic dilemmas that are part and parcel of the modern environmental condition. Nonetheless, the Precautionary Principle has much efficacy because it captures an underlying misgiving over the growing technicalities of environmental management at the expense of ethics and open dialogue; of environmental rights in the face of vulnerability; and of the manipulation of cost-benefit analysis by powerful vested interests supporting development.

So what kind of practical steps does the Precautionary Principle require decision makers to take? To date, precaution provides few, if any, operable guidelines for policy makers nor does it constitute a rigorous analytical schema. Yet, it is accepted by many national governments and supranational entities such as the United Nations (U.N.) and the European Union (EU), for example, as a guiding principle of policy making. It is found in the climate convention and the Rio Declaration (Cameron, 1994). There is little doubt that for the big global issues of climate change, stratospheric ozone depletion, and biodiversity loss, the Precautionary Principle carries much greater legal (and by extension political) weight than it does in day-to-day local environmental issues. What remains desperately unclear is how this high-level interpretation of precaution can be translated to the areas of pollution control, risk management, and waste minimization and assimilation, especially where trans-border policies are required.

In a nutshell, precaution challenges the established scientific method; it tests the application of cost-benefit analysis in those areas where it is undoubtedly weakest (i.e., situations where environmental damage is irreversible, possibly catastrophic, or simply unknown); it calls for changes to established legal principles and practices such as liability, compensation, and burden of proof; and it challenges politicians to begin thinking through longer timeframes than the next election or the immediate economic recession. Precaution also exposes the existing discipline-bound and reductionist organization of academic research and raises difficult issues about the quality of life for future generations and other species. It is profoundly radical and potentially very unpopular. But it is enduring, and it resonates with the anxieties of the modern age.

Precaution reflects the mood of distrust over the introduction of risky technologies, processes, and products that are assumed to be forced on the unknowing and susceptible public by commercial interests, allied to governments, and exerting manipulative, self-interested power over consumers. This is the challenge introduced by the German sociologist Ulrich Beck (Beck, 1992) who argues that society is trapped in a loop whereby risky technologies to which it is inextricably wedded produce socially disruptive and damaging effects such as pollution and habitat loss. Precaution is one of the devices through which that loop can be broken. However, in his characterization of modern society as a struggle to cope with reflexive modernization, Beck does not give the "resistance through precaution" movement as much credit as it deserves.

As precaution becomes increasingly integrated into modern environmentalism, it may well run the risk of following the dangerously successful, epistemologically ambiguous pathway pioneered by sustainability. We say "dangerously successful" because it is precisely the uncritical accumulation of meanings, often contradictory and impractical, that have characterized the success of the sustainability notion in recent years, blunting its uncomfortable message that the existing trajectory of development is socially and environmentally unsustainable.

Lawyers and scientists say the concept is unclear and lacks content. In particular, they argue that it fails to specify where and when precautionary measures should be applied and by whom. We argue below that precaution is a culturally framed concept that has evolved along different pathways and at different rates in different countries. Searching for a single, all-encompassing definition is, therefore, likely to be a fruitless endeavor because individuals will never agree upon what is or is not precautionary in a given situation. Cultural theory tells us that there is no one single context of risk perception. We all "see" the world in a different way, although four broad archetypes can be distinguished (Jordan and O'Riordan, 1997; O'Riordan, Marris, and Langford, 1997). So those who regard the environment as inherently robust and capable of withstanding sustained human impact will tend to be less precautionary than those who regard human impact on nature as unpredictable and potentially calamitous. These value positions are deeply entrenched, and scientists and policy makers need to be more sensitive to this when they communicate risk to the body politic.

We are happy to accept that the Precautionary Principle offers no more than a broad guide to policy makers to move to anticipate problems before they occur, which could involve acting before there is full scientific understanding of the circumstances. Arguably, the precise meaning of precaution will only emerge when stakeholders come together to make a decision in a particular context, trading costs against benefits and determining tolerable levels of damage. After all, what is precautionary in one cultural context may not be regarded as precautionary in another. For example, the introduction of genetically modified organisms (GMOs) into food products is generally tolerated in the United States but is being resisted in Europe. Part of the reason for this difference, which has led to protracted trade disputes between the two areas (Vogel, 1997), is the richer culture of "natural food" preferences among Europeans. Part, too, is the active engagement by high profile NGOs in Europe in what they see as symbolic entry of GMOs into the food chain. Recently, the Austrian Government announced it would oppose cultivation approval in the EU for a genetically modified tomato developed by the British firm Zeneca. The tomato has been cultivated and sold in pureed form in Britain since 1996. Zeneca needs the EU's approval before it can cultivate and market the product in the rest of Europe. The company is confident that the product is safe and believes the public has nothing to fear—sales of the engineered puree in the United Kingdom are apparently double those of the conventional product. Environmental groups in Austria had mounted a last minute campaign to convince the national consumer affairs ministry to block commercialization on the grounds that its safety was unproven. A Greenpeace Austria spokesperson described the tomato as "another example of a genetically modified plant that nobody needs and only serves to optimise the profits of big companies [while] we consumers have to bear the risks" (Environmental Data Services (ENDS) Daily, 1998a). So the political stakes are higher in Europe than in the United States, and the context of risk perception reflects the role of precaution.

The cultural context, therefore, is not just a matter of accounting for the spatial variability of the environment (i.e., the carrying capacity of one locality being greater than another). Rather, people should be provided with the means to work out what precaution means for them in their own localities. We try to identify the principles that should guide that process of searching and discovery. Involving the public in precautionary environmental management is not simply a normative demand for better government. On the contrary, if precaution involves acting before the availability of full scientific information, then grounds other than "good science," such as ethical, moral, or political, are required to legitimate policy decisions. This is why decisions that are imposed without prior consultation are unlikely to be regarded as truly precautionary. Precaution means being honest and open about uncertainty rather than dismissing, ignoring, or downplaying it. It means exploring the worst-case scenario and searching out the ill-informed and possible "losers" from a course of action, asking what they regard as legitimate.

THE HISTORY OF THE PRECAUTIONARY PRINCIPLE

The precautionary principle emerged during the 1970s in the former West Germany at a time of social democratic planning. At the core of early conceptions of precaution (or *Vorsorge*) was the belief that the state should seek to avoid environmental damage by careful forward planning. The word *Vorsorge* means foresight or taking care, although it also incorporates notions of good husbandry and best practice in environmental management even in the absence of risk (von Moltke, 1988). There is no standard statement about its role or its meaning. Albert Weale (1992, 79) quotes one respected German commentator who has identified at least eleven separate meanings. The *Vorsorgeprinzip* (Precautionary Principle) was used by the German government in the early 1980s to justify the implementation of vigorous policies to tackle acid rain, global warming, and pollution of the North Sea. In relation to these problems, *Vorsorge* implied fitting the best available abatement technology in order to minimize polluting emissions at the source.

It is important to understand the context in which precautionary thinking first developed. For Hajer (1995) and Weale (1992, 1993), *Vorsorge* is part of a wider set of ideas or an ideology the Germans label "ecological modernization." This is still a vague notion, but like Brundtland it suggests that the relationship between environmental protection and development is not necessarily antagonistic and can be made to be mutually advantageous. The Precautionary Principle was warmly received in Germany because it seemed to legitimize greener forms of economic growth. One of the fundamental tenets of ecological modernization is that high environmental standards are an opportunity for, rather than a constraint upon, economic growth. High standards of environmental protection were seen to have the potential to spur the development of green technologies, reduce waste, and meet the demands of a more environmentally aware public. Remember that this was a time of unprecedented public concern about environmental damage, particularly the threat of acid rain to natural, coniferous forests to which the Germans are very emotionally and culturally attached. Significantly, the notion of ecological modernization fitted with the dominant path of post-war economic development in Germany, namely the income and employment linked to the export of goods and services with a high value-added technological content. The technology forcing capacity of progressive environmental standards has served the German economy well since the 1970s, encouraging the development of a lucrative clean technology sector that now employs 320,000 people (OECD, 1992). We should expect precaution to receive much less of a clear wind when it sets down limits to economic growth, such as the preservation of a particular habitat or the withholding of a new technology or product such as a genetically engineered crop on the grounds that scientific understanding of its long-term environmental impacts is insufficiently clear.

Both Weale and Hajer show, however, that the precautionary discourse found a less favorable hearing in countries such as the United Kingdom, with a long tradition of scientific corporatism and elitism. The United Kingdom, at that time, exhibited a more secretive and consensual style of environmental management and a deeply institutionalized preference for externalizing waste using long pipes and tall chimneys to make the optimal use of the waste assimilative capacity of the environment. These days of scientific nationalism in the United Kingdom are now over, due largely to the steady incursion of EU laws and policy procedures and the penetration of national policy networks by environmental pressure groups (Jordan, 1998a), although the lesson that precaution flounders in societies that are not instinctively risk averse is an instructive one nonetheless. Openness of information is increasingly the norm in Britain, and regulatory policy has become steadily more pro-active and consensual, though still in a punitive mode. These shifts are part of modernization generally but indicate too that precaution has been opportunistically adapted to these changes.

Precaution, then, emerged in a society experiencing unprecedented levels of support for environmental matters. It was used by German authorities in the early 1980s to justify the unilateral application of technology-based standards to reduce acid rain. Once in place, the Germans pressed the EU to adopt similar standards across the rest of Europe, to prevent its own industries being placed at a competitive disadvantage. This was not enlightened environmentalism at work but the dictates of a competitive market of member states. According to Weale (1998):

> The policy debate was more dominated by competitive considerations rather than environmental concerns, as much of the delay [in adopting measures] was due to fears about comparative costs and benefits of individual states.

The process of negotiating the Large Combustion Plant Directive dragged on for nearly five years, but Germany's conversion to green thinking catalyzed a transition in EU environmental policy from reactive policy making to proactive environmental management the effects of which endure today. It is encapsulated in landmark documents such as the 1992 Fifth Environmental Action Programme and is at work in the EU's progressive position on climate change. Throughout the 1980s, Germany continued to use its political and economic power in the EU to multilateralize precautionary-based environmental policies across Europe (Jordan, 1998b). As Boehmer-Christiansen (1994, p. 30) notes in a comprehensive review of the German experience:

> the precautionary principle therefore helped to lay the conceptual and legal basis for a proactive environmental policy, which, once spread into Europe, was also directed at ensuring "burden sharing" in order that German industry would not lose its competitive edge, but rather gain new markets for its environment-friendly technology and products.

In 1993, precaution was accepted as a fundamental guiding principle of EU environmental policy.

Meanwhile, the Precautionary Principle began to move into other legal and political fora with remarkable speed and stealth. It appears regularly in national legislation, in international statements of policy, and in the texts of international conventions and protocols. More recently, it has been adopted as a guiding principle of environmental policy in both the EU and the UK, and it makes an appearance in the 1992 Rio Declaration—a statement of principles and general obligations to guide the international community toward actions that promote more environmentally sustainable forms of development.

However, the Precautionary Principle still has neither a commonly accepted definition nor a set of criteria to guide its implementation. "There is," Freestone (1991, 30) cogently observes, "a certain paradox in the widespread and rapid adoption of the precautionary principle": While it is applauded as a "good thing," no one is quite sure about what it really means, or how it might be implemented. Advocates foresee precaution developing into "*the* fundamental principle of environmental protection policy at [all] scales" (Cameron and Abouchar, 1991, 27). Sceptics, however, claim its popularity derives from its vagueness; that it fails to bind anyone to anything or resolve any of the deep dilemmas that characterize modern environmental policy making. There are legal scholars, for example, who consider precaution to be too blunt an instrument to act as a regulatory standard or principle of law. Bodansky (1991, 5) is highly suspicious of the Precautionary Principle because it "does not specify how much caution should be taken" in a given situation (see also Bodansky, 1994). It cannot, for example, determine what is an acceptable margin of error or what exact threshold of risk warrants the application of precautionary actions. Nor can it determine when precautionary measures should be taken or define the point at which abatement costs become socially or environmentally "excessive" (*ibid*, 5). Initially, German politicians used it to justify policies that already had high public appeal. In a review of German policy, Boehmer-Christiansen (1994) concludes that it has "little meaning other than that of enabling the policy process to attempt environmentally more ambitious solutions"; Bodansky (1991, 5) suggests that it constitutes little more than "a general approach to environmental issues."

The emergence of the principle has also engendered a very lively debate among scientists and sociologists of science. In the former camp we have writers such as Gray (1990), who deny the principle any role in scientific research, "since by definition it does not have to rely on scientific evidence!" (see also Gray, 1993). He believes that it is, at best, an "environmental philosophy"—purely a matter for administrators and lawyers rather than scientists seeking "objective scientific evidence." This critique is also shared by economists (Pearce, 1994) who worry that there are no proper rules for incorporating economic trade-offs or scientifically manageable uncertainties into precautionary guidelines. Pearce speaks for many economists by claiming that the principle fails the neo-classical tests of marginal utility trade-offs and opportunity cost calculations. For those wedded to economics jargon, this critique argues that too much unexamined and uncritical adoption of precaution may result in costs that cannot be justified by the loss of the use of pre-allocated funds for socially-good purposes. The principle may thus cause the adoption of costs when such monies could be far better deployed improving social well-being elsewhere. The scientific dispute centers on the unwillingness of precaution advocates to accept that there is more use of probabilities in handling uncertain scientific prognoses than precaution is willing to accommodate. The consequence is also that too much effort and money may be directed to dealing with "costs" that are not, at first blush, justified (for example, see Johnston and Simmonds, 1990; and Wynne and Meyer, 1993).

Here is where the solution lies in more participatory and revelatory mechanisms of environmental problem solving of groups representative of the body politic, including elements of legitimate interests. If the process is fair and reliable, it should implement precaution. This could in part be achieved by the explicit acceptance of better liability and compensatory arrangements for dealing with those otherwise disadvantaged by a precautionary act; for example, banning fisherman from fishing waters on the grounds that maximum sustainable catches requires a cut in the number of boats or a reduction in fishing effort. As such socially and economically harsh decisions are strengthened by the application of precaution, so do the rules for compensation become more necessary and politically contentious.

CORE ELEMENTS OF PRECAUTIONARY THINKING

So what does the Precautionary Principle actually mean? At its core lies the intuitively simple idea that decision makers should act in advance of scientific certainty to protect the environment (and with it, the well-being of future generations) from incurring harm. It demands that humans take care of themselves, their descendants, and the life-preserving processes that nurture their existence. Risk avoidance should become an established decision norm where there is reasonable uncertainty regarding possible environmental damage or social deprivation arising out of a proposed course of action. As was indicated in the 1990 Bergen Conference on Sustainable Development, "it is better to be roughly right in due time, bearing in mind the consequences of being very wrong, than to be precisely right too late" (NAVF, 1990, 6). The environment should not be expected to signal pain upon being hurt; it is up to humanity, as a matter of moral principle, to recognize that pain might be imposed and to adopt appropriate avoidance (precautionary) measures.

This in turn suggests that any action likely to result in serious environmental harm is morally wrong and so should be excluded as an option against which other courses of action are to be compared. Thus, a development project that might remove a particularly critical component of life support, say a protective coral reef, simply should not be put up as an option for alternatives to be costed against it. Critical natural habitats such as ancient woodlands, unique wetlands, or other features of the landscape that are judged to be historically, aesthetically, or intrinsically valuable should be left intact. There are strong links here with notions of "inviolability" and "sustainability constraints" (Jacobs, 1991) and, ultimately, social and environmental limits to conventional notions of economic growth. In effect, this means that humans must learn to widen the assimilative capacity of natural systems by deliberately "holding back" from unnecessary and environmentally unsustainable resource use on the grounds that exploitation may prove to be counterproductive, excessively costly, or unfair to future generations. It should be clear from the foregoing that the application of the Precautionary Principle can be both ethically and politically contentious.

From the complex, and at times confusing, debate on the meaning and applicability of the Precautionary Principle, it is possible to abstract a number of commonly occurring themes, such as the following:

- A willingness to take action in advance of formal justification of proof;
- Proportionality of response;
- A preparedness to provide ecological space and margins for error;
- A recognition of the well-being interests of nonhuman entities;
- A shift in the onus of proof onto those who propose change;
- A greater concern for intergenerational impacts on future generations; and
- A recognition of the need to address ecological debts.

These are elaborated on in greater detail below.

By no means are all of these dimensions formally approved of in existing law and common practice. Indeed, the Precautionary Principle will always be slightly protean because it is evolving through distinct pathways, at different rates in different countries.

Pro-Action: A willingness to take action in advance of scientific proof of evidence on the grounds that further delay may prove to be ultimately more costly to society and nature. Precaution is not simply the *prevention* of manifest or predicted risks that have been scientifically proven. Rather, the Precautionary Principle goes *beyond* the notion of prevention in the sense that it insists that policy makers move to anticipate problems before they arise or before scientific proof of harm is established. In practice, the line differentiating precautionary measures from preventative actions may be very blurred given science's inability to "prove" anything. This is the zone of "trans-science" (Weinberg, 1985) that covers matters that are raised by science but cannot be satisfactorily resolved by scientists alone.

Be that as it may, a very strong formulation of precaution demands that all risks to the environment are minimized; that emissions are reduced to as low as possible; and that important (or keystone) ecological assets and processes are maintained. Environmental groups such as Greenpeace, for example, go as far as to say that "the precautionary principle calls for the prohibition of the release of substances which might cause harm to the environment *even if insufficient or inadequate proof exists regarding the causal link*" (emphasis added) (Horsman, 1992, 76). This strong conception offers scientific judgments with a very limited role in decision making. However, in reality, precaution is often linked to some consideration of risks, financial costs, and benefits. In Germany, for example, the application of precaution is tempered by two other general principles of law: proportionality of administrative action and avoidance of excessive costs (von Moltke, 1988, 68).

Although the principle of precaution does not state how these different factors should be traded off, it strongly suggests that a strenuous search be conducted for alternative modes of development that minimize discharges and waste products, regardless of whether they are known to have harmful effects, on the basis that prevention is often, though not always, more cost effective than cure. In general, though, the closer controls are placed to the source of the emission and the earlier environmental considerations are factored into decision making, the more precautionary the overall trajectory of development will be.

Cost-Effectiveness of Action: The application of proportionality of response is designed to show that there should be a regular examination of identifiable social and environmental gains arising from a cause of action that justifies the costs. Economists emphasize that a very strict interpretation of the Precautionary Principle may be very costly to implement especially if the benefits foregone are appreciable and new information reveals that the measures were in fact unwarranted. Furthermore, economists say that it is possible to differentiate between actions taken on a precautionary basis (on the strictest interpretation this would equate with zero emissions) and the optimal level of pollution elucidated by a full risk-benefit or cost-benefit analysis. The difference between the two would represent the social opportunity cost of adopting a precautionary perspective.

Potentially, the net social cost of adopting precautionary measures could be extremely large, especially if the adverse environmental impacts turn out to be less important than predicted. However, cost-benefit analyses presuppose good scientific understanding and sufficient time to undertake them—factors that cannot always be relied upon. This sets up an interesting ethical conundrum. If a possible outcome is potentially destabilizing to the natural order or to social equity, can it truly be regarded as a realistic option to the point where lost benefits ought to constitute a sacrifice? This is the kind of dilemma referred to by Redclift (1993) when he discusses the distortions inherent in seeking to place a monetary yardstick on what essentially are ethical judgments (see also O'Riordan, 1996).

Clearly, there is no easy way of integrating risk, financial considerations, and highly uncertain science into decision making with the appropriate degree of timeliness. Nor does the Precautionary Principle provide a rigorous mechanism for balancing these disparate factors. Environmental economists believe that the realistic course of action is to insist that standard techniques of cost-benefit analysis incorporate the wider social and environmental costs of development. Then, the presumption should be in favor of high environmental quality (e.g., development should only be allowed when the benefits of acting are *much* greater than (rather than simply greater than or equal to) the associated costs) (Pearce, 1994).[1]

Safeguarding Ecological Space: The Precautionary Principle insists that in order for the overall capacity of environmental systems to act as a buffer for human well-being, these environmental systems must be adequately protected: "Any error in risk calculation should be to the advantage of the environment" (Bodansky, 1991, 5). This entails leaving a sufficiently wide natural cushion in the functional equilibria of natural systems. In effect, this means that humans must learn to widen the assimilative capacity of natural systems by deliberately holding back from unnecessary and environmentally unsustainable resource use on the grounds that exploitation may prove to be counterproductive, excessively costly, or simply unfair to future generations. Nature's assimilative capacity cannot always be taken for granted—often, we simply do not know enough about the chaotic behavior of natural systems to be able to identify critical thresholds (points beyond which irreversible phase changes may occur, possibly culminating in the collapse of ecosystems) with any degree of certainty (McGarvin, 1994; Stebbing, 1992).

In more concrete policy terms, the Precautionary Principle offers an explicit challenge to the so-called "dilute and disperse" paradigm (GESAMP, 1986), which has underpinned standards in the United Kingdom for over a century. The paradigm is built on the assumption that science can determine the ability of an ecosystem to assimilate hazardous substances without incurring long-term damage and that individual emission permits can be tailored consonant with these safe margins. The precautionary paradigm is based on the denial of the general validity of these assumptions; that is that science does not always provide the insights with the necessary degree of timeliness or accuracy and that preventing emissions is often, though not always, a better course of action than restoration of damage once it has occurred.

Intrinsic Value and Legitimate Status: The stronger formulations of the Precautionary Principle are consistent with what philosophers term a "bioethic." This states that vulnerable, or critical natural systems, namely those close to thresholds, or whose existence is vital for natural regeneration, should be protected as a matter of moral right. This in turn places a strain both on the application of cost-benefit analysis generally, including the proportionality rule, and the normal practice of considering all options as comparators for decision making. Thus, precaution goes to the heart of the philosophical and political debate on the proper relationship between humans and the nonhuman (natural) world. In promoting a more humble and less rapacious attitude to the environment, the Precautionary Principle presents a profound challenge to some of the unstated assumptions of modern (and particularly Western) societies: material growth, the power and efficacy of scientific reason, and the pre-eminence of human interests over those of other entities. The human race is a colonizing species without an institutional or intellectual capacity for equilibrium, and notions of care, precaution, and restraint strike at the very heart of its common purpose (Jordan and O'Riordan, 1993).[2] The Precautionary Principle lends strength to the notion that natural systems have intrinsic rights and a noninstrumental value that should be accounted for in decision making.

Unfortunately, the Precautionary Principle does not state how much environmental quality should be sacrificed for material growth, nor does it determine how a noninstrumental respect for nature should be incorporated into decision making. However, it does offer a strong presumption in favor of high environmental protection and a justification for treating certain environmental functions or features as inviolable. This is a prospect that usually causes alarm among those who believe that such a concept is an excuse for deep ecology to ride roughshod over sensible forms of development or to impose limits to material growth. The U.S. lawyer Christopher Stone (1987) has sought to allay these fears by indicating that a creative partnership in law can be established to allow nature rights of existence that are not absolute but that require careful deliberation before being set aside.

Shifting the Onus of Proof: The Precautionary Principle suggests that the burden of proof could shift onto the proto-developer to show "no reasonable environmental harm" to such sites or processes, before development of any kind could proceed. This is the reversal of the normal position where it is up to the would-be developer to show likely and unreasonable harm (Cameron and Wade-Gery, 1991). Such a reversal of the liability rule would be truly radical and difficult to implement since it would involve some definition of harmfulness (or if the burden is to be reversed completely, some measure of harmlessness), both of which accord with very strong interpretations of sustainability.

Traditionally, the law has tended to privilege parties accused of degrading the environment rather than the victims of pollution (i.e., those that are forced to bear the external costs). In general, claims for damage in the law are only upheld if the victim can prove that the emission (or the damage) was reasonably foreseeable. "Acts of God" and "accidents" tend to disallow claims for compensation. In this sense, the law offers little inducement to developers or operators of industrial processes to take adequate precautions with regard to the environmental impact of their actions. The introduction of a *strict liability* regime, on the other hand, would only require the victim to prove that the polluter failed to act with due diligence to gain compensation; in the case of *absolute liability*, the victim would merely need to prove that damage had occurred to gain financial restitution. More stringent still, would be to *reverse the burden of proof* entirely (i.e., the burden of proof is placed upon the proto-polluter to prove that emissions are "harmless" before the activity is sanctioned) as in the licensing of new medicines. But again, this presupposes some measurement of harmlessness, which is sometimes equally as difficult to prove as harmfulness.

Finally, reversing the burden of proof entirely would be extremely radical, since it would impose an explicit and legally binding duty of care on those who propose to alter the status quo. Admittedly, this raises profound questions over the degree of freedom to take calculated risks, to innovate, and to compensate for possible losses by building in ameliorative measures. Yet, U.K. and EU law is already moving in this direction with the introduction of formal duties of care, set against the backdrop of a broader, but intensifying, debate on the possible extension of strict liability for environmental damage, no matter how anticipated. A working paper recently issued by the European Commission grapples with some of these problems. The paper, a precursor to a possible directive on waste regulation, proposes that operators be made liable for environmental damage under the polluter pays principle. When damage could have been foreseen (i.e., when it passes the foreseeability test), and it is sufficiently serious, then strict liability would apply. Not surprisingly, industry is reacting strongly to this draft. ENDS Daily (1998b) quotes one Dutch industrialist as saying that the notion of the burden of proof lying with the would-be developer is "totally unacceptable." Although the paper explicitly rules out the application of liability retrospectively (i.e., precaution in reverse), industry quite evidently prefers a much less precautionary regime. In an industrial society, it is unlikely that the stronger interpretation of precaution will prevail.

Meso-Term Planning: The application of precaution extends the scope of environmental policy from certain and known problems that occur in the present to future and more uncertain issues. Precaution urges politicians to act with due care and diligence—to anticipate problems. Precautionary actions could be considered as an investment (or insurance) against unforeseen mishaps or the acceptance of higher costs now in order to guard against dysgenic impacts later. But it also implies committing current resources to investments for the future, the benefits of which may be uncertain or worse nonexistent. Since conclusive scientific evidence of harm or excessive damage in the future may not always be available to justify the commitment of resources to precautionary investments, other grounds for legitimation may need to be present, namely moral, political, ethical, and legal grounds. Cost-benefit analyses rarely take into account the likely costs and benefits of various courses of action during this period. Similarly, legal rules for compensation or obligation to take care are still ill developed. Needless to say, democracy itself is poorly suited to this time scale with its heavy political biases in favor of immediate gratification and gain today rather than tomorrow. Here is an arena where the Precautionary Principle challenges institutional performance and the sense of citizenship that primarily concentrates on the well-being of society today rather than the state of the world tomorrow.

Paying for Ecological Debt: Precaution is essentially forward looking, but there is a case for considering a burden-sharing responsibility for those not being cautious or caring in the past. This is a difficult matter. Responsibility for actions taken in ignorance, or in a climate of opinion that did not regard environmental vulnerability as a serious basis for evaluating options, should not reasonably be placed on those for whom there was no clearly defined duty of taking moral legal care. Nevertheless, shouldering the burden is an important component of precaution. For instance, the notion of common but differentiated responsibility enshrined in the U.N. Framework Convention on Climate Change, and the concept of conducting precaution "according to capabilities" in Principle 15 of the Rio Declaration, reflect an embryonic version of these ideas. Despite all this, "precaution in reverse," while attractive from a moralist or environmentalist viewpoint, could well founder in the courts. Its future, therefore, lies in the political realm.

PUTTING PRECAUTION INTO PERSPECTIVE

Two points should be apparent from the foregoing discussion. First, each of the seven elements requires an institutional mechanism or tool to make it operational. These could be legal (e.g., the introduction of strict liability regimes), economic (i.e., appropriately weighted cost-benefit analyses with citizen juries to consider especially controversial matters), technological (e.g., clean production), or democratic (i.e., participatory visioning based on sharing outcomes and common burdens). Second, precaution works through a continuum within each distinct element. So, there are weak formulations that are relatively protective of the status quo to very strong formulations that predicate the need for much greater social and institutional change. There are, of course, a host of variations in between these poles. The weaker formulations, for example, tend to be restricted to the most toxic and human life-threatening substances or activities. They advocate a role for biased cost-benefit analysis, incorporate some concern for technical feasibility and economic efficiency arguments, and tend to emphasize the importance of basing judgments on the dictates of sound science. These are very much the concerns of the "lighter" greens. The following, a statement from the U.K. government, prescribes a particularly limited role for precaution:

> Where there are *significant* risks of damage to the environment, [we] will be prepared to take precautionary action to limit the use of potentially dangerous materials or the spread of potentially dangerous pollutants, even where scientific knowledge is not conclusive, *if the balance of likely costs and benefits justifies it.* The precautionary principle applies particularly where there are *good grounds* for judging either that action taken promptly at *comparatively low cost* may avoid more costly damage later, or that irreversible effects may follow if action is delayed (emphasis added) (HM. Govt, 1990, 11).

The stronger formulations, on the other hand, have more in common with the "deep green" worldview and ecologism although few political analysts have actually made that link. Dobson (1990, 205), for example, notes that precaution is normally associated with "deep" greenism ("ecologism"):

> ecologism asks that the onus of justification be shifted from those who counsel as little interference as possible with the non-human world, to those who believe that interference is essentially non-problematic.

In the policy domain, examples of strong formulations are more difficult to find. The Third Ministerial Declaration on the North Sea signed by various North Sea states in 1990, for example, states that governments should:

> apply the precautionary principle, that is to take action to avoid potentially damaging impacts of [toxic] substances . . . *even where there is no scientific evidence to prove a causal link between emissions and effects* (emphasis added).

On this conception, science plays little or no role in policy making; administrators go beyond science to address known, but still uncertain, threats to the environment. This interpretation is both promoted by and finds support within environmental pressure groups that challenge the legitimacy of science, such as Greenpeace (Horsman, 1992). The Germans tend also to adopt a fairly strong definition of precaution. Boehmer-Christiansen (1994) for example, quotes the following from a 1984 German federal government report on air quality:

> The principle of precaution commands that the damages done to the natural world . . . should be avoided in advance and in accordance with opportunity and possibility. [Precaution] further means the early detection of dangers to health and environment by comprehensive, synchronized . . . research. . . . [I]t also *means acting when conclusively ascertained understandings by science is not yet available* (emphasis added).

Norton's admonition, namely that meta-concepts must be brought to the heel of pragmatic guidelines, codes of practice, and organizing principles for regulation and valuation, is most apt. The difficulty facing the adherents of precaution is that there is no agreement over how serious the predicament is. At the root of this dilemma lies contrasting positions on the robustness of natural systems to withstand shock, the seemingly bountiful adaptiveness of human societies to cope with change of whatever kind, and the apparently inherent unwillingness to attach much importance to what may or may not happen beyond one's lifetime.

Actual proof of this shift of emphasis remains elusive. The well-known techniques of risk analysis and environmental impact assessment are supposed to convey this role. In practice this is not the case, because few policy-analytical arrangements have incorporated the conditions of "adaptive environmental assessment" associated with Holling and his colleagues (1976) or of participatory (or "civic") science promoted by Lee (1993). Both those contributions bemoan the lack of an adequate institutional arrangement to disentangle the unknown and instead advocate cautious experimentation, seeking to engender interest group support and participation from the bottom up. Both also caution against the use of evaluative procedures that do not explicitly take into account the transitivity and surprise element of ecological and social change in the face of abrupt adjustments.

At this point, we note two common themes that are central to the implementation of the Precautionary Principle:

- A *notion of planetary and social care* that is located in the willingness to accept a measure of ignorance, humility in the face of uncertainty, and the recognition of due democratic procedures based on inclusion, trust, and shared outcomes; and
- *Joint intergenerational and international action* where implementation requires cooperation across nation states and a mutual recognition of obligation to the interests of future generations.

These are fundamental principles of ecocentrism that are sufficiently part of the mainstream to be incorporated into technocentrism at the highest level of cultural acceptance (see O'Riordan, 1996). The steady incursion of environmental and sustainability principles into regulation, business ethics, and management and into decision-making procedures during the latter part of the 1990s, notably in the industrialized nations, is indicative of the degree in which these principles are seeping into the pores of industry worldwide. However, because of globalization of trade and environment responsibilities, nowadays the developing world is no longer immune to such pressures and priorities. It is for these reasons that we feel precaution is coming of age. We may not be able to define it—indeed, it is possibly indefinable at the level sought by some economists and scientists—but its presence is being felt.

Conclusion

The Precautionary Principle is vague enough to be acknowledged by all governments regardless of how well they protect the environment. But the politics of precaution are also powerful and progressive, since they offer a profound critique of the many ways in which the environmental policy is currently determined. Wrapped up in the debate about precaution are forceful new ideas that point the way to a more preventative, source-based, integrated, and biocentric basis for policy. The point about the Precautionary Principle is that it swims against the economic, scientific, and democratic tides. It requires some sort of sacrifice by anyone who cannot see the justification of taking careful avoidance. As we have repeatedly stressed, the strength of the Precautionary Principle lies in beliefs about social or environmental resilience and in the capacity of social groups or political systems to respond to crises. Therefore those who support the notion of resilience and accommodation/adaptation would require precautionary sacrifice as a higher level of cost than those who are more ecocentric on such matters.

The emphasis that governments continue to place on sound science and careful cost-benefit analysis suggests a deep-seated suspicion of the threat that the principle appears to hold to economic growth and rational policy making. Precaution will not explode onto the environmental stage, sweeping away all forms of risk or cost-benefit assessment, careful scientific analysis, and existing legal norms relating to the relative power of polluters and victims. Rather, it will seep through the pores of decision-making institutions and the political consciousness of humanity by stealth. It will do this when, and if, it has the tide of the times behind it.

REFERENCES

Beck, U. 1992. *Risk Society*. Sage, London.

———. 1991. Scientific Uncertainty and the Precautionary Principle. *Environment*, 33(7): 4–5 and 43–45.

Bodansky, D. 1994. The Precautionary Principle: The US experience. In: T. O'Riordan and J. Cameron (eds.) *Interpreting the Precautionary Principle*. Cameron and May: London.

Boehmer-Christiansen, S. 1994. The Precautionary Principle in Germany: Enabling Government. In: T. O'Riordan and J. Cameron (eds.) *Interpreting the Precautionary Principle*. Cameron and May, London, p. 3.

Boulding, K., 1966. The Economics of the Coming Spaceship Earth. In: H. Jarrett (ed.) *Environmental Quality in a Growing Economy*. John Hopkins University, Baltimore.

Cameron, J. 1994. The Status of the Precautionary Principle in International Law. In: J. Cameron and T. O'Riordan (eds.) *Interpreting the Precautionary Principle*. Earthscan, London.

Cameron, J., and J. Abouchar, 1991. The Precautionary Principle. *Boston College International and Comparative Law Review* 14(1): 1–27.

Cameron, J., and W. Wade-Gery, 1991. *Addressing Uncertainty: Law, Policy and the Development of the Precautionary Principle*. CSERGE Working Paper GEC 92-43. CSERGE, UCL and UEA.

Crowards, T. 1997. Non-Use Values and the Environment. *Environmental Values* 6(2): 143–168.

Dobson, A., 1990. *Green Political Thought*. HarperCollins: London.

Environmental Data Services (ENDS) Daily, 1998a. Austria to Oppose Genetically Modified Tomato. *ENDS Daily*, 10 February 1998, 1–2.

———. 1998b. EU Environmental Liability Plan Emerges. *ENDS Daily*, 28 January 1998.

Freestone, D., 1991. The Precautionary Principle. In: D. Freestone, and R. Churchill (eds.) *International Law and Global Climate Change*. Graham and Trotman, London.

GESAMP, 1986. *Environmental Capacity: An Approach to Marine Pollution Prevention*. FAO Report No. 30. FAO, New York.

Gray, J.S., 1990. Statistics and the Precautionary Principle. *Marine Pollution Bulletin* 21(4): 174–176.

———. 1993. A Scientist's Perspective. In: R. Earll (ed.) *The Precautionary Principle: Making it Work in Practice*. Environment Council, London.

Hajer, M., 1995. *The Politics of Environmental Discourse*. Clarendon, Oxford.

HM. Govt, 1990. *This Common Inheritance: Britain's Environmental Strategy*. Cmnd 1200. HMSO, London.

Holling, C.S., 1976. *Adaptive Environmental Planning and Management*. IIASA, Luxembourg.

Horsman, P., 1992. Reduce It, Don't Produce It: The Real Way Forward. In: T. O'Riordan and V. Bowers (eds.) *IPC: A Practical Guide for Managers*. IBC Technical Services, London.

Jacobs, M., 1991. *The Green Economy*. Pluto Press, London.

Johnston, P., and M. Simmonds, 1990. Precautionary Principle. *Marine Pollution Bulletin* 21(12): 402.

Jordan, A. 1998a. 'Private Affluence and Public Squalor?' The Europeanisation of British Coastal Bathing Water. *Policy and Politics* 26(1): 33–54.

———. 1998b. Environmental Policy at 25: The Politics of Multinational Governance. *Environment* 40(1): 14–20 and 39–45.

Jordan, A.J., and T. O'Riordan, 1993. Implementing Sustainable Development: The Political and Institutional Challenge. In: D. Pearce et al. (eds.) *Blueprint Three*. Earthscan, London.

———. 1997. *Social Institutions and Climate Change*. CSERGE Working Paper GEC 97-15. CSERGE, Norwich and London, UK.

Lee, K., 1993. *Compass and Gyroscope: The Role of Science in Environmental Policy Making*. Island Press, Washington, D.C.

McGarvin, G., 1994. The Implications of the Precautionary Principle for Biological Monitoring. In: T. O'Riordan and J. Cameron (eds.) *Interpreting the Precautionary Principle*. Cameron and May, London.

NAVF, 1990. *Sustainable Development, Science and Policy: The Conference Report*. N.A.V.F., Oslo.

Norton, B., 1992. Sustainability, Human Welfare and Ecosystem Health. *Environmental Values* 1: 97–112.

OECD, 1992. *OECD Environmental Performance Reviews: Germany*. OECD: Paris.

O'Riordan, T., 1996. Framework for Choice: Core Beliefs and the Environment. *Environment* 37(8): 4–9 and 25–29.

O'Riordan, T., and H. Voisey, 1998. *The Transition to Sustainability*. Earthscan, London.

O'Riordan, T., C. Marris, and I. Langford, 1997. Images of Science Underlying Public Perceptions of Risk. In: Royal Society (ed.) *Science, Policy and Risk*. Royal Society, London.

O'Riordan, T., A. Jordan, and H. Voisey, 1998. Preparing the Ground for Sustainable Development. In: N. Vig and R. Axelrod (eds.) *International Environmental Law, Institutions and Policies*. Congressional Quarterly Press, Washington, D.C.

Pearce, D. 1994. The Precautionary Principle in Economic Analysis. In: J. Cameron and T. O'Riordan (eds.) *Interpreting the Precautionary Principle*. Cameron and May, London.

Redclift, M., 1993. Sustainable Development: Needs, Values, Rights. *Environmental Values* 2: 3–20.

Stebbing, A., 1992. Environmental Capacity and the Precautionary Principle. *Marine Pollution Bulletin* 24(6): 277–295.

Stone, C., 1987. *The Earth and Other Ethics: The Case for Moral Pluralism*. Harper and Row, New York.

Vogel, D., 1997. *Barriers or Benefits? Regulation in Transatlantic Trade*. Brookings Institution Press, Washington, D.C.

von Moltke, K., 1988. The Vorsorgeprinzip in West German Environmental Policy. In: Royal Commission On Environmental Pollution, *Twelfth Report: Best Practicable Environmental Option*. Cmnd 310. HMSO, London.

Weale, A., 1992. *The New Politics of Pollution*. Manchester University Press: London.

———. 1993. Ecological Modernisation and the Integration of European Environmental Policy. In J.D. Liefferink et al. (eds.) *European Integration and Environmental Policy*. Belhaven Press, London.

———. (ed.), 1998. *European Environmental Governance*. Routledge, London.

Weinberg, A.M., 1985. Science and Its Limits. *Issues in Science and Technology* 2: 59–72.

Wynne, B., and S. Meyer, 1993. How Science Fails the Environment. *New Scientist*, 5 June, 33–35.

NOTES

1. This is consistent with the practice of setting a "'safe minimum standard' to protect important features of the environment" (Crowards, 1997).

2. Boulding (1966) used the term "cowboy" to characterize the human perception of the natural environment as an "open" resource and an infinite sink—a limitless domain to be conquered and exploited.

Chapter 2

✎

THE PRECAUTIONARY PRINCIPLE IN PRACTICE: A MANDATE FOR ANTICIPATORY PREVENTATIVE ACTION

David Santillo, Paul Johnston, and Ruth Stringer

Our ability to protect our environment and to ensure that our exploitation of ecosystems is sustainable depends on our capacity to identify environmental threats and to prevent serious or irreversible harm before it occurs. At the same time, we must recognize that ecosystems are complex entities that can neither be defined explicitly nor described fully, other than through creation of artificial boundaries. This leads effectively to the exclusion of some elements of the ecosystem from further consideration; that is, such elements are "externalized" and are not amenable to description or quantification. Berg and Scheringer (1994) derive the term "overcomplexity" to describe the manner in which the spatial relationships and evolution of ecosystems over time are both unpredictable and impossible to examine to a degree sufficient to reveal their detail and derivative properties. Furthermore, characteristics of undamaged ecosystems (reference conditions) that are meaningful and representative, including variability in time and space and against which ecosystem stresses and damage can be gauged, are likely to be impossible to define.

Scientific research undoubtedly has the ability to improve our understanding of ecosystems, particularly the relationships between organisms and their environment and pathways of energy, nutrient, and contaminant flows.

This may, in turn, lead to a reduction in the level of those uncertainties that have been identified. Nevertheless, all scientific determinations are bound by the largely arbitrary constraints of experimental design and the need to control, as far as possible, all variables that lie outside the study boundaries. Further analysis of a particular system can never address those properties about which we remain ignorant, other than if such properties are identified by chance and are then amenable to analysis. Moreover, natural systems are characterized by complex processes and networks of interaction. It is generally not possible to define precisely where, or even if, chains of cause–effect relationships begin and end. Substantial irreducible uncertainties, or indeterminacies (Dovers and Handmer 1995; Wynne 1992), will always remain as barriers to comprehensive descriptions and predictions of ecosystem function (see figure 2.1).

The Regulator's Dilemma

It is this background against which regulatory decisions are required in order to avoid systematic environmental degradation. With regard to the limitation of damage to ecosystems and the maintenance of their viability through protection or sustainable exploitation, it has long been recognized that "prevention is better than cure" (Bodansky 1991). Clearly the deferral of action until such time as a potentially impacted system can be fully described and the consequences of a particular stressor can be reliably quantified or predicted is not a responsible option, first because such analytical certainty will never be achieved and second because serious or irreversible damage may result in the meantime. Research frequently demonstrates that ecosystems possess greater complexities and are harder to define and predict than previously thought. Overall, while much research funded as part of ongoing regulatory processes is invaluable, it is rarely able to resolve the dilemmas that lead to its commissioning. Environmental regulators are frequently presented with the need to take action to prevent, or avoid the potential for, damage to the environment or human health in the face of considerable uncertainty, an unquantifiable degree of ignorance, and inherent indeterminacies, a situation Bodansky (1991) terms the "regulator's dilemma."

Numerous approaches have emerged in attempts to resolve, or avoid, this dilemma. Approaches based on the assessment and management of risk have perhaps received the greatest attention in recent years, relying essentially on the application of techniques developed in engineering sciences to the forecasting of trends and impacts in more complex and poorly defined natural systems. Risk-based approaches extend from the view that environmental risks can be quantified and managed at sustainable and "acceptable" levels,

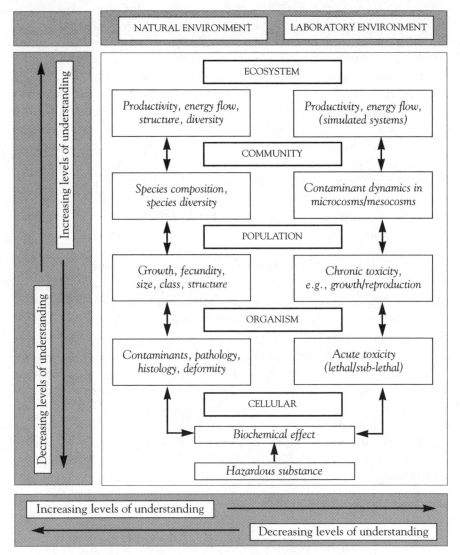

FIGURE 2.1. Diagramatic representation of the relationship between
the level of biological organization of a system, both natural and artificial,
and its amenability to description and understanding.

either in absolute terms or relative to the benefits accrued. In broad terms,
risks are determined from a combination of the intrinsic hazards presented
by a chemical or activity and measurements or estimates of exposure to that
agent. Such approaches assume that it is possible to know enough about the
hazards of, and exposure to, a particular chemical or activity to enable calculation of the risks in a reliable manner.

In practice, risk assessments tend to employ simplistic and subjective assumptions about ecosystem structure and the flows of energy and matter that characterize them. Frequently they lack the breadth and quality of data necessary to facilitate prediction of impacts. Uncertainties and indeterminacies arising from ecosystem complexity are rarely made explicit. Failure to recognize and account for such unknowns can have severe consequences.

It is often assumed that "sound science" will ultimately provide the central basis for environmental legislation. Clearly there is an ongoing need for pure and applied knowledge to facilitate improved understanding of ecosystem function and contaminant behavior (Ducrotoy and Elliot, 1997), but such knowledge will always be incomplete and, in itself, can form only part of responsible policy and management systems. Scientific analyses and deductions undoubtedly serve policy makers with valuable information, and this can be used to identify hazards and to prioritize and guide decisions of a more precautionary nature (Funtowicz and Ravetz, 1994). Fundamentally, however, such assessments cannot replace the decision-making process itself (Power and McCarty 1997; Wynne 1992).

THE NEED FOR PRECAUTION

Recognition of existing and, indeed, inherent limitations to scientific knowledge, coupled with the ongoing necessity to take action, wherever possible, to prevent damage before it has occurred, even in the absence of proven causality, led to the formulation of an alternative approach to environmental protection, which is fundamentally precautionary in nature. The Precautionary Principle, or the Principle of Precautionary Action, undoubtedly has its origins in the German federal government's approach to environmental protection, the *Vosorgeprinzip*, developed in the early 1980s (DoE, 1995; FRG, 1986; McIntyre and Mosedale, 1997). This guiding principle had at its core the recognition that, in order to meet the responsibility of protecting the natural foundations of life for future generations, irreversible damage must be prevented.

The Precautionary Principle not only permits action to be taken in the absence of conclusive evidence of cause–effect relationships, but also stresses that action in anticipation of harm is essential to ensure that it does not occur (Bodansky, 1991). The adoption of such an approach implies a shift in emphasis in favor of a bias toward safety (McIntyre and Mosedale, 1997), ensuring that any errors of judgment made will lead to excess, rather than inadequate, protection.

The principle has perhaps gained its highest profile within the field of marine environmental protection, especially with respect to inputs of hazardous chemicals, although it is equally applicable to other fields of environmental legislation. Ultimately it has application at the science–policy interface not only in relation to the control of the release of harmful chemicals to environmental media, but also to the release of biological agents (including genetically modified organisms), the management of resource exploitation (e.g., fisheries), and, indeed, any field in which human activity might have substantial, far-reaching, or even irreversible impacts.

INCORPORATION OF THE PRECAUTIONARY PRINCIPLE INTO INTERNATIONAL ENVIRONMENTAL LAW

Since its initial formulation, two trends have been apparent in the development of the principle. First, the principle has gained increasing acceptance as a fundamental guiding paradigm within national and international legislative frameworks. It now forms a fundamental component of numerous legislative agreements on the protection of the environment, as discussed below. At the same time, however, the principle has veered away from a strong mandate for precautionary action toward a universal sentiment with little guidance on practical implementation. The principle has been defined in a number of different ways and is increasingly cited without explicit definition. As a result, there is a danger that the initial intentions of the principle may become increasingly diluted or effectively lost during the implementation of some of the legislation in which it is incorporated. There is, therefore, an urgent need to look again at the principle of precautionary action and what it means in terms of practical application.

While the earliest origins of the *Vosorgeprinzip* remain unclear (DoE, 1995; Gray and Bewers, 1996), its most complete definition is probably that given in a German federal government report on the protection of air quality (DoE, 1995). This definition essentially comprises four elements, that damage should, as a priority, be avoided; that scientific research plays an essential role in identifying threats; that action to prevent harm is essential, even in absence of conclusive evidence of causality; and that all technological developments should meet the requirement for progressive reduction of environmental burden. Of these elements, the requirement for action in the absence of analytical or predictive certainty has become the most widely used condensation of the principle.

For example, the Ministerial Declaration, which arose from the First International Conference on the Protection of the North Sea, held in Bremen in 1984, incorporated the statement that North Sea States must not wait for proof of harmful effects before taking action. This commitment was made more explicit in the Ministerial Declarations from the 1987 and 1990

North Sea conferences (MINDEC, 1987, 1990). Although by no means its earliest use, the explicit inclusion of elements of the principle within the 1987 North Sea Ministerial Declaration (MINDEC, 1987) represented a highly significant endorsement, which, undoubtedly, subsequently facilitated its adoption by other regional seas and global marine fora. Notable among these are the Oslo and Paris Commissions (protecting the North East Atlantic) (OSPAR, 1992, 1998), Barcelona Convention (Mediterranean) (UNEP, 1996), and the London Convention on dumping of wastes at sea (LC, 1972, 1996).

It has also gained wider application in legislation designed to prevent environmental degradation, notably the 1987 Montreal Protocol regulating ozone depleting substances, the 1992 Climate Change Convention, the Rio Declaration on Environment and Development (1992), and the 1995 United Nations Agreement on High Seas Fishing (UN, 1995). McIntyre and Mosedale (1997) provided a more comprehensive review.

TOWARD IMPLEMENTATION OF THE PRECAUTIONARY PRINCIPLE

Although the principle has been widely recognized and incorporated into international treaties and conventions, its implementation has been more limited. One of the prerequisites for effective implementation, following agreement and ratification of a particular treaty, is the definitive interpretation of the principle in terms of practical measures. Without such interpretation, the principle would likely remain as a token theoretical ideal that may be acknowledged and subsequently ignored. The degree to which the principle has been interpreted in relation to programs and measures varies greatly between agreements. Perhaps the most transparent and definitive commitments to implementation relate to international agreements for the protection of the North Sea and the North East Atlantic maritime area.

North Sea Ministerial Declarations

The Ministerial Declaration arising from the Second North Sea Conference (MINDEC, 1987) recognized that, in order to safeguard the marine ecosystem, it would be necessary to reduce polluting emissions. Particular focus was placed on substances that possessed the hazard marker properties of persistence, toxicity, and liability to bioaccumulate:

> especially when there is reason to assume that certain damage or harmful effects on the living resources of the sea are likely to be caused by such substances, even when there is no scientific evidence to prove a causal link between emissions and effects (the principle of precautionary action) (MINDEC, 1987).

This interpretation was further strengthened by the Third and Fourth Ministerial Declarations (MINDEC, 1990, 1995), the latter committing to continuous reduction of such inputs with a target of their cessation within one generation. In the sense that this commitment addresses a very broad group of chemicals on the basis of their inherent hazardous properties, without the requirement for individual assessment of causality, it is truly precautionary in nature. Precisely how effectively such measures will translate to precautionary action remains to be seen.

Oslo and Paris Commissions and the OSPAR Convention

The 1992 Convention for the Protection of the Marine Environment of the North East Atlantic (OSPAR, 1992) adopted similar provisions, although the specific timeline for implementation (the "one generation goal") was only adopted at the OSPAR Ministerial meeting in Sintra, Portugal, in July 1998 (OSPAR, 1998). Again, the commitment has been made to make every endeavor to achieve "zero discharge" (cf. discharges, emissions, and losses) for all hazardous substances to the OSPAR maritime region by 2020, with the aim of achieving concentrations close to zero in the environment for synthetic substances and close to background for naturally occurring substances. The convention participants envisioned that implementation would be a staged process, with action on a priority list of chemicals of particular concern within a more limited timeframe. It is also acknowledged that the involvement of industry and other international organizations will form a vital part of effective implementation programs to meet the stated objectives (OSPAR, 1998).

The London Convention and Its 1996 Protocol

The language of the London Convention (1972), a global forum under the auspices of the International Maritime Organization (IMO) that regulates the dumping of wastes at sea, includes a general requirement to take all practicable steps to prevent pollution from dumping operations. The 1996 Protocol to the Convention (LC, 1996) makes explicit reference to the precautionary approach as a general obligation (Article 3), with a practical, though generic, requirement for preventative measures to be taken whenever harm is likely to result from the dumping of wastes, even when conclusive evidence of causality is unavailable. Although such preventative measures are not elaborated further under Article 3, the obligation is also made that action does not simply shift the potential for pollution from one environmental compartment to another.

The current prohibitions on the dumping of both radioactive and industrial waste by contracting parties, resulting from amendments in 1993 to the 1972 convention, could be viewed as essentially precautionary in intent because, again, action is required in the absence of specific assessments of cause and effect. The 1996 protocol, effectively even more restrictive, adopts a reverse list approach within which only a limited number of specified types of waste may be considered for dumping, subject to detailed waste characterization and assessment. For each of the listed wastes (currently including dredge material, fish offal, sewage sludge, inert geological material, vessels, human-made structures at sea, and bulky items comprising iron and steel by small island nations), characterization of the nature and content of the waste is required according to generic guidelines (LC, 1997). Specific guidances for each waste category currently remain under development.

Although such evaluations are essentially case-by-case assessments of hazard and likely impact, the generic guidelines stress the importance of the recognition and consideration of uncertainties in the prediction of impacts in a precautionary manner. One direct interpretation of a precautionary approach in this regard is the provision that

> if a waste is so poorly characterized that proper assessment cannot be made of its potential impacts on human health and the environment, that waste shall not be dumped.

The 1996 protocol remains open for signature. Early ratification is clearly essential in order to move the convention to a more precautionary basis and would be highly significant with respect to the global nature of the treaty.

The Rio Declaration on Environment and Development

The Rio Declaration (1992) provided an opportunity for the Precautionary Principle to gain wider currency in international agreements and national legislation. Nevertheless, the definition of the principle within the declaration is relatively vague, giving little indication as to how it should be applied in practice. The declaration reads:

> Where there are threats of serious or irreversible damage, lack of full scientific certainty shall not be used as a reason for postponing cost-effective measures to prevent environmental degradation (RioDEC, 1992).

This results in part, of course, from the need to reach a consensus declaration among a diverse group of countries. Moreover, the Rio definition introduces to the principle the provision that measures should be cost-effective, which could imply that precautionary action might be subject to cost-benefit analysis; that is, an additional evaluation step in which the advantages and disadvantages of such action are weighed up, primarily in financial terms. In simple terms, it is feasible that if the financial costs of taking necessary precautionary action are deemed too great, such action may be ruled out. This concern is by no means unwarranted. Indeed, the 1994 U.K. strategy for sustainable development uses a definition of the principle on the basis of the Rio definition, making explicit the proviso that precautionary measures should be taken only if

. . . the likely balance of costs and benefits justifies it (DoE, 1995).

Such conditions are becoming more prevalent in the various definitions and interpretations of the precautionary approach and threaten to undermine the fundamental purpose for which the principle itself was developed by subsuming it simply as one "tool" within a risk-based approach. This is discussed further below and, more extensively, by Santillo et al. (1998).

The EC Treaty 1993: Establishing the European Community

While historically and still primarily an economic entity, the EC Treaty (EC 1993) summarizes the European Community approach to the environment under Article 130. Interestingly, while the treaty bases environmental legislation on the Precautionary Principle, it sees the Principle of Preventative Action as an additional and separate obligation. No further definitions are given, and it remains unclear precisely how the Precautionary Principle would be implemented other than through effective, preventative action. Moreover, in practice there would appear to be little room for truly precautionary measures to be implemented within existing directives on environmental issues. Cost-benefit analysis again plays a central role, as does the overriding commitment to the economic and social development of the community and its regions and the maintenance of the single European market. In the regulation of chemicals, the community has clearly adopted a risk-based approach, and the absence of precaution is particularly apparent with respect to current permissive legislation on nonassessed chemicals. The processes by which the production, marketing, and use of chemicals are regulated in the EU and in several member states are currently undergoing review.

United Nations Agreement on High-Seas Fishing

Recognition of the failure of existing management initiatives to ensure sustainable exploitation in numerous fisheries around the world (e.g., North

Sea—Cook et al., 1997; Serchuk et al., 1996—Canadian cod stocks; Myers and Mertz, 1998) initiated the development of a more precautionary approach to the exploitation of fish stocks (Stephenson and Lane, 1995). Such approaches differ fundamentally from those aimed at controlling or eliminating pollution, as those relating to fisheries work from the assumption that some level of continued exploitation is ultimately sustainable. They also differ in a number of fundamental ways from traditional approaches to fishery management, primarily in their recognition of the underlying importance of species conservation and the mechanisms they incorporate to address uncertainties and indeterminacies (Dayton, 1998).

The 1995 United Nations Agreement on the Conservation and Management of Straddling Stocks and Highly Migratory Fish Stocks (UN, 1995) includes specific reference to the need to ensure long-term sustainability through adoption of a precautionary approach. The agreement goes on to describe, in general terms at least, how such an approach would be applied, including improved science-based decision making, development of techniques to address and account for uncertainties in stock size and productivity, and the implementation of methods to reduce by-catch of nontarget species (i.e., unwanted catch of other organisms, including seabirds and sharks, resulting from fishing efforts directed at the target species).

The agreement has yet to be ratified and implemented, despite the urgency imposed by the very poor state of many fisheries on a global basis. Meanwhile, the development of thinking within the precautionary approach to fisheries continues. Recently, Myers and Mertz (1998) suggested allowing each fish cohort, or year group, to spawn at least once before they are subject to commercial fishing as an additional practical and more precautionary measure to safeguard long-term sustainability. Nevertheless, the interpretation of the principles in this field still lag considerably behind their application to the control of chemical contaminants.

SCIENCE AND THE PRECAUTIONARY PRINCIPLE

It has been argued (Bewers, 1995; Gray, 1990) that the Precautionary Principle is, in essence, unscientific as it promotes preventative action even in the absence of proof of causality. Risk-based approaches are commonly presented as the science-based alternative. Such views do not appear to recognize that the principle is founded on the use of comprehensive, co-ordinated research in order to guide precautionary action. The fundamental difference between risk and precautionary approaches is not that one uses science while the other does not, but simply the way in which scientific evidence is employed for decision making at the science–policy interface. The precautionary approach is, to a degree, less prescriptive in its evaluation of the need for action, in that it does not rely on a need explicitly to define and quantify risks, but rather on the more general application of scientific research as a means for the early detection of dangers to human or wildlife health or to the environment as a whole. The commitments within the North Sea and OSPAR processes, for example, clearly have a firm basis in science, as they rely on scientific research to identify those properties and, thereby, sub-stances or groups of substances that are of concern. Nevertheless, in the requirement to address all substances with those properties, the legislation is also clearly precautionary in nature. It is in this manner that science can continue to play a central role in the formulation and implementation of effective environmental legislation without the need for a risk-based approach.

The Precautionary Principle is, in its own right, a crucial scientific tool to mitigate threats to the environment (Johnston and Simmonds, 1990, 1991). Clearly it is not intended as a substitute for a scientific approach but rather as an overarching principle to guide decision making in the absence of ana-lytical or predictive certainty. It provides a mechanism to compensate for inherent uncertainty and indeterminacy in natural systems and a central paradigm for responsible, timely, and definitive preventative action.

Gray and Bewers (1996) suggested that, in the context of the North Sea Ministerial agreements, the Precautionary Principle should be implemented through the employment of pessimistic assumptions in standard risk-assess-ment procedures. Such an approach captures neither the spirit nor the pro-visions intended for the principle and threatens to undermine its utility by subjecting it to the self-same limitations of risk assessment and management procedures. Their arguments are challenged in more detail by Santillo et al. (1998).

In short, the Precautionary Principle cannot and should not be subsumed under a risk assessment mechanism, as is also currently implied within guid-ance for risk assessment and management in the United Kingdom (DoE, 1995), to be invoked only when an risk assessment is judged to have failed (Brown, 1998). Neither should risk assessment be seen as a means of imple-menting the Precautionary Principle, as one tool in the full suite of risk assessment methodologies. Contrary to Brown (1998), the Precautionary Principle should operate at all times in recognition of the fact that assess-ments of hazards, exposure, and risk, despite their apparent objectivity, can never alone ensure an adequate level of environmental protection. Indeed, if the principle is not operational at all times, its effectiveness is greatly diminished.

IMPLEMENTATION INTO THE FUTURE

If the Precautionary Principle is to act as a truly effective means of ensuring that serious and irreversible environmental damage is avoided, continued development of its interpretation and implementation within international treaties and conventions is essential, in addition, of course, to the timely ratification of the treaties themselves. As noted by McIntyre and Mosedale (1997), the principle is now very much a norm of customary international environmental law, but it is essential that its incorporation as such results in more than "lip service" to precautionary measures. As one example, the obligation for North East Atlantic States to meet the "one generation goal" with respect to discharges, emissions, and losses of hazardous substances will require urgent implementation of precautionary action.

In order to facilitate the translation of the principle from theory to practice, it may be necessary to revisit and reaffirm the necessity for, and initial intentions of, the precautionary paradigm (see also Santillo et al. 1998). The Precautionary Principle or, more definitively, the Principle of Precautionary Action, could be defined in terms of the four elements below, based largely on the early formulations of the principle in German federal law (FRG, 1986).

Implementation of the Principle of Precautionary Action demands that:

1. Serious or irreversible damage to ecosystems must be avoided in advance, both by preventing harm and by avoiding the potential for harm;
2. High-quality scientific research is employed as a key mechanism for the early detection of actual or potential impacts;
3. Action to protect ecosystems is *necessary*, not simply possible, even in the presence of uncertainty, ignorance, and irreducible indeterminacy; and
4. All future technical, social, and economic developments implement a progressive reduction in environmental burden.

Such an interpretation would reaffirm the principle as a mandate for anticipatory action of a preventative nature.

Bodansky (1991) argued that the choice faced by environmental regulators will always be between one risk and another and that the precautionary withdrawal of one process may simply lead to the transfer of the problem to other media. In this regard, it is essential to recognize that the Precautionary Principle is not intended to be applied in a simple one-sided approach to decision making, without consideration of the potential hazards of alternatives. For action to be truly precautionary, it must also ensure that the fundamental objective of the reduction of overall environmental burden is strictly observed. If this requirement is not observed, then the goal of environmental legislation would not be guaranteed.

In order to meet this objective, it must be recognized that a decision, for example, to prevent the use or discharge of a certain chemical may require a fundamental reevaluation of societal need for that product and may not always imply simple substitution with an alternative. For example, if the commitments within the North Sea Ministerial process (MINDEC, 1995) and, more recently, under OSPAR (1998) are to be met, particularly to achieve zero discharge, emissions, and losses of hazardous substances to the North Sea and North East Atlantic regions respectively, changes to industrial practice, process, and even products will undoubtedly be necessary.

CONCLUSION

During the ten to fifteen years since its early formulation and development, the Precautionary Principle has, therefore, been progressively incorporated as a guiding paradigm in treaties and conventions designed to protect the environment. At the same time, however, there has been relatively little focus on the development of mechanisms by which precaution may be effectively implemented.

The initial development of the Precautionary Principle stemmed from the necessity for a mechanism to address uncertainties and limitations to scientific knowledge at the science–policy interface. It is of fundamental importance, therefore, that the principle should not be weakened to a point at which it is seen merely as an ideal to be noted but ignored. Furthermore, attempts to make the principle subject to cost-benefit analysis or to reduce its status to one of a suite of tools within risk-based approaches must be strongly resisted. Such changes threaten to prevent the principle from serving the essential role for which it was designed.

The North Sea Ministerial process and the OSPAR Convention provide some of the strongest bases yet for the practical implementation of the Precautionary Principle as it applies to the protection of the marine environment, particularly with regard to hazardous substances. It is now essential to ensure that the principle is strictly observed during the further development of these agreements, and particularly during the development and application of practical programs and measures to address threats to the marine environment. Moreover, it is important that similar provisions are extended to provide similar levels of protection to other compartments of the environment. It is only through adopting mechanisms that enable and, indeed, require precautionary action that we will be capable of ensuring that environmental damage and threats to human health can, wherever possible, be avoided in advance.

REFERENCES

Berg, M., and M. Scheringer, (1994). Problems in Environmental Risk Assessment and the Need for Proxy Measures. *Fresenius Environmental Bulletin* 3: 487–492

Bewers, J.M. (1995). The Declining Influence of Science on Marine Environmental Policy. *Chemistry and Ecology* 10: 9–23.

Bodansky, D. (1991). Scientific uncertainty and the Precautionary Principle. *Environment* 33(7): 4–5 and 43–44

Brown, D. (1998). Environmental Risk Assessment and Management of Chemicals. In: *Risk Assessment and Risk Management*, R.E. Hester and R.M. Harrison (eds.) *Issues in Environmental Science and Technology* 9, The Royal Society of Chemistry, Cambridge, UK, pp. 91–111.

Cook, R.M., A. Sinclair, and G. Stefansson (1997). Potential Collapse of North Sea Cod Stocks. *Nature* 385: 521–522.

Dayton, P.K. (1998). Reversal of Burden of Proof in Fisheries Management. *Science* 279: 821–822.

DoE (1995). *A Guide to Risk Assessment and Risk Management for Environmental Protection*. Department of the Environment, UK, 92 pp.

Dovers, S.R., and J.W. Handmer (1995). Ignorance, the Precautionary Principle and Sustainability. *Ambio* 24(2): 92–97.

Ducrotoy, J.-P., and M. Elliott (1997). Interrelations Between Science and Policy-Making: The North Sea Example. *Marine Pollution Bulletin* 34: 686–701.

EC (1993). Treaty Establishing the European Community, Article 130: 297–299.

FRG (1986). Umweltpolitik: *Guidelines on Anticipatory Environmental Protection*. Federal Ministry for the Environment, Nature Conservation and Nuclear Safety, 43 pp.

Funtowicz, S.O., and J.R. Ravetz (1994). Uncertainty, Complexity and Post-Normal Science. *Environmental Toxicology and Chemistry* 13(12): 1881–1885.

Gray, J.S., (1990). Statistics and the Precautionary Principle. *Marine Pollution Bulletin* 21, 174–176.

Gray, J.S., and J.M. Bewers (1996). Towards a Scientific Definition of the Precautionary Principle. *Marine Pollution Bulletin* 32(11): 768–771.

Johnston, P. and M. Simmonds (1990). Precautionary Principle (Letter). *Marine Pollution Bulletin* 21: 402.

———. (1991). Green Light for Precautionary Science. *New Scientist*, 3 August 1991, p. 4.

Johnston, P.A., D. Santillo, and R.L. Stringer (1996). Risk Assessment and Reality: Recognizing the Limitations. In: *Environmental Impact of Chemicals: Assessment and Control*. M. Quint, R. Purchase, and D. Taylor (eds.) Royal Society of Chemistry, Cambridge, pp. 223–239.

LC (1972). Convention on the Prevention of Marine Pollution by Dumping of Wastes and Other Matter, 1972, The London Convention, LC.2/Circ.380, IMO.

———. (1996). 1996 Protocol to the Convention on the Prevention of Marine Pollution by Dumping of Wastes and Other Matter, LC.2/Circ.380, IMO.

———. (1997). Guidelines for the Assessment of Wastes of Other Matter That May be Considered for Dumping, LC 19/10, Annex 2., IMO.

McIntyre, O., and T. Mosedale (1997). The Precautionary Principle as a Norm of Customary International Law. *Journal of Environmental Law* 9(2): 221–241

MINDEC (1987). Ministerial Declaration of the Second International Conference on the Protection of the North Sea. 24–25 November 1987, London, UK.

———. (1990). Ministerial Declaration of the Third International Conference on the Protection of the North Sea. 7–8 March 1990, The Hague, Netherlands.

———. (1995). Ministerial Declaration of the Fourth International Conference on the Protection of the North Sea. 8–9 June 1995, Esbjerg, Denmark.

Myers, R.A., and G. Mertz (1998). The Limits of Exploitation: A Precautionary Approach. *Ecological Applications* 8(1) Suppl.: S165–S169.

OSPAR. (1992). Final Declaration of the Ministerial Meetings of the Oslo and Paris Commissions. Oslo and Paris Conventions for the Prevention of Marine Pollution, Paris, 21–22 September 1992.

———. (1998). The Sintra Statement, Final Declaration of the Ministerial Meeting of the Oslo and Paris Commissions. Oslo and Paris Conventions for the Prevention of Marine Pollution, Sintra, 20–24 July 1998.

Power, M., and L.S. McCarty (1997). Fallacies in Ecological Risk Assessment Practices. *Environmental Science and Technology* 31: 370A–375A.

RioDEC (1992). Rio Declaration on Environment and Development, ISBN 9-21-100509-4, 1992.

Santillo, D., R.L. Stringer, P.A. Johnston, and J. Tickner (1998). The Precautionary Principle: Protecting Against Failures of Scientific Method and Risk Assessment. *Marine Pollution Bulletin*.

Serchuk, F.M., E. Kirkegaard, and N. Daan (1996). Status and Trends of Major Roundfish, Flatfish and Pelagic Fish Stocks in The North Sea: Thirty Year Overview. *ICES Journal of Marine Science* 53: 1130–1145.

Stephenson, R.L., and D.E. Lane (1995). Fisheries Management Science: A Plea for Conceptual Change. *Canadian Journal of Fisheries and Aquatic Science* 52(9): 2051–2056.

UN (1995). Agreement for the Implementation of the Provisions of the United Nations Convention on the Law of the Sea of 10 December 1982 Relating to the Conservation and Management of Straddling Fish Stocks and Highly Migratory Fish Stocks, The Earth Summit Agreement on High Seas Fishing. United Nations, New York, November 1995: 33 pp.

UNEP (1996). Final Act of the Conference of the Plenipotentiaries on the Ammendments to the Convention for the Protection of the Mediterranean Sea Against Pollution from Land-Based Sources. UNEP(OCA)/MED 7/4, United Nations Environment Programme, March 1996.

Wynne, B. (1992). Uncertainty and Environmental Learning: Reconceiving Science and Policy in the Preventative Paradigm. *Global Environmental Change*, June 1992: 111–127.

Chapter 3

~

THE PRECAUTIONARY
APPROACH TO CHEMICALS MANAGEMENT:
A SWEDISH PERSPECTIVE

Bo Wahlström

The Precautionary Principle has always held a strong position in the Nordic countries and Germany. The origins of the principle in environmental legislation may be traced back to the late 1960s. This chapter explores several examples of the application of the Precautionary Principle in Sweden, as well as examples when it should have been used. This chapter discusses the future of chemicals policy in Sweden as well as the concept of chemical sunsetting. The chapter concludes by noting the need for generic approaches toward unwanted chemicals as well as the need to pay attention to chemicals in products. Criteria such as persistence and bioaccumulation should be used to identify those chemicals that should be phased out in the future. In the long term, perspective global agreements on unwanted chemicals are also needed.

HISTORICAL BACKGROUND OF THE
PRECAUTIONARY PRINCIPLE IN SWEDEN

The written history of the Precautionary Principle in Swedish legislation can be traced back to the Government Commission on Environmental Management, established in 1969, which delivered its report to the Swedish government in the Spring 1972. On the basis of the report, the government proposed a new act, the Act on Products Hazardous to Man or the Environment (Parliament of Sweden, 1973). Article 5 in the act states:

> Anyone handling or importing products hazardous to man or the environment shall take such steps and otherwise observe such precautions as are needed to prevent or minimize damage to man or the environment. Particularly anyone manufacturing or importing such a product must carefully investigate the composition of the product and its properties from the perspective of health and environmental protection. The product shall be clearly labelled with data of importance from the point of view of protecting health and the environment.

This article was slightly reformulated by the Chemicals Commission who in 1984 proposed a new Act on Chemical Products, which became law in 1985 (Parliament of Sweden, 1985). In 1990 the Swedish government added the substitution principle to the article, which now reads:

> Anyone handling or importing a chemical product must take such steps and otherwise observe such precautions as are needed to prevent or minimize harm to man or the environment. *This includes avoiding chemical products for which less hazardous substitutes are available* (emphasis added).

In the present legislation, violation of government or agency regulations established with reference to Article 5 may be penalized. If there are no specific regulations on an issue, the general rule in the article applies. In such a case, to be penalized, there must be gross or conscious violations of the article.

The recent Swedish Chemical Policy Committee discussed the application of the Precautionary Principle in its report "Towards a Sustainable Chemicals Policy" (1997) but did not propose any changes in the existing legislation.

The Swedish Environmental Code Committee has proposed that the Precautionary Principle should be one of the guiding principles for the new Swedish Environmental Code, in which the Act on Chemical Products will be a chapter.

THE PRECAUTIONARY APPROACH IN SWEDISH LAW AND ADMINISTRATION

In the Swedish system of government, the preambles to Acts of Parliament constitute an important part of the acts. In the preambles, which form part of the Bill from the Government, principles and objectives of the law are explained and interpreted in detail. Both the Act on Products Hazardous to Man or the Environment and the Act on Chemical Products are framework laws that need to be complemented with government ordinances and agency regulations on specific issues such as notification, classification, and labeling.

What is said about the Precautionary Principle in the preambles to the Act on Products Hazardous to Man or the Environment has been quoted with approval by later commissions, including the recent Chemicals Policy Commission, as well as by the various administrations since then. In essence, therefore, it still stands as the fundamental precept of Swedish chemicals management and control.

In the preamble from 1973, there is a section devoted to the precautionary approach "that should be applied by anyone who produces or handles hazardous products." First of all there is a general obligation to investigate, as far as possible, the effects of the product on man and the environment, using to the fullest extent possible existing scientific knowledge. There is also an obligation to label the product and to inform all users as carefully as possible on measures to be taken to protect man and the environment during use and disposal of the product. The preamble regards all protective measures as being part of a precautionary approach.

It is further stated that the responsibility to take precautionary measures includes such things as reducing or eliminating to the extent possible substances in the product that make it hazardous to man or the environment. When assessing the appropriateness to market a product that contains a hazardous substance, the possibility to substitute this substance with a nonhazardous or less hazardous substance shall be considered. A product that contains a substance that entails health or environmental risks shall not be accepted, if there exists, or, if it is possible to produce, acceptable alternatives with smaller risks.

In discussing measures that might need to be taken against chemical products, the preamble concludes that a scientifically based suspicion of risk shall constitute sufficient grounds for measures, from the producer as well as from the government. The authorities need not abstain from measures until harm has been done. To avoid restrictions on his or her product, the producer must show, beyond reasonable doubt, based on existing scientific knowledge and principles, that the suspicion is unfounded. Otherwise he or she will have to accept that the product is treated as being hazardous by the authorities. The uncertainty still residing about the hazard of a substance shall not be borne by the public but by whomever wants to market the product. This is often referred to as the reversed burden of proof. This is an important part of the Precautionary Principle.

PRACTICAL APPLICATIONS OF THE PRECAUTIONARY APPROACH IN SWEDEN

In general, it may be said that all ordinances and agency regulations under the Swedish Act on Chemical Products are applications of the Precautionary Principle (e.g., regulations on classification and labeling, storage, and child-protective packaging). It has, however, also been applied in special cases against individual chemicals and groups of chemicals. A generic application of the principle is the general restriction for public use of preparations containing substances classified as carcinogenic, mutagenic, or reproductive toxicants. The following discussion includes specific examples illustrating both appropriate and less appropriate uses of the principle.

Pesticides

The area shows the most evident, earliest, and well-established use of the principle in Sweden is the pesticide approval system. Many pesticides that were cancelled or withdrawn in the 1970s and early 1980s had, *at the time of the decision,* limited data sets that would not in any modern sense of the word be considered sufficient for risk assessment. Rather it was the inherent toxicity and the *potential* risk of substances such as aldrin, dieldrin, and endrin that caused the Swedish authorities, first the Poison Board and later the Products Control Board, to cancel their registration. When DDT was cancelled for most uses in 1970, the reason was suspected effects in the environment, including eggshell thinning of birds of prey, but the direct link to DDT was weak and was questioned by the chemical and forestry industries. Also, the health effects were not prominent, and well-designed studies on chronic effects in animals were not available until much later. The decision was not unanimous. Opponents pointed to the great economic importance of DDT in forestry and to the great costs that would entail from a ban on its use. Since forestry-based industries (e.g., the timber, pulp, and paper) contribute to a major share of Sweden's export income, this was an important argument. Years later, when the use of aerial spraying of herbicides in forestry was first restricted and then finally banned, the same catastrophic scenario of major export industries going bankrupt was raised by industry, as well as by some diehard herbicide fans in the scientific community. In neither case was there any measurable impact on the economy of individual companies, let alone the national economy.

Several pesticides have been cancelled because less hazardous alternatives were available. The insecticides aldrin, dieldrin, and endrin, commonly called the "drins," are good examples. Another is the insecticide parathion and its derivatives, which were cancelled since a less toxic alternative, fenitrothion, was available. Similar arguments that less toxic pesticides were available were used for the cancellation of the herbicides dinoseb and amitrol in 1972.

An interesting case was the ban on the herbicide 2,4,5-T in 1977. In the wake of the ICMESA accident and the concomitant dioxin release to the nearby village of Seveso in Northern Italy in the summer of 1976, there was a heated discussion in Sweden on the continued use of herbicides containing chlorinated dibenso-para-dioxins as a result of the production process. There was intensive public pressure to ban the phenoxy-acid based herbicides. The Products Control Board lingered in its decision. There were voices advocating a corpses on the table approach, in some cases driven by a benevolent attitude toward pesticides. The government then stepped in, in a fashion most unusual for Sweden, and changed the Pesticide Ordinance overnight to include an outright ban on 2,4,5-T as an active ingredient in pesticides. The Products Control Board then had to execute a withdrawal of all registrations for such products.

In 1978, the Products Control Board behaved more in line with precaution. All products based on chlorinated phenols, including pentachlorophenol (PCP) mainly used for impregnating wood and textiles, were banned because they were suspected to contain chlorinated dioxins, or such substances that may be produced upon incineration of the impregnated materials. There was limited scientific proof as to the hazard of the treated materials, but by analogy with the substances released during the Seveso accident the Products Control Board decided to cancel the registration.

A different way of using the Precautionary Principle was the decision of the Products Control Board in 1978 to cancel the registration of nine pesticide formulations for which the documentation was mainly based on reports of tests performed by the Industrial BioTest Laboratories (IBT) in the United States. The reason given was that in view of the uncertainty related to reports from this laboratory, these formulations could not be said to be backed by sufficient data for a proper assessment of their hazards and risks and thus no correct advice as to their proper handling, use, and disposal could be given.

Halving the Use of Pesticides in Agriculture

In the parliamentary election campaign in 1985, the environment was one of the issues in focus. The sitting government proposed a program for halving the use of agricultural pesticides. Since the elections were won, the new National Chemicals Inspectorate (KemI), established in January 1986, was given the task to set up this program together with the Board of Agriculture, the National Food Administration, and the Swedish Environment Protection Agency. The role of KemI was to ensure that old, hazardous pesticides were systematically substituted, whenever possible, by more modern, well-documented, and less hazardous pesticides (Bernson and Ekström, 1991). The program was very successful and reduced pesticide use in agriculture by 65 percent in ten years. In the process, KemI developed a set of principles for identifying unacceptable pesticides that were clearly precautionary in character. These principles were published as a separate report (KemI, 1992). In that report KemI described a stepwise procedure for the approval process including cut-off values for unwanted properties. KemI also systematically applied a method of comparative assessment, which implied that if a new, substantially less hazardous pesticide was approved, the registrations of old, less acceptable pesticides were withdrawn. The comparative assessment can be seen as a practical application of the substitution principle in Article 5, Act on Chemical Products.

Industrial Chemicals

There is a major difference between pesticides on the one hand and industrial and/or consumer chemicals on the other. While the former are not allowed to be used unless approved by central agencies, in most countries industrial and consumer chemicals are put on the market without formal approval. Management and control by the government take place after their introduction on the market. There is generally much less data on industrial and consumer chemicals, with the exception of a few compounds that for various reasons have come into focus (e.g., CFCs and PCBs). Therefore, a much greater onus is placed on the producer to apply the Precautionary Principle before placing a chemical on the market. As the history of chemical incidents shows again and again that responsibility fails. On the following pages are some examples of when the government has applied the precautionary principle, and when it has not.

CFCs

An early case of applying the Precautionary Principle to industrial chemicals in Sweden is the ban on the use of chlorofluorocarbons (CFCs) in spray cans, which occurred in 1977. At that time there was only limited evidence of ozone depletion, and the theory was still under heavy attack from many quarters, including scientists in the field. Obviously, this was a case of acting on a limited scientific database because of the serious consequences involved. There were alternatives available (e.g., butane/propane mixtures), although from an acute risk perspective this meant a step back from nonflammable to flammable solvents, which actually may lead to more accidents.

PCBs

PCB was discovered and identified in fat from seals in the Baltic by Sören Jensen in 1966. At the time of the first act on PCBs (Parliament of Sweden, 1971), the persistence and bioaccumulative capacity of PCBs were known. There were suspicions linking PCBs to some effects on reproduction in seals, otters, and birds of prey, but hard scientific evidence establishing a cause and effect link was lacking. The government, however, took action, which slowly started the process toward a complete phase out, occurring recently. The exposure from historical uses of PCBs is still ongoing.

Cadmium

With cadmium the situation was different. The first regulations in 1974 concerned cadmium containing solders, where there was a risk of substantial direct exposure from water taps soldered with cadmium. Another early source to be regulated was kitchenware, for which maximum permissible extraction levels were issued in 1975. These regulations were protective although not necessarily precautionary in character. A quite different story was the Swedish ban on cadmium in electroplating, pigments, and stabilizers in 1981. Although the scientific knowledge has increased since then, there existed a substantial body of scientific literature on the effects of cadmium on man and the environment at that time. There was a growing insight into the importance for the population exposure of the diffuse uses of this metal (e.g., in products such as pigments and fertilizers). However, the extrapolation of the effects from massive occupational exposure or high-dose animal experiments to the low doses that the general population was exposed to remained to a large extent conjectural. At the time of the ban there was not so much stress on an existing alarming situation with respect to health. Rather, the fundamental concern was for a projected future situation that was considered unacceptable unless the increase of total cadmium burdens of Swedish agricultural soil was stopped. The Cadmibel study (Buchet et al., 1990) has strengthened this line of argument, but the threshold dose levels for probable adverse effects are decreasing.

Cadmium is still a hotly debated issue. At an Organization for Economic Cooperation and Development (OECD) workshop on Risk Reduction of Cadmium held in Stockholm in September 1995, some participants questioned the assumption that the level of cadmium in soils is increasing. Others opposed the view that present cadmium extraction and use eventually lead to unacceptable exposures. Sweden still needs to fight a battle within the European Union to be able to uphold its restrictions on cadmium (KemI, 1997).

Mercury

The discovery that inorganic mercury as well as phenyl mercury released into the environment may be metabolized by bacteria in sediments to organic methyl mercury, which is then readily absorbed in the gills of fish or in the intestinal tract of birds and mammals, established a clear link between industrial and agricultural activities on the one hand and high levels of methyl mercury in predatory fish, seed eating, or predatory birds on the other. A risk assessment performed in Sweden in the mid-1960s showed that the margin of safety was only a factor of two for effects to occur in the growing human fetus. Still, the decision of the Poison Board in 1966 to ban methyl mercury as a seed dressing agent was taken unwillingly, and under pressure from the public, including some foresighted scientists. In the case of methyl mercury, we were very close to a "corpses on the table" situation for man. For the environment, and particularly for birds such as the osprey, the king eagle, the sea eagle, and the peregrine falcon, the dead embryos were already there for all to see.

Lead

The first evidence that lead at low levels could affect children's learning capacity was strongly opposed by some vocal leading Swedish experts on metal toxicology. At a public meeting arranged by the Swedish Academy of Science in Stockholm in 1978, two scientists from Denmark and the United States, Drs. Grandjean and Smythe, respectively, were rudely attacked by some Swedish professors because they insisted that levels of lead in the blood (not far from those present) in many children living in urban areas could lead to mental retardation. The studies of Grandjean indicated that a major source of exposure was lead in gasoline. Another source for child exposure, particular to the United States and some Southern European countries, was lead from paint eaten in so-called pica behavior. Later studies have confirmed these early results and decreased the no-observed adverse effects level.

In 1978, Sweden had already taken some steps to decrease the level of lead in petrol. Volvo and Saab started selling cars with catalysts for the U.S. market a year later. At home, however, they successfully lobbied the Swedish government to postpone the introduction of catalysts in cars and thus lead-free petrol until 1989. In 1991 a government-commissioned report from KemI and the Swedish EPA proposed a complete phaseout of lead and lead products in the long term (KemI, 1991). They also proposed a differential tax on lead in petrol, which later became law. Within a couple of years, the alternatives had swamped the market, with the exception of some limited uses for veteran cars and small aircraft petrol sold in Sweden today.

EARLIER APPLICATIONS OF THE PRECAUTIONARY PRINCIPLE

This chapter has discussed examples of proper and not so proper uses of the Precautionary Principle. There is certainly a wealth of material for social science to study the way controversial environmental issues are resolved.

In many of the examples, the time necessary to bring the issues to conclusion is in the order of decades. For many chemicals (e.g., PCBs), most of the risk reduction work still remains to be done at the international level. To illustrate this point, it recently became clear that there is de novo production of PCB within the Russian Federation, and that the Russian government intends to continue production for at least a decade. Only a handful of compounds have, in any sense of the word, been permanently put aside during the last twenty years. For the persistent, bioaccumulating organic pollutants, an international process is just about to begin. Many chemicals (e.g., the metals) are coming back into focus again and again. Sweden still has to fight hard to uphold precautionary measures taken ten to twenty years ago against cadmium, pentachlorophenol, mercury, and various pesticides because these measures are stricter than regulations within the European Union (KemI, 1997).

At the same time there is extensive experience to show that for the chemicals mentioned previously precaution was necessary. New data have in general shifted the levels for no-observed adverse effect levels (NOAELs) progressively lower or have uncovered new effects for those chemicals. Their continued use at present levels or, in some cases, at all, is not safe for man, woman, or child nor for the environment. The accumulated experience from the last decades tells us that we should not be so overcautious in applying the Precautionary Principle. Nine times out of ten we will not err on the side of caution, but later realize that precaution was necessary.

Compared to other countries, even those that adhere to the Precautionary Principle in one form or another, Sweden has often been in the forefront in regulating environmental pollution issues related to chemicals. This has sometimes caused problems with other countries and in particular with the earlier European Community (e.g., for cadmium and chlorinated solvents). So far, there is no substantiated evidence that the Swedish industry or the economy in general has suffered as a result of Sweden regulating ahead of other countries or differently from other countries.

Being in the forefront on contentious issues also means that other countries, particularly in the developing world, tend to put Sweden on a pedestal to be admired, while adversaries (e.g., in Swedish or European industry), use Sweden as an example to be abhorred. For the sake of argument, Swedish solutions are normally distorted by both sides in such a discussion.

SCIENTIFIC UNCERTAINTY AND THE REVERSED BURDEN OF PROOF

In several of the cases previously discussed, the main reason for action was a perceived hazard or risk, based on scientific data. Normally, however, existing data do not permit comprehensive risk assessments of all possible exposure situations. Therefore, a certain amount of scientific uncertainty is by necessity inherent in most situations. The Precautionary Principle makes one behave as if these risks were true. Hopefully, in most instances we will not have the answer. If we had such an answer, by decreased morbidity, increased biological diversity, or some other parameter we would not have applied the principle correctly. At the same time this is what makes it so hard for some sectors of society (e.g., industry) to accept some of its applications. You will never, or rarely, be able to put a discrete figure on what you have saved.

The invoking of the reversed burden of proof is in principle not so very different from acting on scientific uncertainty. Rather, it is the measure itself that differs. Instead of banning something on a scientifically founded suspicion, you require that someone else, usually the producer, convince you that a chemical, product, or process is safe. Unless that is done, you will take action. In most legislatures, pesticide approval is based on the reversed burden of proof, although this is seldom realized. Until you are convinced that a pesticide might be safely used, it will not be allowed. The data requirements to fulfill the burden of proof have increased enormously over the last two decades.

For industrial chemicals, it has proven to be very difficult to use the reversed burden of proof in a practical way. In many legislatures, including Sweden, the legal tools are weak or missing. It has been tried in Sweden, in following up the proposals in the report Risk Reduction of Chemicals (KemI, 1991). For most of the chemicals in that report, the Swedish government chose the nonregulatory option for reaching the targets. For some of the chemicals, where the report actually stated that more data were necessary for a first hazard or risk assessment, the agency has tried to persuade the companies that they should do more testing. In most cases there were diverging views in industry on whether there is a potential risk or not, and in general the data have not been forthcoming.

THE PRECAUTIONARY PRINCIPLE IN USE

Notwithstanding what has been discussed earlier, it is far easier to list examples of situations where the Precautionary Principle should be used or should have been used than to find cases where it has been applied successfully. I will just mention a few cases of nonapplication of the principle. One may believe that the use of phthalates in plastics resulting in the continuous dissipation of such substances during the life cycle of the product is not defensible because of the lingering uncertainty as to their health and environmental effects. Neither is it in line with the Precautionary Principle to continue using nonyl phenol ethoxylates in detergents, when some of them have been shown to have endocrine disrupting properties in laboratory experiments and it has not been disproved that the effects are characteristic of all members of this group of chemicals. Many wetland birds are found with their craws full of lead shots, which they have picked up from the lake bottom probably believing them to be gravel. The Substitution Principle should have been applied by the producers of shotgun ammunition. The concentrations of some metals are increasing in the upper soil in and around cities because of their diffuse spread through use in metal containing products. There should be ample reason for both producers and users of these products to apply the Precautionary Principle as well as the Substitution Principle.

In October 1997, the use of metolylacrylamide containing 1–4 percent acrylamide for grouting in a railway tunnel construction project was underway in Southern Sweden. Acrylamide is a known carcinogen, genotoxic mutagen, reproductive toxicant, and a strong irritant and sensitizer. Despite knowledge about the uneven quality of the rock, with a lot of fissures and cracks, and defying warnings from expert geologists about the improper techniques used for sealing the rocks, the company in question, partly misled by the French producer of the chemical, went ahead on a big scale. Hundreds of tons of the product were injected into the rock together with a polymerizing agent with the objective to produce a polymer in situ to seal off the tunnel from water leaking from above and the sides. As one might imagine, the polymerizing process did not keep pace with the penetration of the monomer through the cracks and fissures of the rock. However, it was not until a number of cows, drinking water from a nearby stream polluted by outlet water from the tunnel, developed neurological symptoms and died did the breadth of the catastrophe start to dawn on those responsible. Available information shows that a similar grouting product containing acrylamide was banned in Japan in 1974 following a careless application near a well that led to several cases of acrylamide poisoning.

It is evident that an appropriate information search and an assessment of available data would have led to a decision not to use such a risky technique with a very hazardous chemical. Corroborating factors that led to the wrong decision in this case were that the alternatives were, in the short term, more expensive and that the tunnel had been severely delayed (at least three years) by several earlier mistakes in the operation of the work. Obviously, the construction company felt pressed to use a technique that, although risky, was cheap and quick. So much for the Precautionary Principle.

THE SUNSET CONCEPT

Whichever way the Precautionary Principle has been used in the examples stated earlier, the striking similarity is that it has been one chemical in focus each time. The assessment of hazards and risks has been on a case-by-case basis, and enormous resources have been spent on assessing and reassessing these individual chemicals in one country after another. What if we try to find another solution, more focused on internationally accepted generic criteria for unwanted chemicals, ambitious but long-term targets, that gives early warnings to industry as well as lead time to develop new processes and products and ends in a sunset (i.e., a phaseout) or ban? Where is an international organization that could make this happen? Such were the thoughts that led to a letter to the editor of *Nature* magazine (Wahlström, 1989). In that letter, an international procedure was outlined that started with identifying a list of essentially multiproblem chemicals, advertising the list with deadlines for phaseout, inviting industry to report annually on its efforts in phasing out those chemicals, and ending the process in a ban or sunset. An important aspect of the process was the openness and the lead times that should be long enough for industry to develop alternatives. To give sufficient impetus to the work it was proposed to start work within an international organization (e.g., OECD) where most of the multinational industry is located.

The OECD Risk Reduction Pilot Project

The letter to the editor of *Nature* was also the start of the OECD Risk Reduction Pilot Project. As a national support for the OECD activity, Sweden decided to start the KemI Sunset Project. But, first a few words about the OECD project. At first, everything ran smoothly. There was an ad hoc meeting in Stockholm, Sweden, in March 1990 to propose chemicals for the pilot project. Lead, mercury, and cadmium were selected together with methylene chloride and brominated flame retardants. For all of the chemicals, there were countries volunteering to be so-called "lead countries" for the OECD work. The motto was "learning by doing." Work started quickly on all chemicals. In May 1991, there was a second meeting in Stockholm to set criteria for selecting further substances. Workshops were held to gather data, compare experiences, and discuss risk reduction options for the individual substances. The original lead countries for each chemical were quickly joined by others. The newcomers were not so intent on concerted OECD-wide risk reduction. Some of them in fact questioned the whole idea of the OECD risk reduction work (e.g., Canada and Australia). A couple of countries, both major metal producers, went as far as trying to bury the whole project. In 1995 the pilot project was put to an end. Since not much was actually "done," there was not so much to "learn." An OECD Advisory Group on Risk Management rose out of the ashes. Its track record with the present participation is so far one of discussing tentative plans on what might be an outline of a program for OECD. Recently, the OECD Chemicals Group has agreed on the dissolution of the Advisory Group on Risk Management. Most likely the new impulses for work on sunseting will arise outside of the OECD.

The KemI Sunset Project

Concomitantly with the OECD pilot project, an activity was started within KemI, to support the international work and to provide a database from which different sets of chemicals for risk reduction could be selected using criteria that could be easily tailored to specific needs. This was probably the first serious attempt to practically apply the OECD "Blue Book," *Existing Chemicals: Systematic Investigation* (OECD, 1986). The basic concept was to identify, in a stepwise process of increasing complexity, what was termed as *multi-problem chemicals* (i.e., chemicals that presented problems both from an environmental and a health perspective and for which there was exposure, measured or estimated). To add an international angle to it, all possible lists of priority chemicals established by national, regional, international organizations were collected. The lists were categorized separately for health and the environment and in the first stage chemicals were ranked according to the number of categories in which they appeared. The database contained altogether 7,000 chemical entities, coming out of 70 lists (see also KemI, 1994a). In the first stage, chemicals were selected if they appeared on a number of health and environmental priority lists. This narrowed down the number of chemicals from 7,000 to about 500.

KemI developed a scoring system to select those chemicals with the highest scores. For this, some qualitative parameters had to be normalized on numerical scales. This is described in another report (KemI, 1995a). The advantage with the database and the priority setting scheme is that it is very flexible. Anyone using the database may select their own criteria based on other priorities and get another listing of candidate chemicals suited to their purpose (e.g., chemicals hazardous to the aquatic environment or chemicals with occupational exposure). In the scoring exercise, the number of chemicals was narrowed down from 500 to 100.

Thus, by a stepwise process the number of candidates was narrowed down from 7,000 to 100. In the third and final assessment step, chemicals for which international work was already ongoing, or which had already been banned or phased out in several countries, were excluded. The remaining chemicals (less than 50) were subjected to a hazard analysis coupled with a preliminary exposure analysis to find out if they would be of interest for further risk reduction. This last assessment step was the most time-consuming, and it focused on national priorities for pragmatic reasons. The overall outcome of the project is summarized in the report *Selecting Multi-Problem Chemicals for Risk Reduction* (KemI, 1994b). The hazard profiles for the final selected chemicals have been published separately (KemI, 1995a).

The UNEP Global Information Network on Chemicals (GINC), supported by the Japanese National Institute of Health and Safety, has shown great interest in the Sunset project. The database is available through the home page of GINC and, according to sources close to GINC, is one of the most frequently visited. Unfortunately, so far KemI has not been able to allocate resources to update the database.

THE KEMI OBSERVATION LIST

In addition to the Sunset project, KemI has tried to further promote industry's use of the Precautionary Principle by publishing a so-called "observation list," in cooperation which the Swedish Environment Protection Agency and the Swedish Occupational Health and Safety Board. The list contains more than 200 chemicals or groups of chemicals selected by criteria for health and environmental hazard. The intent of the list, as explained in the Introduction, was to focus attention on a relatively large number of chemicals that might be candidates for risk reduction by companies.

The list was primarily aimed at the individual producers and users but could also be used by regional and local authorities in their enforcement of the chemicals regulation. The observation list was favorably received by the domestic industry, but it has been used by purchasers as a banning list, which creates problems for individual companies. At the European industry level, there has been less understanding. The objective and procedure have been questioned by European industry representatives on the grounds that Sweden—and other Nordic countries who have tried similar approaches—are trying to circumvent notification of restrictions according to EU directives while at the same time aiming at restrictions or even bans.

TOWARD A SUSTAINABLE CHEMICALS POLICY: A RECENT GOVERNMENT COMMITTEE PROPOSAL

In 1996 the Swedish government commissioned a committee, the Swedish Chemicals Policy Committee, to evaluate the chemical policies of the past ten years and to propose new policies for the future. The Swedish Chemicals Policy Committee delivered its report "Towards a Sustainable Chemicals Policy" to the government in June 1998. The report is a political document by the members of the committee.

The Swedish Chemicals Policy Committee concluded that the present way of managing chemicals one by one is too slow, that present policies have too little environmental profile, that the Precautionary Principle is not applied as much as it should be, and that the Esbjerg Declaration from the Fourth Conference on the Protection of the North Sea necessitates a new chemical policy. The committee showed a strong preference for generic methods for risk reduction (e.g., criteria for persistent), bioaccumulating chemicals, and working product groups.

In discussing measures at the national level, the committee proposed a Chemicals Management Model, with the following steps:

1. Try to use simple and clean products whenever possible.
2. Additives should not be mobile.
3. Articles should not contain "stop" substances (i.e., human-made organic substances that are persistent and bioaccumulating, or mercury, cadmium, or lead, or substances that may give rise to "stop" substances). Products should not contain humanmade substances that are carcinogenic, mutagenic, or endocrine disrupting including reproductive toxicants.
4. "Stop" substances should not be used in production processes. Such substances may exist transiently in production processes.
5. No releases of "stop" substances or carcinogenic, mutagenic, or reproductive toxicants should occur from production processes.
6. For substances that are used or arise during the life cycle of the product, companies shall show their properties and effects and prove that the product is safe to use in the short and long term.
7. Hazardous substances shall be substituted with better alternatives from

a health and environmental perspective as often as possible based on existing knowledge.

The committee's report has been sent for comments to a wide range of stakeholders. On the basis of the report and the comments, the Swedish government will forward an environmental bill to the Parliament in the spring of 1999.

Is There an Ongoing Devaluation of the Precautionary Principle?

The original concept of the Precautionary Principle, as given in Swedish law, is very broad. It encompasses all kinds of measures to reduce or prevent harm (e.g., minimizing exposure, minimizing dose, substituting alternatives with equal function, substituting alternatives with limited function, and restricting use of a chemical or even eliminating its use).

There is, however, a move toward different interpretations that is less inclusive (e.g., measures shall be taken when there is a reasonable suspicion of *serious* effects on man or the environment, or the reversed burden of proof shall be applied when *serious* harm is suspected). Such language is used, for example, in Principle 15 in the Rio Declaration.

Principle 15 in the Rio Declaration should be interpreted as follows: The first part covers in a broad sense all precautionary measures, while the second part refers to a particular situation (i.e., a specific application of the precautionary approach). The way some stakeholders refer to Principle 15, however, indicates that the specific case of limited scientific data indicating *serious or irreversible damage* becomes the principle. The impracticality of taking *cost-effective measures* at a stage when there is a lack of full scientific certainty, and hence, limited possibilities to perform a comprehensive risk assessment on which to base such measures, begs the question of whether the second part of Principle 15 is a practical tool for environmental protection or not. It is evident that most precautionary measures taken by governments might be contested on the grounds of cost-effectiveness because of the uncertainty on which they are inherently based.

Recent Proposals to Weaken the Substitution Principle Rebutted

Industry has criticized the codification of the Substitution Principle in the Act on Chemical Products, and particularly the possibility to initiate court action against potential violations of this principle. Several local and regional authorities have established "black" lists of products and tried to persuade shops not to sell them and companies not to market them locally or regionally. In some cases they have approached consumers with information intended to influence their choice of products in an environmentally suitable direction. Some of these actions have been less clearly thought out and have created great resistance and anger among the chemical companies concerned.

In the recent proposal for a new Swedish Environmental Code, the penalization of violating the Substitution Principle had been left out. This recent attempt to weaken the chemicals legislation was criticized by several commenters on the proposal (e.g., local and regional authorities) and could clearly be seen as a step backward at the national level. The final bill to the Parliament of the Swedish Goverment has changed the proposal from the Environmental Code Committee back to the status quo (i.e, violations of the Substitution Principle may still be penalized).

THE PRECAUTIONARY PRINCIPLE IN THE FUTURE?

There is a need to defend as well as to develop the concept of the Precautionary Principle. There are several ways to do this.

1. The Precautionary Principle should be used more in the generic sense. The Swedish Chemicals Policy Committee set the target that products should not contain persistent and bioaccumulating substances by the year 2007. In doing this, the committee considered that substances that are persistent and bioaccumulating will, regardless of their place of origin, redistribute themselves after use, ending up in species at higher altitudes, latitudes, and trophic levels, humans included, and remain there long after their use has been discontinued. We do not know enough about all the possible effects of these chemicals. Experience shows that there are surprises in store for us also in the future in the shape of new and unexpected effects. Thus, as long as such substances continue to be used there is no real chemical safety. Once we know that they are unsafe, there is not much we can do about it but wait for the substances to disappear over a generation or more. If they affect the reproductive system, then the residual period of a generation is too long. The Swedish Chemicals Policy Committee has described this by saying that persistent, bioaccumulating chemicals have a braking distance that is too long.

    Needless to say, even if persistence and bioaccumulation are considered sufficient properties for taking measures to reduce use and emissions of a chemical, there is certainly room for toxicologists in the future. Toxicity would be very useful to rank chemicals that need to be phased out and to set priorities.

2. There must be more attention paid to chemicals in products. Many of the problems that we see on the horizon today are related to products containing unsustainable chemicals (e.g., brominated flame retardants and rare earth metals in electronics, heavy metals in pigments, and stabilizers). We must strive toward clean products, where the unavoidable hazardous components stay in the product for its life cycle. There must be a continuous process of product development to find new, less hazardous substances which can substitute for the unacceptable ones.

3. There is a need to collect and analyze cases where the Precautionary Principle has been invoked or used in the past. One needs to look at *the data available at the time of decision* and to compare them with later, more extensive knowledge. Based on my own experience I would say that early suspicions about a chemical are generally confirmed or strengthened by later data. So far, however, there has been no systematic attempt to build a case, based on existing experience, to show that as a rule precaution is justified from a health and environmental perspective.

4. There must be a global outlook. The world trade in chemicals and articles is doubling in a decade. We are approaching the limits of our planet to tolerate a growing population using more and more commodities. The rapid penetration of new products in all markets around the globe makes the margin for mistakes more narrow. Before too long it will become necessary to start discussions on a global framework agreement or instrument on chemicals. In such an agreement, the Precautionary Principle and Substitution Principle as well as other principles (e.g., the polluter pays principle) should be amalgamated into a philosophy of pollution prevention that will guide us toward a responsible and sustainable use of chemicals to the benefit of humanity and our environment in the future.

REFERENCES

Bernson, V., and G. Ekström. 1991. Swedish Policy to Reduce Pesticide Use. *Pesticide Outlook*, vol. 2, p. 3.

Buchet, J.P., et al. 1990. Renal Effects of Cadmium Body Burden of the General Population. *Lancet*, vol. 336, pp. 699–702.

Myers, R.A., Hutchings, J.A., and Barrowman, N.J. 1996. "Hypotheses for the Decline of Cod in the North Atlantic. *Marine Ecology Progress Series*, 138, 293–308.

Organization for Economic Cooperation and Development (OECD). 1986. *Existing Chemicals*. Systematic Investigation. Paris.

Parliament of Sweden. 1971. Act on PCBs. *Swedish Book of Statutes*, p. 385.

————. 1973 Act on Products Hazardous to Health or the Environment. *Swedish Book of Statutes*, p. 329.

————. 1985. Act on Chemical Products. Swedish Book of Statutes, p. 426. In: Swedish National Chemicals Inspectorate. 1997. *Swedish Legislation on Chemicals*. Stockholm.

Swedish Chemicals Policy Committee (Ministry of the Environment of Sweden). 1997. Toward a Sustainable Chemicals Policy. Stockholm, *Government Official Reports*, 1997, 84.

Swedish National Chemical Inspectorate (KemI). 1990. *Cadmium—An Analysis of Swedish Regulatory Experience*. Stockholm, KemI Report 6/90.

————. 1991. *Risk Reduction of Chemicals*. A Government Commission Report. Stockholm, KemI Report 1/91.

————. 1992. *Principles for Identifying Unacceptable Pesticides*. Stockholm, KemI Report 4/92.

————. 1994a. *Chemical Substances Lists*. Stockholm, KemI Report 10/94.

————. 1994b. *Selecting Multi-Problem Chemicals for Risk Reduction*. Stockholm, KemI Report 13/94.

————. 1995a. *A Priority Setting Scheme for Scoring Hazardous Properties*. Stockholm, KemI Report 2/95.

————. 1995b. *Hazard Assessments*. Stockholm, KemI Report 12/95.

————. 1996. *Supplement to Report 13/94, Observation List*. Stockholm.

————. 1997. *Swedish Restrictions Benefit the Environment Reduced Threats from Cadmium, Arsenic, PCP, and Oganotin Compounds*. October.

United Nations Conference on Environment and Development (UNCED). 1992. *Agenda 21. The United Nations Programme of Action from Rio*. New York.

Uvurderede kemiske stoffer. *Teknologiraadets rapporter 1996/2*.

Wahlström, B. 1989. *Nature*, vol. 341, p. 276.

Part II

LAW AND THEORY

Current U.S. science, law, and policy protect economic well-being over public health and the environment not as a matter of malfeasance, but as an outcome of science and legal history. Modern science has been built on the foundations of Descartes and Bacon, who promoted a separation of values and science. Science constructed out of reductionism, replicability, and quantifiablity was well-suited to problems that could be taken into the laboratory and solved. Cartesian science is not well-suited to problems that are global or multigenerational in scale (climate change) and that have a broad scope, crossing many disciplinary boundaries (endocrine disruption); nor is it suited to problems that appear generations after the harm occurred and are diffuse, and unpredictable given current scientific models. Consequently, we lack a robust philosophy of science that can address the uncertainties and ignorance embedded in problems that require the Precautionary Principle.

Similarly, law, built on centuries of burdens and standards of proof that presume innocence and that are constructed on the notion of reasonability, requires reformation in order to restore justice to those harmed or to protect from injustice those in future generations. We need to move toward a philosophy of law and policy that fosters justice and does not simply maximize utility.

The outcome of these legal and scientific regimes is that we have an explicit U.S. policy that protects chemicals and potentially harmful activities. For example, trade treaties like GATT and NAFTA support global trade and promote the role of science in protecting exports. The force of science and law advances the notion that what is good for Monsanto, Mattel, or McDonalds is good for the country: Research leads to new products; new products lead to increased trade; and trade enhances the gross national product (GNP).

Accordingly we have a regulatory system that encourages the entire train of science into law that will support a strong GNP. Unfortunately, that regulatory system has balanced public health and environmental degradation with economic increase. Public health and the environment have fallen short on the scale. We are beginning to realize that weighing diffuse future risks against current economic benefits underestimates harm to this generation and our offspring.

This section considers these fundamental assumptions and offers new ways of approaching the philosophies underlying the kinds of information and the ways we use information to make decisions. Read broadly, these chapters address the question of "How do we react to scientific uncertainty?" How did we come to put the burden for proving harm (reducing uncertainty) on the public? Isn't respect, rather than reason, a more appropriate response to uncertainty? What kind of mistakes or errors in science do we tolerate as a society? Is causality the central question of environmental and public health science?

Carl F. Cranor opens this part with an examination of asymmetric information about benefits and costs of production and about standards and burdens of proof. In essence, asymmetrical information refers to the fact that society has much better information about the benefits of production and products than the development of waste and pollution and their subsequent impacts on health and the environment. This inadvertently favors keeping substances in commerce or permitting them into use. Cranor proposes ways of reducing standards and shifting burdens of proof in such a way that we can act precautiously.

David Ozonoff presents the counterintuitive notion that the environmental community should prevent false positive results in screening for environmental harm because false positives may lead us to use scarce resources to address nonexistent problems. He argues from the metaphor of medical screening, which has high costs associated with false positives.

Barrett and Raffensperger counter that idea with a discussion about the values of an entirely different model of science that promotes contextual thinking, deemphasizes reductionism, and argues that a precautionary science prefers false positives over false negatives because it promotes more science and gives better information. This chapter on precautionary science focuses on the new problems facing environmentalists and suggests that they require a new philosophy of science that fosters a deep respect for the scope and scale of problems that are now challenging scientists.

Michael M'Gonigle offers an important perspective on the political economy of the Precautionary Principle by presenting regulation as embedded in the larger political and economic world. He defines political economy as "the study of society's way of organizing both economic production and the political processes that affect it and are affected by it." This paper argues for a fundamental change in the way society organizes production and economics so that we can incorporate ecological concerns under the rubric of precaution.

Finally, Anita Bernstein approaches the Precautionary Principle sideways, coming at it through the unlikely means of sexual harassment. Bernstein proposes a new standard, the Respectful Person, for judging sexual harassment. In this chapter, she compares the Respectful Person with the Precautionary Principle and underscores their value for environmental legal reform.

These chapters all share common threads: the belief that law can reflect something truly good; the profound affirmation of the public's role in civic life; and the assumption that the philosophy of science must be reformed in order to fulfill its promise in environmental and public health problems. Should society adopt these precepts, the Precautionary Principle may actually forestall environmental and public health catastrophe.

Chapter 4

Carl F. Cranor

ASYMMETRIC INFORMATION, THE PRECAUTIONARY PRINCIPLE, AND BURDENS OF PROOF

Carl F. Cranor

What implications does Principle 15 of the Rio Declaration on Environment and Development of the United Nations Agenda 21—the Precautionary Principle—have for countries to implement Agenda 21 within existing legal systems? In addressing this question, I focus on toxic substances posing threats to human health, using carcinogens as the main example. Analogs to ecological or broader environmental harms can be easily developed. This chapter considers as a preliminary issue some areas in which the Precautionary Principle needs further specification and interpretation. Next, the chapter highlights some difficulties in acquiring information about environmental problems, particular problems with toxic substances, and some of the asymmetric informational problems and asymmetric political forces that militate against acquisition of the appropriate knowledge about harms from potentially toxic substances. (Lawyers, economists, and others should be forewarned that I use the notion of "asymmetric information" differently than the term commonly used in other fields. Often it is applied to different kinds and amounts of information available to different people. I refer to different kinds and amount of information we might have about particular substances, or perhaps processes, that might turn out to be toxic.)

This chapter analyzes further the Precautionary Principle, restating it in forms commensurate with the assignment of burdens of proof and the stringent standards of proof associated with those burdens in a legal system to help provide protections against toxic substances. The Precautionary Principle suggests strategies for reducing or mitigating the stringent scientific standards of proof for legal purposes in order to take action to protect the environment and human health. Finally, I discuss several ideas in the spirit of the Precautionary Principle for changing the burdens and standards of proof in different legal venues in order to facilitate protection of the environment and human health.

INTERPRETATIVE ISSUES

There are many statements of the Precautionary Principle not one of which is canonical. The Precautionary Principle in Chapter 15 of the Rio Declaration on Environment and Development, which has widespread endorsement (e.g., by the United Nations and the vast majority of its member countries) states:

> Where there are threats of serious or irreversible damage, lack of scientific certainty shall not be used as a reason for postponing cost-effective measures to prevent environmental degradation. (U.N. Agenda 21, 1992, 10)

Several elements of any version of the principle, including this one, need clarification and specification:

(1) A reasonably precise statement of the desired environmental goal and the environmental condition that justifies invoking the Precautionary Principle;
(2) An identification of the jurisdictional scope of the agreed precautionary obligations under the principle;
(3) A specification of those human activities for which precautionary measures are required; and
(4) A clear statement of the precautionary measures that must be undertaken before engaging in a covered activity.[1]

In addition to these necessary specifications, there is the "cost-effectiveness" clause. This can also interact in subtle ways with the rest of the principle to render the principle either plausible or vacuous.[2]

The Precautionary Principle is usually used in international law agreements. However, with appropriate specifications it could be used to guide both international and domestic laws. This chapter mainly addresses the "uncertainty" phrase of the principle and how existing legal systems might utilize the principle, suitably specified, to provide a reason for modifying legal presumptions, burdens of proof, and standards of proof within a legal system to better serve the aims of the Precautionary Principle. There are, however, some other preliminary issues that merit discussion.

First, a reasonable interpretation of the principle suggests that those to whom it is addressed not refrain from *anticipating* threats of serious damage and taking steps to *prevent* it from occurring even though they do not have full scientific certainty that this will occur. The idea of anticipating damage and taking steps to prevent it is suggested by the idea that a decision maker is aware of threats of serious damage, that the threats are judged quite serious, and that since the threats are sufficiently affecting something highly valued, lack of full scientific certainty should not constitute sufficient reason to refrain from taking preventive action. What seems to underlie this is a judgment about the great value of something that is threatened.[3] Such value commitments are not different from our approach to protecting other highly valued people and things around us (e.g., children or other loved ones, precious art objects, and the like). Moreover, the principle is silent about standards for judging what is so important that it warrants someone taking an anticipatory and preventive approach toward it apart from unspecified "serious or irreversible threats of damage to the environment." This is another way in which the principle would need specification before implementation.

Second, some have suggested that the Precautionary Principle does not merely address action under conditions of uncertainty but has implications for precautionary and anticipatory action even when decision makers are not faced with considerable uncertainty.[4] To the extent that decision makers are *certain* of serious or irreversible risks of damage to human health or the environment, the anticipatory and preventive implications of the principle seem correct. If decision makers should not refrain from taking precautionary steps to prevent risks of damage from materializing when some scientific uncertainty obtains, *ceteris paribus* they should not refrain from taking similar steps when they are certain damage will occur. However, since the certainty of decision makers about risks to health and the environment seems a much rarer circumstance than their having some (or considerable) uncertainty about such risks even when they exist, I consider the latter case, because this is both the more typical and the harder case to prove.

Third, it is important to clarify what the principle requires of decision makers. Most statements of Precautionary Principle state that "lack of scientific certainty shall not be used *as a reason* for postponing cost-effective measures to prevent environmental degradation" (emphasis added). Reading the principle strictly merely removes one reason for postponing a course of action to prevent damage from occurring. It does not actually recommend a course of action; this is part of the needed specification of the principle. The strong suggestion is that a highly valued environmental state is so threatened with serious damage that the mere fact that one does not have full scientific evidence that the threat will materialize should still provide sufficient reason for taking cost-effective precautionary steps to prevent the damage from occurring. Moreover, the Precautionary Principle focuses attention on a particularly powerful reason often cited by decision makers for not acting—lack of full scientific certainty about the threatened damage. Thus, there may be good reasons for acting even when there is some scientific uncertainty with respect to the threat of serious damage to a highly valued environmental resource.

Fourth, the principle indicates that "lack of full scientific certainty shall not constitute *a reason* for postponing cost-effective measures to prevent environmental degradation." The principle is probably better stated as requiring that "lack of full scientific certainty shall not constitute *a sufficient reason* for. . . ." For one thing, there are circumstances in which lack of full scientific certainty would constitute *a reason* for postponing cost-effective action, for example, when two or more cost-effective actions otherwise identical in their protection of the environment could be pursued and one was supported with greater scientific certainty than the other, the action supported by less certainty would provide a reason for choosing between them. That is, if everything else is equal (which is unlikely to be the case except in rare circumstances) and both actions cannot be pursued, then the action supported by greater scientific certainty would have priority.[5] In addition, the lack of full scientific certainty should not be a *sufficient* or *decisive* reason for not taking cost-effective action. Lack of full certainty might be *a consideration* along with a number of other considerations in deciding on a course of action, just not a decisive one. It appears to be the sufficiency of such considerations that is the difficult point, not whether it would constitute "a reason" in some circumstances.

ASYMMETRIC INFORMATION AND KNOWLEDGE ACCUMULATION

Acquiring the relevant scientific information about toxic substances is difficult. For example, carcinogens have long latency periods (from five to forty years), operate by obscure causal mechanisms, rarely leave causal "signatures," and inflict diseases that are typically causally overdetermined.[6] These features of carcinogens all increase the difficulty of acquiring adverse health effect data about them. Other toxic substances will have many of the same properties (e.g., endocrine disrupters and some other reproductive toxins), while some may not. Moreover, different substances cause different kinds of harm by different mechanisms; there are typically few generalizations from one substance to another (e.g., compare the mechanisms of reproductive toxins or neurotoxins with carcinogens and compare carcinogenic initiators and promoters with each other and with other toxins). Furthermore, many of the scientific fields, in themselves or in application, on which we must rely for assessing the risks from toxic substances—epidemiological studies, animal studies, various short-term studies indicating toxicity mechanisms—are in their infancy. Thus, even if the mechanisms of toxicity were not difficult to determine in themselves, the means by which scientists might discover the health and other harms they would cause are also not as well developed now as they might be at some point in the future. However, we should not be overly optimistic about the future, in any event, because depending on what our stock of knowledge about toxic substances is at any given time, we will always face questions about the toxicity of new substances and if this raises new questions about mechanisms, effects, or distinguishing the source of causally overdetermined harms, we may face many of the same problems.

In addition, the fact that much of the information about potentially toxic substances is asymmetrical information about the benefits of products, by-products of production, or pollutants, compared with information about the typical health effects of those same substances, inadvertently favors keeping the substances in commerce or permitting them in commerce. The generic nonmonetary benefits of either products or pollutants, as well as the monetary benefits of keeping them in commerce (or at least not regulating them or banning them from commerce), tend to be better known than the adverse health effects. For example, firms develop products because they understand their potential benefits and believe that people will pay to have them, and they develop manufacturing processes in ways that benefit their goals and often their existing facilities. Thus, firms are well aware of the monetary costs of not having the products, changing the processes, or being forced to reduce their pollutants. Firms do not appear to be as aggressive in identifying adverse health effects.

Political constituencies asymmetrically favor protections for substances. Products, their contaminants, or the by-products of production have obvious constituencies, namely, the firms that manufacture and sell the products or those that benefit from their use. Potential victims typically are much more diffuse and less well organized, because of the nature of the diseases—the causal path from exposure to disease may be obscure and diseases may be overdetermined. Moreover, large populations that move frequently may make the etiology even more difficult to discern. Such collections of individuals, unlike industry associations, will not be formed into effective polit-ical constituencies unless and until those in the collection can identify themselves as victims of a common toxic exposure, but this is very difficult to do. Indeed, in most cases collections of potential victims do not consti-tute a constituency at all. For more diffuse environmental harms, it will be even more difficult to find constituencies.

The asymmetries in acquiring information about toxic substances are reinforced in various ways by the norms of scientific epistemology and evi-dentiary practices. First, scientific inquiry about potentially toxic substances begins with a *presumption* that substances have no properties in particular until they have been established by appropriate studies. Second, scientists express a concern to avoid one kind of mistake, false positives (e.g., their procedures mistakenly showing that a substance is associated with a health harm when it is not). They appear to have a much lesser concern to avoid false negatives (e.g., their procedures mistakenly showing that a substance is not associated with a health harm when it is), perhaps in the view that if there is sufficient investigation of a substance, the harm will ultimately be identified. (A caveat to this point is that the relative concern to prevent false positives and false negatives varies by field, e.g., compare various pub-lic health fields that should have a concern to prevent false negatives with many other fields that do not face this particular moral issue.) Third, the burdens of proof and the standards of proof typically followed in science and designed largely to protect against false positives reinforce protections for potentially toxic substances. By the assignment of *burdens of proof*, I mean the party or scientists who must take the argumentative initiative in per-suading other scientists that the evidentiary or knowledge status quo should be changed. Typically, the burden falls on one who would argue that we are mistaken about what is currently known; on the one who would change the knowledge status quo. By *standards of proof*, I mean the degree of certainty required to substantiate a claim, as established by scientific evidentiary norms and practices, that a person who would argue to change the status quo ante would have to meet.

The burdens of proof and standards of proof in science are there for good reasons. They aim to protect against mistakenly overturning the hard-earned epistemic status quo and mistakenly adding to the stock of scientific knowl-edge and to protect against making *certain* kinds of inferential mistakes.[7] These discourage individual scientists from enthusiastically advocating their own ideas and from wasting their own research efforts, and they help to pre-vent whole fields from chasing research chimeras and wasting collective efforts and resources.

It is difficult to specify exactly what constitutes appropriate scientific certainty. It is reasonable to suppose that the degree of certainty varies with the field in question; that is, much greater certainty is probably needed before justifying claims about subtle effects in subatomic physics than is needed for claims about changes in weather patterns. Moreover, elsewhere I have argued that the degree of scientific certainty needed depends on the context in question and answering—this is not at all a scientific question.[8] Finally, it is difficult to find information about the appropriate degree of certainty needed to support a scientific conclusion, and queries to scientists interested in methodological questions have not resulted in appropriate literature on the subject. Thus, in understanding the issue, it has been necessary to rely on anecdotal references in the literature and occasional debates to provide suggestions of the stringency of this standard. Nonetheless, satisfying scientific standards of proof appears to be quite difficult—it is reinforced by considerable skepticism, inferential caution, and reluctance to suggest that scientific conclusions are known with considerable confidence. The following examples support this conclusion.

First, in research articles the authors always emphasize what is not known, further research that needs to be done to reinforce suggested conclusions, and in general the high degree of tentativeness present in the conclusions. Second, in debates between epidemiologists about the extent to which health professionals should be prepared to act on the basis of epidemiological studies, some participants suggested that the requisite degree of scientific certainty was equivalent to the "beyond a reasonable doubt" standard of proof typically used in the criminal law (compared with others concerned about public health who suggested the need to take action on the basis of less than fully certain evidence).[9] Finally, I have found revealing remarks of Arthur Furst, a past President of the American College of Toxicology. Before he would classify something as a human carcinogen for *scientific* research purposes, he would require confirmation by multiple epidemiological tests, multiple animal studies subject to stringent experimental conditions, and a variety of short-term tests all positive and pointing toward the same conclusion.[10] His are stringent requirements indeed.

However, the inferential caution and skepticism protecting the knowledge status quo, which is present in scientific research for its own sake, can have quite different results when used in different contexts. In research where we seek carefully to add to our knowledge, skepticism helps to protect the epistemic status quo and protects against making certain kinds of inferential mistakes. In the regulatory setting or in tort law, such skeptical attitudes can have substantial unintended consequences in general and scientific skepticism and reinforce the knowledge and *legal* status quo. Under a postmarket regulatory statute (or in the tort, personal injury, law), where the burden of proof is on the government (or plaintiffs) to show that a substance is harmful, skepticism toward conclusions about toxicity keeps a substance in commerce until the skepticism is overcome. Under a premarket regulatory statute, where the burden of proof is typically on the manufacturer or regis-trant of a substance to show that it is safe, skepticism about its safety pre-vents a substance from entering commerce until the skepticism is overcome. Thus, the actual legal effect of scientific inferential caution and skepticism depends on the context and the legal burdens and standards of proof estab-lished for that venue and context. However, much of our regulation of toxic substances proceeds by means of either postmarket regulatory statutes or the tort law. The regulation of carcinogens, for example, is in large part by means of post-market regulatory laws (e.g., the Clean Air Act, the Clean Water Act, the Safe Drinking Water Act, the Consumer Product Safety Act, Resource Conservation and Recovery Act, aspects of the Toxic Substances Control Act, the Comprehensive Environmental Response, Compensation and Liability Act (Superfund), the Occupational Safety and Health Act, aspects of the Food, Drug and Cosmetic Act, and even aspects of the Federal Insecticide and Rodenticide Act). In a few cases, premarket regulatory statutes also address carcinogens (aspects of the Food, Drug and Cosmetic Act—notably the Delaney Clause—the Toxic Substances Control Act, and aspects of the Federal Insecticide, Fungicide and Rodenticide Act).[11] Where the regulation of toxic substances is by means of postmarket regulatory statutes or the tort law, however, the applicable legal burdens and standards of proof reinforce the scientific burdens and standards of proof.

To sum up, the difficulty establishing information about, informational asymmetry and asymmetrical political constituencies favoring potentially toxic substances are all further reinforced scientifically by presumptions, bur-dens of proof, and standards of proof, and legally (in the vast majority of cases) by *postmarket* regulatory statutes and the tort law that imposes legal burdens of proof on the government or on victims to demonstrate sufficient harm or risk of harm before legal remedies are available. The Precautionary Principle envisions a setting much like that obtained in post-market regula-tory statutes. It envisions a serious threat to a valued environment, and it assumes that some would react by postponing action until the scientific uncertainty is removed, much as postmarket statutes may tempt decision makers to do. The Precautionary Principle does, however, remind us that in order to take appropriate precautionary action, decision makers might well shift burdens of proof or mitigate or attenuate the standards of certainty that must be satisfied before taking action in order to try to prevent serious or irreversible harms from materializing.

The previously stated reasons add up to providing substantial considerations for adopting a precautionary approach are a mixture of different kinds of considerations:

(1) Some are epistemic, having to do with difficulty of acquiring knowledge and information about the health effects of substances because of the nature of the thing being assessed.

(2) Some have to do with asymmetric acquisition of knowledge about substances because those who develop them tend to focus on their benefits and perhaps less on their potential harms.

(3) Some have to do with the politics of chemical regulation—small well-organized groups have considerable advantage over nonexistent or much less well-organized groups.

(4) Some have to do with the conventions and internal mandates of research science that tend to focus less on preventing false negatives than on preventing false positives.

Reasons 2–3 are the result of easily altered social practices (some of which the Precautionary Principle aims to change), while Reason 4 results from social practices in the scientific community that could be modified (but is better justified and deeply embedded in the norms of scientific inquiry). Reason 1 results from difficulties in the nature of the harms caused. These different contingent reasons, some of which could be more or less easily changed, all tend to support the precautionary and preventative approach of Precautionary Principle. However, perhaps the most important reason is one more difficult to articulate and that will vary case by case, the value of protecting human health and greatly threatened environmental resources of value—the value or importance of what is at stake. This, as noted above, is one of the important specifications of using the principle.

VARIATIONS ON THE PRECAUTIONARY PRINCIPLE

This section reformulates the Precautionary Principles in order explicitly to bring in analogs of legal standards of proof to help clarify the vague notion of "scientific proof" and to facilitate an understanding between two different areas of inquiry: science and the law. There are several legal standards of proof with which we might be familiar from the most stringent to the least stringent that the moving party—the party whom the law requires to argue for its position—typically the state in the criminal law or a plaintiff in the tort law,[12] must satisfy in order to establish his or her case. Placed in order from the most to the least demanding, they are that the moving party must establish his or her claims "beyond a shadow of a doubt" or "beyond a reasonable doubt" or by "clear and convincing evidence" or by a "preponderance of the evidence." The most demanding consists of a finding for the defendant when there is even a "shadow" of doubt about the moving party's case. The least demanding requires a finding for the defendant if the balance of probabilities and the quality of the evidence do not favor the moving party (this is the standard typically followed in most civil litigation, including toxic tort litigation). The second most demanding standard—that the state must establish its claims "beyond a reasonable doubt"—is the one most typically used in the criminal law. The criminal law makes an interesting foil because there are views about why the state must overcome such a heavy burden in making its case before a jury. The state typically possesses much greater resources and expertise than are available to the defense (except in extremely unusual cases involving wealthy defendants), so part of the rationale for burdening the state is to help balance the scales of justice. The more important reason, however, is that it is at least ten times worse for an innocent person to be found guilty than for a guilty person to go free.[13] In the language of science, an innocent person being found guilty is a legal "false positive," and a guilty person going free is a legal "false negative." In the actual world, either a person is guilty or not, but our procedures may or may not result in a correct outcome, thus we need to recognize that our procedures may result in mistakes and design our institutions accordingly.[14]

With this information as background, consider reformulations of the "uncertainty" phrase of the Precautionary Principle that explicitly utilize a standard of proof term from the law in place of "scientific certainty." This will make the Precautionary Principle somewhat more precise (or suggest various formulations of it) or at least more familiar. Consider three formulations from the most to the least demanding, with the differences indicated in italic.

Precautionary Principle #1:
- Where there are threats of serious or irreversible damage, *failure to establish the threat of damage beyond a shadow of a doubt* shall not be used as a sufficient reason for postponing cost-effective measures to prevent environmental degradation.

Precautionary Principle #2:
- Where there are threats of serious or irreversible damage, *failure to establish the threat of damage beyond a reasonable doubt* shall not be used as a sufficient reason for postponing cost-effective measures to prevent environmental degradation.

Precautionary Principle #3:

- Where there are threats of serious or irreversible damage, *failure to estab-lish the threat of damage by a preponderance of the evidence* shall not be used as a sufficient reason for postponing cost-effective measures to prevent environmental degradation.

Finally, there is an even stronger version of the Precautionary Principle that helps relieve some of its limitations. A weak scientific standard of proof might be that a scientist has a "rational basis for a threat of damage based on empirical data even though other scientists might evaluate the evidence dif-ferently." One may consider this to be a minimal kind of empirical evidence with which a scientist would identify a possible area of concern—it is based on empirical evidence, but not necessarily sufficient evidence to command near scientific unanimity or high degrees of confidence. Moreover, we might think of this as one of the most minimal scientific standards of proof. Incor-porating it into the Precautionary Principle results in the Precautionary Principle #4.

Precautionary Principle #4

- Where there are threats of serious or irreversible damage, *failure to have a rational basis for the threat of damage based upon empirical data* shall not be used as a sufficient reason for postponing cost-effective measures to pre-vent environmental degradation.

Precautionary Principles #1 and #2 explicitly incorporate standards of proof from the criminal law, which typically make it quite difficult for the party making a claim to establish it to the satisfaction of a fact finder. Pre-cautionary Principle #3 incorporates the civil law's standard of proof for the burden of persuasion, and Precautionary Principle #4 incorporates a modifi-cation of a "rational basis" principle that is used in some scientific and legal contexts.[15] Precautionary Principles #1 and #2 are quite plausible, even innocuous, versions of the Precautionary Principle. That is, even if one can-not establish beyond a shadow of doubt or beyond a reasonable doubt that a *threat* of serious or irreversible damage to an environmental resource exists, this should not be used as "a reason" for postponing cost-effective measures to prevent environmental degradation. Put more positively, we may have reasons for acting to prevent certain serious kinds of damage even when we do not have the highest degree of certainty that it will occur. This makes eminently good sense, since in many areas of our lives we act on the basis of reasons that are short of being supported by the highest degree of certainty. If the damage threatened to a valued environmental resource is serious enough, especially if there are cost-effective ways to avoiding the harm, then the fact that we cannot establish the threat with the highest degree of certainty does not provide us a reason against acting.

Precautionary Principles #3 and #4, however, pose greater problems. Do we want to subscribe to a principle that says that failure to establish the threat of harm as being more likely than not should not be used as a sufficient reason for postponing cost-effective measures to address it? What this version of the principle suggests is that failure to establish a threat of harm by using the balance of the evidence available should not provide a reason for postponing cost-effective action. Thus, if one has some evidence of a threat of serious damage, but the balance of the evidence does not yet show such a threat, one should nonetheless not postpone cost-effective action. Precautionary Principle #4 as stated would be tantamount to foregoing a scientific basis for the presence of a threat of serious or irreversible damage. Stated as Precautionary Principles #3 and #4, the Precautionary Principle may initially generate considerable opposition. Nonetheless, it is quite possible that there may be things so precious that we are prepared to take precautionary actions to protect them, even when the balance of the evidence does not show that they are threatened[16] or there is little or no rational basis in empirical evidence that they are threatened.[17] If there are such things, we need to identify what they are and how much evidence is needed to take cost-effective precautionary actions to reduce or remove the threat of serious injury. Considering Precautionary Principles #3 and #4 helps us to analyze the Precautionary Principle and forces us to think through which versions we adopt for which purposes.

The pertinence of Precautionary Principles #1 and #3 for the legal system is that the most stringent scientific standards of proof need not be met when valued environment resources are threatened with serious damage. This in turn suggests that the laws protecting such resources should not demand full scientific certainty (either proof beyond a shadow of a doubt or beyond a reasonable doubt) before legal action protecting the environment should be taken. What might such recommendations mean in a legal system such as ours?

There is another element of the Precautionary Principle that is not explicitly stated, but which seems clearly present insofar as the Precautionary Principle aims at protecting highly valued environmental resources and human health. That is, if we anticipate that serious damage may occur to human health or the environment and we seek to prevent it, one of the best ways of doing so is to shift the burden of proof to those whose actions pose the threat. That is, the agents who can do much to reduce such threats are those whose actions contribute to or pose the threats. Thus, if legally they have the burden of proof to show to some standard of certainty that their actions will not pose the threat of concern, this provides them with *a stake in reducing* or *an incentive to reduce* the threats from their activities. In short, it gives them a stronger reason than they ordinarily would have without that burden of proof to minimize any potential damage. Moreover, shifting the burden of proof moves toward ameliorating some of the knowledge asymmetries concerning toxic substances, because it would place the onus of producing information or generating knowledge on the manufacturer who is probably in the best position to address issues about the effects of its substances. The manufacturer then has an incentive to produce the needed information or suffer loss of the substance from commerce or at least have it regulated more stringently.

Two different generic strategies have been suggested in this section, depending on the actors in question, for preventing serious threats of damage to human health or the environment. One is that decision makers (these are likely to be governmental decision makers, although they need not be) should be prepared to act to prevent harms from occurring even though they do not have full scientific certainty that the threats will materialize. The second is for governments to shift onto the contributors of the threat some of the burden of proof to show that threats of damage will not occur. The burden of proof must fall on those whose actions can directly affect whether the threats materialize and onto those who can produce information about the substance. If such contributors cannot show that their actions would not cause serious damage, the legal effect would be that they must change their actions (e.g., not produce a pollutant or not expose humans or ecological systems to the harmful substances or actions in question). Both strategies foster action in furtherance of the Precautionary Principle.

Shifting Burdens of Proof and Reducing Standards of Proof to Implement the Precautionary Principle

In implementing a suitable version of the Precautionary Principle in legal systems, at least three different normative circumstances should be discussed: How might the Precautionary Principle be implemented within the existing legal system, given decisions that have been made by courts and current legislation that has been passed by legislatures and interpreted by the courts? In this scenario, one holds all of the current legal system constant and looks for ways to implement the Precautionary Principle within that context. Second, one might ask how the Precautionary Principle could be implemented within the current legal system and existing laws by suggesting how minor but crucial aspects of the law, such as leading court decisions, might have been modified to help achieve the aims of the Precautionary Principle. This holds constant somewhat less of an existing legal system and suggests changes that might have been made. The suggestions considered here are not terribly radical, but they do focus on some critical Supreme Court cases that might have been decided in other ways, and on decisions that had the potential to move in a more protective direction in the spirit of the Precautionary Principle. Had those cases been decided somewhat differently, it seems to be the case that even the current U.S. legal system would be more favorable toward the Precautionary Principle seems to be the case. Finally, one could consider how the Precautionary Principle might be implemented within a legal system where one was not constrained by existing legislation, court decisions, and the like. While lack of space limits discussion of this possibility, one should be aware of it and even consider various radical alternatives to existing law in order to have models and ideals to guide changes in the law.

Implementing the Precautionary Principle within Existing Law

There are several ways in which scientific standards of proof can be changed for legal purposes to ease the burden of discovering adverse health effects. These subscribe at least implicitly to Precautionary Principles #1 and #2. That is, both regulatory agencies and toxic tort judges should permit legal action on the basis of evidence that falls short of scientific proof beyond a reasonable doubt concerning the regulation of toxic substances, but which evidence is nonetheless appropriate for the venue in question. The modest changes described on the following pages should be seen as attempts to attenuate or mitigate the cumulative effect of a large number of epistemic, political, and legal considerations that asymmetrically tend to preclude discovery of adverse health effects from toxic substances and, thus, to hinder preventive action.

Prior to the early 1990s, potency assessments of carcinogens in California had been quite slow, taking from .5 to 5 person years per substance. Yet the diagnosis of the sluggish pace of this one step in risk assessment suggested that there were policy, research, and administrative practices that accounted for the problems, changes in which would greatly expedite the procedures. Instead of state toxicologists researching the toxic properties of a substance de novo in the literature as a good research scientist might, they could utilize existing and quite respectable databases on the carcinogenicity of a substance to provide the basic toxicological data. Moreover, these toxicologists could utilize regulatory policies for choosing the appropriate tumor data and subject the data to the same high-dose-to-low-dose and rodent-to-human extrapolation models that would be used in any case according to state policy for data derived de novo from the literature. Finally, they could dispense with many of the write-up requirements that had been used in previous assessment documents. Utilizing these changes, state officials were able to expedite potency assessments, which resulted in about 200 carcinogens being assessed in one year (at a rate of about 1 per day compared with .5 to 5 person years per substance in the past).[18] The potency assessments are accurate, provide a consistent regulatory framework, are health-protective, and appear to save millions of dollars in governmental, public health, and overall social costs.[19]

Second, the *identification* of carcinogens has also been slow. Whether one considers the entire universe of substances registered for use in commerce (about 100,000 substances) or only several thousand of the most high-use chemicals, there appear to be thousands of substances whose toxicity may not have been identified and whose toxicity remains largely unassessed (at least according to the public record).[20] Thus, it seems desirable to find quicker administrative procedures or scientific approximations to identify carcinogens. For example, quick, inexpensive, short-term tests (STTs) (e.g., mutagenicity tests and structure-activity tests) seem promising compared with slow and costly animal bioassays or human epidemiological studies. Their use could provide an approximate characterization of the risks posed by substances and result in public health benefits (because toxic substances would be identified earlier). A National Science Foundation (NSF) funded workshop evaluated the accuracy of approximately fifteen different expedited identification procedures for accuracy against the results of animal bioassays.[21] Validating them proved difficult, but nevertheless several procedures appeared to be sufficiently accurate for identification purposes. In addition, modeling procedures similar to those used by environmental economists suggest that even STTs that are less than fully accurate—with false negative and false positive rates higher than .05—may have uses both in premarket and postmarket regulation, depending on some facts about the world and what is at stake.[22] For example, if the percentage of carcinogens in the chemical universe is 7.5 percent or greater, then STTs appear to be justified compared with reliance on the results of animal bioassays for providing preliminary identification of carcinogens. Such procedures will result in mistakes, both false positives and false negatives, but the balance of the social costs of such mistakes favors using faster, even if somewhat less accurate, identification procedures rather than slow, science-intensive methods. The greater the percentage of carcinogens in the chemical universe, the better the case for utilizing STTs.[23]

Third, susceptible subpopulations have not been well protected by environmental regulations (e.g., children, the elderly, the genetically susceptible, those whose health is already compromised). Yet there are good legal and moral principles that support the view that they should be. In addressing this problem, there may be a well-motivated, but unfortunately misguided, temptation on the part of scientifically trained people to want to identify susceptible subgroups and to try to quantify on the basis of empirical evidence the extent of the susceptibility. When we have world enough and time, these studies should be done. However, for administrative health law purposes, giving in to such temptations in the present state of regulation seems almost perverse. Rather, agencies should adopt, as many currently do, default safety factors or high upper confidence extrapolation models to serve as placeholders for variations in susceptible subpopulations until substantial evidence is provided to remove some of the uncertainty and change the default position; otherwise, research on the issue will further slow the already sluggish assessment of risks.[24] However, preliminary evidence suggests that the defaults should perhaps be higher than the typical tenfold factor used for such purposes in the past.[25]

Fourth, in the tort law, insensitive adoption of scientists' standards of proof and pragmatic rules about the use of evidence in order to establish causation will distort the accuracy, policy aims, and balance of interests between plaintiffs and defendants. Yet some courts and many commentators urge this, citing the aphorism "science is science wherever you find it." Judges, however, should resist the scientific and political pressure to go down the above path in their boundary-drawing around what counts as science for tort law purposes. Courts need to develop a more sensitive understanding of the science involved—sensitivity to the subtlety, complexity, strengths, and weaknesses of different kinds of scientific evidence—and not issue overly simple rules for admitting or barring available evidence. But they must also learn not to demand the same high degree of certainty for inferences used by scientists or else they must return to more relaxed standards for admitting scientific evidence. Moreover, all of the evidence on which scientists rely when making judgments about causation should be admissible in toxic tort cases, clinical studies, epidemiological studies, case studies, animal studies, structure-activity relationships, and other short-term tests (some courts currently preclude, dismiss, or denigrate some of these kinds of evidence).[26] Admissibility rules should not explicitly or implicitly change the legal standards of proof so dramatically that plaintiffs must establish a piece of scientific evidence to a very high level of certainty, approaching the criminal law's "beyond a reasonable doubt" burden of persuasion, to satisfy admissibility conditions. Failure of courts to be sensitive to the contexts for use of scientific evidence in question will tend to undermine the corrective justice and deterrence goals of the tort law.[27] It will then function less effectively to help control toxic substances.

Unless something like the four suggestions indicated previously are followed, we risk having the epistemology implicit in scientific standards and burdens of proof preempt other social and legal goals of our existing legal system. Were this to occur, our legal system would also become more hostile to the values implicit in the Precautionary Principle. We would postpone action on serious threat of damage to the environment because we lacked scientific certainty. In short, stringent epistemology puts our other social and legal goals at risk.

The earlier recommendations, however, call for relatively minor changes in the standards of proof in scientific risk-assessment procedures or in legal procedures in order to permit courts or governmental agencies to take action on the basis of somewhat less evidence than some of the most stringent scientific evidentiary requirements and practices would have it. Some areas of the law exemplify more radical legal strategies that involve shifting who has the burdens of proof on an issue. For example, under various sections of the Food, Drug and Cosmetic Act (FDCA), some categories of substances must be shown to be "safe" by the manufacturer before they are permitted into commerce and some are not permitted into commerce if they are found to cause cancer in humans or laboratory animals (under the Delaney Clause). Similar premarket screening procedures are part of U.S. pesticide legislation, but the substantive health protections in the statute tend to be less protective (at least on the surface) than the procedures under the FDCA.[28] Under the Toxic Substances Control Act, substances proposed for manufacture must be submitted to the U.S. EPA together with certain minimal toxicity data (which may or may not be sufficient for an adequate assessment of the health and environmental effects of substances). The EPA then has 90 days to review this data to try to determine whether the substance may pose a threat that must be investigated for further toxicity data. If the EPA does not act within the statutory period, the firm may proceed anyway.[29] These are two models of premarket screening that attempt to reverse some of the legal burdens of proof and place them on one who might threaten human health or the environment—the manufacturer or registrant of a product—who then has the burden of showing its safety before entering commerce (instead of the government showing its damage once it is in commerce). Both premarket models legally require potential contributors to damage so as to have a stake in identifying and reducing adverse effects of their products. This provides them with important reasons for action.

A third example of shifting the burden of proof is California's Proposition 65, which requires the governor to list substances known to be reproductive or carcinogenic toxins. For listed substances, firms that expose the public to them (at some predetermined concentration level) have the burden of proof to show that exposure does not pose a risk beyond some legally specified level. This combines postmarket circumstances—the substances are already in commerce and people are exposed—with placing the burden, not on the government as is typical under most postmarket regulatory statutes, but on the person or firm whose substances might pose risks to others. It too aims to give them a legal stake in preventing damage from occurring.

Implementing the Precautionary Principle in the Legal System with Modest Changes

I have suggested that modifications in burdens or standards of proof are either part of the existing legal system (in the referenced statutes) or are easily consistent with current laws. However, reformers interested in implementing the Precautionary Principle in a more thorough way might want to know its implications in somewhat changed circumstances. One such normative scenario is to consider how our existing legal system might be slightly altered in ways that would promote greater assimilation of the goals of the Precautionary Principle into the legal system. Consider court decisions in two areas of law that might have been somewhat different and might have served the goals of the Precautionary Principle better than at present.

In 1980, the U.S. Supreme Court in *Industrial Union, AFL-CIO v. American Petroleum Inst.*,[30] (the *Benzene* case) held that the Occupational Safety and Health Administration (OSHA), before issuing occupational health standards, must show that workers face a "significant health risk" at existing levels of exposure and that the agency must show this by a preponderance of the evidence.[31] On the face of it, this principle might not seem particularly onerous. However, given subsequent interpretations of it and agency reactions to it (not only OSHA, but other agencies as well), it is arguable that this had a substantial chilling effect on regulatory activity by OSHA and other agencies, hindering OSHA from issuing more health-protective regulations. The point for this essay is that the original case need not have been decided this way at all. The court could have permitted the regulation that triggered the review to stand and not have issued such a chilling decision. Or, the court could have selected different aspects of the statute to emphasize—such as the fact that OSHA is required to protect workers from "material impairment of health or functional capacity" even if they are exposed to a toxic substance for a "working lifetime" and that Congress recognized that OSHA must work at the "frontiers of scientific knowledge"—which would have resulted in a more health-protective interpretation of the statute. In addition, agencies and subsequent reviewing courts in interpreting the *Ben-*

zene decision need not have insisted upon such substantial scientific evidence to support a regulation. Such modest changes in a highly visible and influential court decision could easily have made quite a difference in our actual legal system and led agencies to issue, and reviewing courts to permit, much more health-protective regulations than perhaps has tended to be the case. This might have made a substantial difference in U.S. regulatory law and invited much greater health protective regulation in the spirit of Precautionary Principle. It is arguable that the Benzene decision greatly increased the legal pressure on agencies to perform detailed, science-intensive risk assessments in support of regulations, which has resulted in many fewer substances being considered for regulation by OSHA or by other agencies and in agencies being legally barred by reviewing courts from considering analogs of expedited regulations described above.[32]

In the tort law, two recent Supreme Court decisions could easily have been decided in more health-protective ways, but under present interpretations they are hardly in the spirit of the Precautionary Principle. The Supreme Court, in Daubert v. Merrell Dow Pharmaceuticals, Inc.,[33] rejected existing admissibility rules for scientific evidence in toxic tort cases, the so-called Frye rule,[34] writing "[I]n order to qualify as 'scientific knowledge,' an inference or assertion must be derived by the scientific method. Proposed expert testimony must be supported by appropriate validation—i.e., 'good grounds,' based on what is known."[35] It further held that the Federal Rules of Evidence contemplate "some degree of regulation of the subjects and theories about which an expert may testify."[36] Thus, a trial judge "must ensure that any and all scientific testimony or evidence admitted is not only relevant, but reliable."[37]

Much of this decision is salutary in requiring appropriate scientific methodology to support the testimony of an expert and an improvement on the Frye rule. A critical choice in the decision, however, was how strongly to insist that testimony based on scientific evidence had to be based on a good scientific foundation. The court had choices in its decision and opted for more rather than less restrictive rules for admitting scientific evidence. Nonetheless, this decision itself might not be incompatible with Precautionary Principle provided other reviewing courts did not insist on high degrees of scientific certainty in order to get evidence admitted on a plaintiff's or defendant's behalf. Although it is early in the debate, initial trends suggest that many courts will require, and even more commentators are recommending, that the scientific evidence on which experts testify must satisfy fairly stringent standards of proof.[38] Moreover, the Supreme Court has revisited this topic in General Electric v. Joiner,[39] a case mainly based on arcane procedural issues. The *Joiner* decision suggests that it will be even more difficult for plaintiffs to introduce scientific evidence into toxic tort cases.[40] As a result of these two decisions, toxic tort law not only risks being even less health protective than previously, but also, despite having a preponderance-of-evidence standard of proof that victims must provide in order to receive compensation for injuries suffered, will depart even further from the goals of the Precautionary Pinciple. With these decisions the court is approaching a requirement that plaintiffs must establish their conclusions with considerable scientific certainty before they can even have their evidence admitted into court. This will frustrate recovery of damages in the tort law and reduce deterrent effects on defendant's actions in exposing the public to toxic substances. While the law in this area is far from settled, preliminary indications are not reassuring.

A More Thorough Implementation of the Precautionary Principle

In considering possible alternatives to existing legal burdens and standards of proof, consider several different models beginning with familiar and moving to less familiar possibilities.

First, the standards of proof for eliminating/controlling problematic substances under postmarket statutes could be reduced in order to make it easier to regulate, or ban from commerce, substances that pose great threats of damage to human health and the environment. Under current statutes and especially under the procedures that agencies have adopted to review substances and perform risk assessments on them, agencies tend to adopt science-intensive procedures to be sure that they are not wrongly regulating the risks of the substance in question. For the most part, such changes would not involve statutory amendments to current law, only changes in what agencies require by way of scientific evidence and the minimum evidence needed to show that substances should be regulated. As we considered in the previous section, reviewing courts could permit agency regulations to survive court review even if they were not based on the most highly certain evidence so that court review itself did not act as a significant deterrent to action.

Second, more legislation could have premarket protections analogous to the food additive section of the Food, Drug, and Cosmetic Act, which in principle requires a fairly careful screening of substances before they ever enter commerce. Such legislation would shift the burdens of proof onto the potential contributors of harm in order to give them a legal reason to take precautions. That is, before a company introduced a substance into commerce, it would have to persuade an agency, in accordance with some appropriate standard of proof, that, for example, the substance was "safe" or posed

"no serious risk of harm." There are limitations to such statutes. They probably work best for substances that are part of products and other substances
that are deliberately introduced into commerce or the workplace. They may
work less well for pollution or for some of the unanticipated by-products of
production that find their way into the environment.[41]

Third, we should look for ways to shift the burdens of proof onto those
who expose people to toxic substances and who contaminate the environment. For substances that are already in commerce, subject to postmarket
regulation, more of our laws could be modeled on something like California's
Proposition 65. Such legislation not only gives producers of potential toxins
a stake in taking precautionary action, but also reverses incentives to produce information, placing regulations on a polluter unless the firm can produce experimental data to exonerate itself.

Fourth, there is a modification of both premarket and postmarket statutes
that would permit decision makers to gain some control over currently problematic substances. One would first compile a list of substances that were
known to be toxic. For premarket statutes, firms would not be permitted to
introduce substances chemically similar to those on the list without proof
beyond a reasonable doubt that they were not toxic.[42] For post-market
statutes, one could compile a list of toxic substances and require them to be
phased out over time or firms could avoid phase-out only if they could show
beyond a reasonable doubt that any exposures posed no risk of harm or posed
no threats of serious damage to health or the environment. A more stringent
standard of proof is suggested before permitting such substances to enter or
to remain in commerce, because they have properties likely to cause serious
threats to human health or the environment. This does not seem unreasonable given the recent history concerning bioaccumulating or persistent substances, such as polychlorinated biphenols, that have been found highly
concentrated in the body fat of Arctic animals and in the tissues of native
human populations.

More interestingly, one could compile lists of *characteristics* of substances
(e.g., their tendency to persist, to bioaccumulate, or to be mutagenic) and
then establish a presumption under a premarket statute that no substances
with such properties be permitted into commerce without scientific proof
beyond a reasonable doubt that they would not be problematic. Under a
postmarket statute such substances would be presumed to be hazardous to
human health or the environment and subject to removal, unless the firm
could show beyond a reasonable doubt that they were not. The presumptions
suggested here would be founded on policy decisions based on historical evidence that toxic substances or particularly problematic properties justified

different legal strategies than those for completely unknown substances and properties.[43]

Fifth, however, one could take Proposition 65 analogs a step further. That is, one could design legislation so that anyone—a research scientist, a public interest group, a union—who claimed to have *credible* evidence that a substance posed threats of serious or irreversible damage to persons or the environment could initiate a hearing by an impartial third party, such as a governmental agency, into the credibility of the evidence. If such evidence were found to be credible, the burden of proof would then fall on the manufacturer to show to some required level of certainty that the substance did not pose the threat alleged. In effect, such a law would create a legal mechanism for raising warning flags about threats of serious damage and would then shift the burden of proof—to provide information and to exonerate the substance or activity in question—to the manufacturer of the substance once a credible warning had been raised. The standard of proof such a polluter would have to satisfy in order to carry the burden would need specification: Would they only have to show that it is "more likely than not" that their exposures did not pose threats of serious damage or would they have to demonstrate it "beyond a reasonable doubt"?

Finally, we might look to Sweden as a model for how some of these suggestions could be implemented. Many of the burden-shifting and standard-of-proof-reducing strategies suggested have been adopted in Sweden as attempts to reduce the damage to human health and the environment. Generic strategies include the following: "a general obligation to investigate, as far as possible, the effects of the product on man and the environment, . . . to label the product and to inform all users as carefully as possible on measures to be taken to protect man and the environment during use and disposal of the product, . . . to [reduce or eliminate] to the extent possible substances in the product that make it hazardous, . . . [to find substitutes for hazardous substances and to utilize them] if it is possible to produce acceptable alternatives with smaller risks."[44] Once there is a scientifically based suspicion of risks from a product, the producer has the burden of proof to "show beyond a reasonable doubt, based on existing scientific knowledge and principles, that the suspicion is unfounded. . . ." and any remaining uncertainty about its hazardous properties must be borne by the manufacturer who wants to market the product, not the public.[45] While Sweden is a different country with different traditions, attitudes toward the environment, and political constituencies, it may serve as a useful concrete (vs. a mere theoretical) model for how some of the problems of serious threats to human health and the environment should be addressed.

These suggestions are only a beginning for thinking about implementing the Precautionary Principle. We should continue to develop and implement alternative ways of thinking about some of the difficult problems of addressing toxic substances in the environment. Such strategies should help overcome asymmetric information about the toxic properties of substances, asymmetric political forces favoring their introduction, and many of the scientific and legal burdens and standards of proof that currently handicap action on substances that pose serious or irreversible threats of damage to human health or the environment.

NOTES

1. James E. Hickey, Jr. and Vern R. Walker, "Refining the Precautionary Principle in International Environmental Law, *Virginia Environmental Law Journal* Vol. 14, p. 426 (1995).
2. Scott Altman of the University of Southern California Law School pointed this out.
3. Different countries and different parts of the legal system may adopt a more or less preventive strategy at present as I discuss in Section III.
4. Nicholas Ashford, "A Conceptual Framework for the Use of the Precautionary Principle in Law," delivered at the Wingspread Conference on Implementing the Precautionary Principle, January 23–25, 1998.
5. Frances Kamm and Michael Outsuka of the UCLA Philosophy Department made this point.
6. Carl F. Cranor, *Regulating Toxic Substances* (New York: Oxford University Press, 1993), p. 3.
7. I emphasize the point about inferences because of what appears to be scientists' asymmetric concerns to prevent false positives and their lesser concern to prevent false negatives.
8. Cranor, *Regulating Toxic Substances*.
9. The first view was endorsed by H. J. Eysenck, "Were We Really Wrong?" *American Journal of Epidemiology* Vol. 133(5), p. 432 (1991). For a contrary view that has a more health protective orientation, see Sander Greenland, "Invited Commentary Science versus Public Health Action: Those Who Were Wrong Are Still Wrong," *American Journal of Epidemiology* Vol. 133(5), pp. 435–436 (1991).
10. Arthur Furst, "Yes, But Is it a Human Carcinogen?" *Journal of the American College of Toxicology* Vol. 9, pp. 1–18 (1990).
11. These laws are described in more detail in the U.S. Congress, Office of Technology Assessment, *Identifying and Regulating Carcinogens* (Washington, D.C.: Government Printing Office, 1987), pp. 199–220.
12. The government also has the burden of proof in typical postmarket *regulatory* statutes, except that governmental agencies do not face typical standards of proof in regulatory settings like those in the criminal law or in the tort law. Typically, under a postmarket regulatory statute, the government must marshal

sufficient evidence and implement appropriate policies so that it does not "act arbitrarily or capriciously" or so that it enacts the policy that is "based upon substantial evidence in the record." The quoted phrases are standards of judicial review to which agencies would be subject by appellate courts should they choose to enact a regulation under two distinct sets of regulatory procedures. Agencies must marshal sufficient evidence and act in accordance with procedures so that they are not subject to having their regulations overturned.

13. Alexander Volokh, "Guilty Men," 46 *U. Pa. L. Rev.* 173–216 (1997).

14. A myriad of factors, some having to do with the values at stake in the law, some having to do with more pragmatic matters, shape the assignment of burdens of proof and the stringency of standards of proof. Fleming James, Jr., and Geoffrey C. Hazard, *Civil Procedure,* 2d. Edition (Boston: Little, Brown and Company 1977), pp. 249–262.

15. Hickey and Walker, p. 449.

16. Most of us would take precautions if we knew that there was a 40 percent chance of being killed or that a certain airplane had a 25 percent chance of crashing if it takes off. In neither case does a preponderance of evidence support the risk in question. I owe this point to David Dolinko of the UCLA Law School.

17. Precautionary action to protect our own or other's children might fall into the second category. While we may have little or no empirical evidence of a threat, we may nonetheless take precautionary actions because of the great value we attach to protecting them.

18. Calif. Code of Regulations, Title 22, Sec. 12705.

19. Sara M. Hoover, Lauren Zeise, William S. Pease, Louise E. Lee, Mark P. Henning, Laura B. Weiss, and Carl Cranor, "Improving the Regulation of Carcinogens by Expediting Cancer Potency Estimation," *Risk Analysis* Vol. 15, No. 2, April, 1995, pp. 267–280, and Carl F. Cranor, "The Social Benefits of Expedited Risk Assessment," *Risk Analysis* Vol. 15, No. 4, June 1995, pp. 353–358.

20. In 1984, 78 percent of chemicals in the United States with production volume greater than one million pounds per year lacked even "minimal toxicity information." National Research Council, *Toxicity Testing* (Washington, D.C.: National Academy Press, 1984), p. 84. Little has changed in thirteen years. Environmental Defense Fund, *Toxic Ignorance* (1997).

21. "Science and Policy Issues in the Application of Alternative Carcinogen Identification Methods," (1995) (funded by NSF Grant SBR-93107995, C. F. Cranor, PI).

22. Lester Lave and Gilbert S. Omenn, "Cost-Effectiveness of Short-Term Tests for Carcinogenicity," *Nature* Vol. 324, p. 29 (1986); Talbot Page, "A Framework of Unreasonable Risk in the Toxic Substances Control Act," in *Annals of the New York Academy of Science Management of Assessed Risk for Carcinogens,* p. 145 (W. J. Nicholson, ed. 1981). Also the author used a similar modeling method in "The Social Benefits of Expedited Risk Assessment," *Risk Analysis* Vol. 15, p. 353 (1995).

23. Carl F. Cranor, "The Normative Nature of Risk Assessment Features and Pos-

sibilities," *Risk, Health, Safety and Environment* Vol. 8, pp. 123–136 (Spring 1997).

24. Carl F. Cranor, "Eggshell Skulls and Loss of Hair from Fright: Some Moral and Legal Principles which Protect Susceptible Subpopulations," *Environmental Toxicology and Pharmacology* Vol. 4, pp. 239–245 (1997).

25. Dale Hattis, "Variability in Susceptibility: How Big, How Often, for What Responses to What Agents?" *Journal of Toxicology and Environmental Pharmacology* Vol. 4, pp. 195–208 (1997).

26. Some courts currently adopt screening rules that would seem to preclude consideration of scientific evidence on which scientists themselves would and have relied upon in order to draw conclusions about the carcinogenicity of substances. For example, IARC classified ethylene oxide as a *known* human carcinogen on the basis of mixed epidemiological studies, animal studies, and mechanistic studies, an evidentiary basis that would not be permitted in many jurisdictions.

27. Carl F. Cranor, John G. Fischer, and David A. Eastmond, "Judicial Boundary-Drawing and the Need for Context-Sensitive Science in Toxic Torts after *Daubert v. Merrell-Dow Pharmaceutical,*" *The Virginia Environmental Law Journal* Vol. 16, pp. 1–77 (1996), and Carl F. Cranor, "Discerning the Effects of Toxic Substances Using Science without Distorting the Law," *Jurimetrics* (forthcoming).

28. Under the Federal Fungicide, Insecticide and Rodenticide Act, the EPA must find that a substance registered for use must not pose "unreasonable risks to health or the environment," whereas the FDCA requires, for example, that food additives be found "safe." At least on the face of it, the latter seems like a more health protective standard than the former.

29. It appears that this piece of legislation was designed to avoid delays in the pre-market review process that may occur under legislation governing food additives, drugs, and cosmetics or legislation governing pesticides.

30. *Industrial Union, AFL-CIO v. American Petroleum Inst.*, 448 U.S. 607 (1980)

31. *Id.* at 642–644, 653.

32. *Occupational Health and Safety Letter* Vol. 22, No. 15, p. 112. In 1989 OSHA issued a regulation that set standards for 212 airborne toxicants based upon the best information available and largely in accordance with industry consent standards already in place, but the 11th Circuit Court of Appeals invalidated the entire regulation. Needless to say, this threw additional ice water on agency attempts to address a modest number of the hundreds or thousands of toxic substances to which people are exposed in the workplace.

33. 509 U.S. 579 (1993).

34. *See Daubert,* 509 U.S. at 587.

35. *Id.* at 590 (emphasis added).

36. *Id.* at 589. The dissenting opinion concurred in rejecting the Frye rule and also concurred that scientific testimony must be relevant but argued that the majority had gone too far in arguing for the "reliability" of evidence as part of Rule 702. Id. at 599–600 (Rehnquist, C.J., dissenting).

37. *Id.* at 589.
38. Cranor et al., "Judicial Boundary Drawing."
39. *General Electric v. Joiner,* 1997 Westlaw 764563.
40. The court decision appears to permit District Court judges to critically evaluate each piece of evidence on which plaintiffs would base their cases and to exclude those pieces of evidence that do not clearly support plaintiff's conclusions, even if the evidence considered as a whole might provide a plausible basis for plaintiff's conclusions. Justice Stevens was so concerned about this aspect of the opinion that he dissented from it. *General Electric v. Joiner* (slip opinion) pp. 5–8.
41. It may be more difficult to anticipate what the pollutants or by-products of production might be especially when these are emitted in small amounts. However, for large and consistent releases into the environment (e.g., especially the air or water) firms might be required before they are permitted to go into operation to establish which pollutants would pose serious threats of damage and then to eliminate or greatly reduce them.
42. The U.S. EPA may be following a procedure somewhat similar to this under premarket screening provisions of the Toxic Substances Control Act. Chemical structure and chemical activity are two features of substances that are checked under these provisions. Moreover, they are in the process of developing an elaborate computer program that mimics expert decisions about toxicity properties of substances in order to have a faster way of identifying problematic substances.
43. This would be much like legislative or judicial strategies for creating legal presumptions.
44. Bo Wahlström, "The Precautionary Approach to Chemical Management—Swedish Views and Experiences," p. 3, presentation to the Wingspread International Conference on Implementing the Precautionary Principle, Racine, Wisconsin, January 23–25, 1998.
45. Wahlström, p. 4.

Chapter 5

~

THE PRECAUTIONARY PRINCIPLE
AS A SCREENING DEVICE

David Ozonoff

Medicine has had its own Precautionary Principle since the days of the Hippocratic Oath, whose injunction, *Primum non nocere*, "First Do No Harm," has been its most quoted phrase. The Precautionary Principle is a kind of population and ecological analog to this. This chapter exploits another similarity with a common maneuver in clinical medicine, the medical screening test. Such tests (e.g., a routine chest X-ray or a cervical Pap smear) aim to cast a wide net to catch potential problems, whose early detection might result in better prognosis or limit to harm. On the basis of the screening test results, it is usually necessary to take additional steps (more specific and usually more invasive diagnostic tests like a biopsy). There is a cost to the confirmatory tests, both in economic costs and in terms of health risk. These costs must be weighed against the benefits of making an earlier diagnosis, say in economic terms or in improved prognosis or lessened disability (i.e., in quality of life).

Similarly, the Precautionary Principle employs available scientific evidence to make a judgment concerning the possibility of environmental or societal harm and suggests (in vague terms) that if such a possibility exists (threshold unspecified), additional "precautions" should be taken. Usually such precautions will impose costs of some kind on someone. These raise difficult problems of distributive justice, but I will not discuss such problems here, concentrating instead on other unexpected consequences of the analogy.

WHAT IS A "GOOD" SCREENING TEST?

Traditional criteria look at a test's reliability and accuracy. Rather than treating the Precautionary Principle and screening tests separately, let me just translate the language of screening tests into that of the Precautionary Principle. Reliability is another word for repeatability; that is, if another person were to apply the scientific method to "the evidence," how likely would she be to arrive at the same decision? Scientific interpretation involves a process not unlike assembling a large jigsaw puzzle with many pieces missing. The gaps are large, and it usually is not clear if a particular piece ("evidence") is a centrally important one, an insignificant one on the periphery of the picture, or perhaps not a true piece at all. Different scientists will assign different weights to the available pieces (and even their selection of appropriate pieces might differ) with resulting different pictures. The weighing of the various puzzle pieces can legitimately be done in so many different ways and be affected by so many different factors (many of them social and economic) that it is exceptional (and not of interest in this context) to have identical or even similar outcomes across the board. If we consider the evaluation or interpretation of scientific evidence as a screening tool it fails on the criterion of reliability for a decision instrument. By this measure, science as a method to make important societal decisions would be discarded outright. But of course we do not use reliability as our only criterion here, any more than we do so in other areas, like grading students, deciding college admissions, or even the routine clinical maneuver in the doctors office (like reading a chest X-ray).

One might even argue that the weakness in reliability is in reality a strength, allowing judgment and extrascientific factors to find their way into the mix. Most of us would be uncomfortable with having a strict algorithm (that could be done by a computer) govern important societal decisions. On the other hand, the large amount of "play" in the decision process opens the way for mischief of various sorts to affect important decisions. In response, environmentalists and public health advocates have tried at one time or another to increase reliability by stipulating certain rebuttable presumptions that a hazard exists. An example was the "generic carcinogen standard" of the late 1970s, which set generic criteria for when a chemical would be considered a carcinogen in an effort to avoid the uneven and time-consuming process of doing so on a chemical-by-chemical basis.

The lack of reliability also means that decision rules or tendencies can be crafted to lean more in one direction or another as a general principle. The Precautionary Principle is one such example. While phrased in terms of "scientific uncertainty," it is really a reflection of the lack of reliability in scientific evaluation.

The other measure of a screening tool's worth is its accuracy. Accuracy is how close the judgment comes to the "truth." It has two quantitative dimensions: sensitivity and specificity. Sensitivity is the proportion (or probability) that you will correctly identify a hazard, given it truly is a hazard. Specificity is the probability that you will correctly identify a nonhazard, given that it truly is a nonhazard. Unfortunately there is usually no "gold standard" to determine if something is "truly" a hazard. This has several consequences. Consider the following table classifying the outcome of a decision and the "real truth":

	judged a hazard	*judged not a hazard*
truly a hazard	a	b
truly not a hazard	c	d

In making a decision, one would obviously like to be in boxes *a* or *d*, that is, make the "right" decision. But we don't know which "row" we are in. We only control which column. The Precautionary Principle would say that if the first row is possible (i.e., there is some chance a hazard exists), the decision rule is to be in the first column. Most of the discussion of this problem in medicine or for social decisions deals with the costs to the patient or to society of the wrong guesses, (boxes *b* and *c*). Too little has dealt with the consequences to the decision maker of being in one column or another, although it is the calculation of this cost that can have a profound effect on the decision: Whichever column we elect (and for whatever reasons), no one can tell if we have arrived at the right answer (it is unknown), so the consequences to us, the decision makers, inhere in the consequences (to us) of the judgment.

For example, if we decide that a local facility poses a threat to the community, or that there is a cancer cluster in the town, we will certainly cause an upset in the local legislator (who now must "do something"), the Departments of Health and Environmental Protection (who also must take some action), local real estate interests (who don't want to have their town known as cancer alley), the governor's office (who must answer publicly for a plan of action), a potential responsible party (who then hires opposing experts to make us look foolish and incompetent), and we may be interviewed by the news media who will ask difficult questions we don't know how to answer. On the other hand, if we opt for "no problem" (column 2) the consequences are usually much less severe and dramatic. We may incur the wrath of a small group of vocal activists who have pushed for an investigation for many years, perhaps irritating town officials, neighbors, and others in the process, but most everyone else will be relieved. Soon, experience teaches us, the problem will go away. The personal calculus of the average decision maker is clear, especially as he or she will never be "caught" since the true state of affairs will remain indeterminate. Of course we cannot jump into column 2 arbitrarily. We need to construct a defensible argument to justify this, but this is normally not difficult within the bounds of current scientific practice. Indeed much current exercise of risk assessment or hazard evaluation by public officials can safely be put under the heading of decision justification.

Thus, if the Precautionary Principle is operationalized as "choosing a column," one must consider the incentives and disincentives for choosing one column over the other. It might seem that if the incentives for choosing the first column could be increased and disincentives decreased, and *mutatis mutandis* for the second column, we could achieve a better implementation of the Precautionary Principle. Unfortunately it is not quite so straightforward, even in principle. We would like a decision rule that is perfect in reflecting the true state of nature, that is, a rule with 100 percent sensitivity and 100 percent specificity. Usually we don't have this, although we can always choose to make a decision rule 100 percent sensitive (i.e., never miss a true hazard) by saying everything is a hazard (always in column 1). We will then identify all those that are truly hazards, but of course also misidentify many that aren't (so-called false positives). This is the extreme case of the Precautionary Principle. Likewise we could make a rule 100 percent specific by never saying something is a hazard (always in column 2), at the expense of many false negatives. This is the "trade-association decision rule." The flexibility noted in the discussion of reliability means that we can do some trade-offs of sensitivity for specificity according to social or philosophical factors without violating the underlying science. Realistic implementations of the Precautionary Principle will require some negotiation here. Indeed most of us would not be willing to say that we should always take precautions no matter how unlikely it is that harm will result. This is because most decisions carry some risk either way, so there is no risk-free ride (if there is no cost to the precaution, then the decision rule is easy but not very interesting).

Medicine's Precautionary Principle ("first do no harm") has not prevented it from making some costly and terrible mistakes. We need not always trade sensitivity for specificity, or vice versa, however. Our goal should be, by using the "best science," to attain both maximum sensitivity and specificity. Suppose we succeed in arriving at a rule that has both very high sensitivity and very high specificity (i.e., high overall accuracy). Now consider the question of how likely one is to be correct given that we decide there might be a hazard. Note that this is the reverse of the sensitivity question (which is how likely are we to say there is a hazard given that there really is one?).

This probability is called "the positive predictive value of the decision rule," and it turns out that even for very high sensitivity and specificity (say 95 percent sensitive and 95 percent specific), if the proportion of things that are truly hazards is low (i.e., if most things are not hazards), our positive predictive value will also be low. Said another way, if most things are not hazards, even highly accurate decision rules will get it wrong most of the time when they declare something to be a hazard. To use a clinical example, suppose we have a screening test for cancer that is highly accurate, say 95 percent sensitive and 95 percent specific, and the prevalence of a disease in the screened population is 1 percent (most tests or instruments are nowhere near this accurate). Then the probability that we truly do have cancer, given that the highly accurate screening says we might, turns out to be only one in six. The positive predictive value improves either with the accuracy of the test (and here it is specificity that has the most influence) or more importantly, with the prevalence of the disease.

Here is an environmental example. Suppose that of all waste sites, one in ten pose serious threats. This is perhaps not an unrealistic estimate since many waste sites involve little if any exposure. Suppose we weigh our decision procedure heavily to "err on the side of safety," thereby boosting our sensitivity to 99 percent, that is, correctly identifying something as a hazardous site when it truly is such. In the process of increasing our sensitivity, however, we could easily drop our specificity to 20 percent or so, meaning that when something is not a hazard, we only correctly identify it as such 20 percent of the time. In such a circumstance, we will wind up taking "precautions" six times out of seven when it is not necessary (arithmetic not shown, but straightforward).

The lesson for the Precautionary Principle is that you should apply your scientific evaluations to situations where the proportion of cases that are hazards is high and/or use methods with high specificity, that is, that correctly identify "no-hazard" situations. For many of us who have "bought into" the Precautionary Principle, this seems counterintuitive, especially the latter requirement. We are usually more interested in high sensitivity (if there is the slightest chance there is a hazard, take further action) rather than high specificity, but a quantitative analysis suggests that to disregard this will result in unacceptably low payoffs for high cost, not a situation that will encourage widespread implementation of the Precautionary Principle and a weak basis upon which to argue.

CONCLUSION

Thinking of the Precautionary Principle as akin to a screening test reveals interesting and unexpected features. First, we observe that as a decision "instrument" the scientific evaluation upon which it is grounded has low reliability. As the software makers have learned, this can either be viewed as a "bug" or a "feature," depending on your point of view. Second, the philosophical preference for "erring on the side of safety," which is integral to the principle, can lead to costs in certain circumstances that would make the Precautionary Principle a hard sell politically and socially.

REFERENCES

Herman, P., Hessel, S., Gerson, D., Blesser, B., and Ozonoff, D. "Disagreement Analysis as a Measure of Radiologic Performance," *Chest* 66:278–282, 1975.

Ozonoff, D., and Boden, L., "Truth and Consequences: Health Department Responses to Environmental Problems," *Science, Technology and Human Values*, 12:70–77, 1987.

Chapter 6

◁ ~◁ ▷

PRECAUTIONARY SCIENCE

Katherine Barrett and Carolyn Raffensperger

Scientific uncertainty is widely recognized as a condition for invoking the
Precautionary Principle. Although definitions and interpretations of the
principle vary, precautionary measures have been invoked in environmental
treaties and legislation at least since the 1970s.[1] A typical and frequently
cited formulation of the Precautionary Principle is the Bergen Ministerial
Declaration of 1990: "Where there are threats of serious or irreversible damage, lack of full scientific certainty should not be used as a reason for postponing measures to prevent environmental degradation." Although precaution is intuitively appealing, implementation of the Precautionary Principle
in a consistent and broadly acceptable manner has been fraught with philosophical, legal, political, and scientific problems. This is, perhaps, because
the principle incorporates all of these elements in potentially (but not necessarily) antagonistic ways.

This chapter addresses the question of whether the Precautionary Principle is antagonistic to science. Some critics claim that precautionary measures are anti-science because they incite us to take action without waiting
for definitive scientific proof of harm. Such action may be construed as a disregard for scientific knowledge and irreverence toward scientific authority.
Consequently, the Precautionary Principle may be perceived as a threat to

current "science-based" environmental and health regulations because the legitimatory power of "sound science" would be severely weakened if action was justified prior to scientific certainty or consensus. The problem, in other words, is that regulators, the scientific community, and society in general may be reluctant to embrace the Precautionary Principle if such an approach is seen to undermine, or be completely devoid of, scientific knowledge.

However, as we discuss in this chapter, science and the Precautionary Principle are not inherently antagonistic. On the contrary, science can (and should) play a vital role in precautionary approaches. However, we must pay close attention to the variable constructions of science used in health and environmental regulations. In other words, what is at issue is the kind of science that is used to support the Precautionary Principle, and consequently the way in which "sound science" is characterized for the purposes of environmental regulation. We address the problem of "precautionary science" by comparing two ideals of "good" or "sound" science, namely the mechanistic and precautionary models. We suggest that while the mechanistic model lends itself to current public policies based on cost- or risk-benefit calculations, the precautionary model is an alternative, rigorous scientific method and a more explicit scientific ethic that can provide effective support for precautionary policies. Although our paper focuses on a specific case study, the release of genetically engineered organisms, our arguments may contribute to an alternative decision-making process for other indeterminate, potentially hazardous technologies.

SOUND SCIENCE

"Sound science," or "good science," functions as an authoritative basis for many current environmental policies. For example, national and international regulations for the release of genetically engineered organisms (GEOs) are often explicitly grounded in sound science or science-based frameworks. This emphasis on the authority of science often coincides with a marked move toward cost-benefit and/or risk-benefit policy frameworks.[2]

It is worth examining the concept of sound science more closely. Because sound science is rarely defined in policy documents, we must consider how it is employed in policy making. Liora Salter's concept of "mandated science" provides valuable insight in this regard.[3] Mandated science is all scientific research "produced and/or interpreted for the purposes of public policy." Through case studies of chemical exposure standards, Salter and co-authors found that certain types of scientific knowledge are more likely to be mandated than others. Specifically, . . . "the science [policy makers] seek is one that is capable of being justified and explained to a wide variety of publics. . . . It must facilitate clear choices. It must represent a body of evidence on which decisions can rest and be seen to be rational." To this end, science used in policy is often idealized. For example, science is frequently portrayed as a value-free and disinterested endeavor that strictly follows the scientific method and is publicly validated through open debate. Salter describes this ideal science as a "protective shield" that may be used to justify policy decisions. Building on Salter's argument, if we wish to implement the Precautionary Principle in a manner that gives an appropriate role to scientific research, we must try to unearth and understand the scientific ideals that support specific policy trends.

TWO MODELS OF SCIENCE IN ENVIRONMENTAL POLICY

We propose that current environmental policies that are based on risk and/or cost-benefit calculations require, and therefore mandate, a particular scientific ideal, which we will call "mechanistic science."[4] The mechanistic ideal lends support to policies that are based on risk assessment because it overstates the accuracy and authority of calculations and predictions required for such assessments. Moreover, the mechanistic model tends to frame scientific uncertainty either in terms of a temporary and surmountable lack of knowledge or as inexorable and extrascientific. Under this view of science, strong precautionary measures may appear alarmist, defeatist, or antagonistic to the power of scientific inquiry. The first part of this chapter outlines characteristics of mechanistic science that lend authority and legitimacy to risk-based policies.

In the second part of the chapter, we will compare the mechanistic ideal with an alternative model of science, namely a precautionary model. We suggest that precautionary science can provide a solid basis for the goals and values of the Precautionary Principle. The precautionary model incorporates a broader perspective than mechanistic science, a perspective more consistent with the complexity of the environmental and health hazards we are currently facing. In doing so, precautionary science adopts different methods, raises new questions, and provides an active role for scientific inquiry.

We fully acknowledge that both the mechanistic and precautionary models are ideal types of science and that no research endeavor will, nor should it, fall squarely into one category or the other. However, these models are an effective tool for examining the kind of science that is currently promulgated in environmental policy, and the effects of this policy trend. Furthermore, our outline of precautionary science suggests an alternative ideal toward which science might strive in an effort to better address complex environmental and health issues in a manner consistent with the Precautionary

TABLE 6.1. Characteristics of Mechanistic and Precautionary Science

	Mechanistic Science	Precautionary Science
Authority of Science/Scientists	• separation of science from social issues • exclusive peer review system • closure and consensus	• multidisciplinary approaches • inclusive peer review • co-problem solving • open-ended dialogue
Definitions of Harm	• direct harm measured by few variables	• disruption of biological, ecological, or social systems
Points of Reference	• molecular or organismic time • human	• ecological or evolutionary time and multigenerational • nature • all species
Error and Burden of Proof	• Type I minimized • Type II maximized (fewer false positives) • burden on public • explanations in terms of causality	• Type II minimized (fewer false negatives) • burden on proponents/ producers • explanations in terms of pattern and association
Evidence and Data	• empirical, experimental • quantitative • replicable • deductive	• analytical, experiential, empirical, and experimental • qualitative and quantitative • inductive and deductive
Uncertainty	• lack of data or extrascientific	• indeterminacy

Principle. Table 6.1 summarizes the characteristics of mechanistic and precautionary sciences.

MECHANISTIC SCIENCE

In very general terms, mechanistic science is reductionist science, which ignores the larger context of the research subject. This approach is typified by molecular or toxicological studies. However, mechanistic approaches are by no means limited to these fields, nor are these fields necessarily limited to mechanistic approaches.

In philosophical terms, the mechanistic model embodies many assumptions of positivist science in which methods are based primarily on deduction, experiments are replicable, theories are predictive, and the scientific endeavor is considered to be value free. Although many such characteristics are recognized as naively idealistic (both within and outside the scientific community), the notion of empirically verifiable and value-free science retains authority in policy making. Significantly, we believe this is also the model of science taught in many school and university curricula.

We propose five specific characteristics of the mechanistic model on the following pages. These characteristics are a convenient way to show how mechanistic science can be effectively and productively applied to very specific scientific questions. However, we also aim to show that this approach to science is inadequate for addressing long-term, complex, and heavily value-laden environmental and health issues.

Authority and Autonomy

The mechanistic model portrays good science as simultaneously objective and disinterested and thereby demarcates science from value-laden ethical or political activities. According to this view, highly specialized scientific knowledge is developed and distributed without reference to political, social, or cultural contexts. This autonomous knowledge is then deemed "expert knowledge" and scientists are the "experts." Autonomy is further established through rigid boundaries among scientific disciplines, collection of data from isolated methodological or cognitive frameworks, and validation of research through exclusive peer review.

Separation of scientific and normative issues is politically convenient, if somewhat ironic given the blatant and pervasive role of science in policy making. Attempts to harmonize trade regulations under the World Trade Organization (WTO), for example, illustrated the potential power of a "neutral" science.

A larger and perhaps more firmly entrenched distinction lies between scientific risk assessment and political risk management. Despite meticulous documentation of the interplay between risk assessment and risk management, and the growing recognition that value judgments permeate both processes, separation of scientific and political activities remains a powerful source of credibility in policy making.[5]

In drawing strict boundaries around valid scientific questions,[6] the mechanistic model sets aside social, cultural, and political contingencies that permeate the production and use of scientific knowledge. This is convenient from a political point of view because the irrefutable, yet irreducible, uncertainty arising from interactions between science and society does not seem to compromise scientific authority.

Definitions of Hazard

The mechanistic model defines "hazard" as direct and measurable impacts on individuals or single components of a system, an approach largely adopted from toxicological and carcinogen studies. At present, the primary protocol for testing a public health hazard under risk assessment is the rodent bioassay for carcinogenicity.[7] Other protocols are being developed, including a screening and testing program for endocrine disruption. But generally, we test a single chemical for its cancer-causing properties. While it is widely recognized that there are other "harms," we have few settled methods for testing how, or even whether, a chemical causes endocrine disruption or ecological disturbances.

Definitions of environmental hazard in GEO regulations are also frequently limited to a well-defined "event," such as the creation of an uncontrollable pest.[8] Under this framework, the problem is simplified to the introduction of a new gene and "hazard" is limited to events that can be foreseen, but are not yet controlled. Consequently, risk assessments that are confined to short-term laboratory or field trials appear to adequately address the situation. Several unexpected adverse effects have been reported for GEOs released into the environment. These events challenge the adequacy of limited, linear definitions of harm.[9]

Points of Reference

When mechanistic science is mandated for political ends, expediency and short-term goals often assume paramount importance. Consequently, mechanistic science is often restricted in temporal and geographic scope, and studies tend to focus at the molecular and/or organismic temporal levels, which are more easily manipulated within short time periods and in confined areas. For example, risk assessments for GEOs are generally conducted on a small scale in several respects. Methodologically, assessments are limited to small geographic areas (lab, greenhouse, field trials) and small time scales (several years at most). Epistemologically, assessments are focused on genes or organisms (questions such as the effect of a new gene or differences between engineered and parent organisms, for example). In terms of ethics, broad and difficult questions about future generations, inherent value, or impact on other cultures often remain subordinate to short-term goals of efficiency and productivity.[10]

Error

Under the mechanistic model, scientific experiments are designed to avoid false positives. Indeed, it is considered "better" science to erroneously claim there is no effect (of a toxin, for example) than to erroneously claim there is an effect. In the language of statistics, good science, under the mechanistic model, aims to minimize Type I errors and consequently to maximize Type II errors. By this model, we would, for example, avoid claiming that GEOs are hazardous when they are actually safe. It would be more scientifically "sound" to claim that GEOs are safe when they are actually hazardous. In other words, science errs on the side of "no effect" and, therefore, requires stringent standards of experimentation and replication to prove there is an effect.

In one sense, this is a judicious approach for some scientific questions; discretion is used in claiming certainty. However, such an approach provides support for regulatory policies that valorize scientific certainty because lack of certainty (and, to reiterate, mechanistic science is conservative in claiming certainty) may be used as a rationale for proceeding with questionable technologies. The argument is that science has not proven harm, therefore, we can legitimately charge ahead. Clearly, there are ethical problems with this standard of proof. As Kristin Shrader-Frechette has well argued, avoidance of Type I errors places the public at risk, while protecting producers or proponents of hazardous technologies.[11] What may appear good for "pure" science, may prescribe disastrous policy, potentially resulting in serious harm to the environment and its inhabitants.

Validity of Data

Mechanistic science relies heavily on data that are empirically verifiable and quantitative. For example, since the mid-1980s, scientific and policy literature on GEOs has emphasized the need for accurate measures of probability and empirical evidence of "an event." A frequently stated scientific rationale for continuing with large-scale GEO release is that no such evidence of hazard has been observed in confined field trials.[12] In the GEO case, negative evidence, or a "no data observed" argument, is used to support the hypothesis that GEOs are safe. Following this line of reasoning, scientific uncertainty arises from a temporary gap between current observations and theory; it is uncertain whether or not the observations support the theory. Ideally, this discrepancy could be strengthened through further experimentation.[13]

However, there are several problems with relying on further empirical evidence to resolve uncertainty. Many environmental and health hazards are considered "low probability/high consequence hazards." Because the chances of observing a low frequency event are very small by definition, a null hypothesis can easily be supported by negative data. But how many "safe" assessments do we need to conclude that a technology is "safe enough" or that a risk is "acceptable"? Gathering positive evidence of hazard, on the other hand, is relatively difficult. Taken to the extreme, definitive proof of hazard may require an environmental disaster to occur. In fact, this situation presents a paradox for research on broad-scale environmental hazards: Negative data tell us little about possible effects; yet seeking positive data may condone using "society as a laboratory."[14]

These problems are exemplified by current procedures used to conduct GEO field trials. As discussed previously, field trials are limited in geography and time and in the types of harm they aim to test. In many cases, field trials are designed to examine the efficiency of the genetically modified trait, rather than ecological hazards per se. Through limited frames of reference and use of confinement measures that prevent gene escape, it becomes relatively easy to support the hypothesis that GEOs are safe. Major assumptions have not been challenged in the technological leap between controlled small-scale trials and uncontrolled commercial releases in many different environments.

Uncertainty

We have, so far, discussed two related definitions of uncertainty used in the mechanistic model. In the Authority and Autonomy section, we saw that some types of uncertainty are considered beyond the boundaries of scientific activity and responsibility. This type of uncertainty is portrayed as an uncontrollable, inexorable function of ethical, social, or political factors. In everyday language we might say "Nothing is risk-free" or "Life requires us to take some risks." Under the mechanistic model then, this type of uncertainty is seen neither to arise from nor impinge on scientific inquiry. It is, in other words, extrascientific.

We then discussed a second type of uncertainty. In contrast to the first definition, uncertainty defined as a temporary lack of data is considered to be within the boundaries of the mechanistic model, and, therefore, a valid, albeit manageable, scientific problem. This definition is based on the idealistic notion that all scientific uncertainties can be accounted for and controlled—a view premised on a world that is determinate (ultimately knowable) or probabilistic (calculable).

Narrow frames of uncertainty raise significant philosophical and ethical problems, as we have discussed. These problems highlight a further consideration that is crucial for effective implementation of the Precautionary Principle. We should be careful in assuming that acknowledging scientific uncertainty will necessarily lead to precautionary regulations, or, in Salter's terms, that recognizing uncertainty removes the "protective shield" of mechanistic science. There are many definitions of uncertainty, ranging from statistical to stochastic to inherent ignorance.[15] This flexibility may be used to imbue policies with cautionary language, while failing to embrace the values of the Precautionary Principle. Herein lies the danger: We may condone regulations which have only a thin gloss of precautionary language, but which, at a practical level, rely on balancing risks, costs, and benefits. This is dangerous because the accuracy with which these calculations can be made has been exaggerated, and it is the environment and society that stand to lose. To illustrate this point, we describe three types of regulations that have such a precautionary gloss.

First, regulations that are enacted in advance of a specific adverse event may appear to be, or even claim to be, precautionary yet the actual policy framework may rely heavily on risk or cost-benefit trade-offs. Any regulations that are installed prior to a disaster are precautionary in some sense. However, to base subsequent actions and policies on the outcome of cost/risk calculations fails to appreciate the inherent limits of scientific certainty—limits that invoked the "precautionary" policies in question. A useful distinction might be drawn between development of policies in a precautionary manner (i.e., before a disaster, many policies fall into this category) and policies that advocate a precautionary approach to the development of technology. The former concerns the act of policy making and may be consistent with the risk/benefit framework. The latter concerns the content of policies and is more consistent with a genuinely precautionary approach to hazardous technologies.

Second, some regulatory frameworks explicitly contain both precautionary and cost-benefit elements, for example, the "not entailing excessive cost" (NEEC) part of "best available techniques not entailing excessive cost" (BATNEEC). This framework entails an initial cost-or-risk benefit assessment, then proceeds on the side of caution.

Third, the step-by-step approach advocates taking safety "precautions" while the technology proceeds. Each step provides more (negative) scientific evidence to rationalize proceeding to the next step. This approach has been strongly advocated for GEO regulation[16] and seems more akin to trial and error. Reactionary measures to prevent further harm may be adopted only after something goes wrong.

The above examples are "cautionary" in that they aim to control or mitigate known adverse effects. However, such decision-making frameworks do not take us very far toward a genuinely *precautionary* approach because they fail to question the ideals and assumptions that underlie the mechanistic model and therefore fail to *anticipate* unknown and uncontrollable effects. As a result, the legitimatory power of "sound science" remains unchallenged and the mechanistic ideal is further entrenched in policy decisions.

To summarize, by idealizing the accuracy of scientific assessment and prediction, the mechanistic model adds legitimacy to policies that are based on risk-cost-benefit assessments (but rarely require assessment of benefits), place the burden of proof on the risk averse, consider a narrow range of expertise, and favor the status quo through incremental regulatory changes. Policies founded on the Precautionary Principle require a more radical approach: We must reconsider current concepts of good science.

PRECAUTIONARY SCIENCE

We have suggested that the mechanistic ideal of science is inappropriate and inadequate to account for the complexities of current environmental and health issues. Nevertheless, mechanistic science provides strong support for risk-based policies and, in doing so, glosses over the need for strong precautionary measures. As a result, the Precautionary Principle may appear unnecessary or anti-science. To address this problem, we present precautionary science as an alternative ideal toward which scientific research might aspire, in an effort to confront uncertainty and complexity "head-on." In doing so, precautionary science can make a vital contribution toward implementing the Precautionary Principle.

Authority and Autonomy

In the precautionary model, scientists act as co-problem solvers in a broad community of peers. This community extends not only beyond the boundaries of individual disciplines but also beyond the traditional boundaries of the scientific community. There are both methodological and ethical reasons for broadening the scientific peer community to include a wide range of expertise.

As discussed previously, the environmental and health problems that precautionary science aims to address are often far-reaching in their geographic and temporal effects. Such methodological and ethical complexity lies beyond the scope of a single scientific discipline. Shrader-Frechette and McCoy,[17] for example, argue that research methods, theories, and empirical bases in ecology, as well as in more reductionist sciences, are underdetermined. As a result, isolated scientific disciplines cannot provide a strong basis for environmental policy. For such reasons, scientists must participate in research that is multidisciplinary (e.g., incorporates social sciences), multilevel (e.g., considers networks and relationships), and community based (e.g. includes many different value judgments). For example, in making decisions about GEO technology, farmers, First Peoples, and nongovernment organizations could contribute an informed perspective particularly when and where they have observed the natural world for long periods of time and understand the local agricultural methods. Precautionary science recognizes, in other words, that research priorities, data, and conclusions are shaped by social context and values. A more inclusive peer community should therefore contribute to all stages of the research project, from formulating questions to reviewing outcomes.

Environmental and health issues, in other words, are societal questions as well as scientific questions. Research priorities and policy decisions will affect how people live and the choices available to future generations. Ulrich Beck, for instance, has argued that environmental hazards are democratic because affiliation with a particular group or class does not guarantee safety.[18] Other researchers have suggested that marginalized groups tend to bear more environmental and health risks than other groups (such as proponents of the technologies). In either case, it is clear that the public has a vital interest in the outcome of environmental and health decisions. For such ethical reasons, the problem-solving community ought to be extended beyond the narrow peer community of the mechanistic model.

Definitions of Hazard

The definitions of hazard used in mechanistic science could be expanded in at least two respects. First, while mechanistic science focuses on direct effects, precautionary science seeks indirect, secondary, cumulative, and synergistic interactions. Second, as discussed earlier, precautionary science considers the context in which these effects are, or may be, manifested. Framing the scientific questions in broader terms necessitates biology and technology assessments that consider local-and-global ecological and social conditions. As a result, precautionary approaches often raise prior questions of appropriate or alternative technologies, rather than attempting to establish an acceptable level of risk at the outset. For example, the potential hazard of releasing GEOs might include: effects on food webs and soil structure; changes in agricultural practices and concomitant changes in lifestyle and quality (e.g., loss of biological control options or increased exposure to chemical herbicides); effects of technology transfer; and the outcomes of community choice.

Points of Reference

A precautionary model of science adopts an ecological, evolutionary, or multigenerational timeframe or a combination of these timeframes. Our comparison with mechanistic points of reference is based on E.O. Wilson's four timeframes: biochemical time, organismic time, ecological time, and evolutionary time.[19] Precautionary research programs encompass ecological, if not evolutionary, time because the consequences of introducing a new chemical or technology into the environment can rarely be observed in bio-chemical or even organismic time. Endocrine disruption provides one of the best examples of the need to study multigenerations.[20]

While we affirm the value of biochemical timeframes and other reduc-tionist methods for some research problems, we would like to emphasize the limitations of these approaches. Reductionism has been defined as "the breaking down of things into smaller and smaller parts, so that if you under-stand how the small parts work that will in principle tell you how the big thing works."[21] Reductionism makes an inquiry manageable. But a precau-tionary model requires going beyond reductionism to a full understanding of the whole as well as the parts and suggests that the whole is more than the parts. Ecological or public health problems rarely lend themselves to facile reductionism: They require contextual thinking. As noted earlier, a GEO will respond differently in a controlled field experiment than under broad commercial application with different weather conditions and in a different ecological system.

Error

As explained previously, science under the mechanistic model avoids find-ing that there is an effect when there is really no effect. That is, the researcher is less likely to claim that a false positive is significant. In con-trast, precautionary science errs in the opposite direction. Under this model, it is better to claim there is an effect when there is none than to falsely claim there is no effect. For example, it would be better to erroneously find that technologies such as GEOs are hazardous than to erroneously claim they are safe. In fact, following false positives may generate more research questions and ultimately yield more information than following false negatives. Was the positive correct? What contextual factors led to a positive result in this particular instance? A false negative, on the other hand, may lead to a sci-entific dead end and a premature conclusion that no further research on the potential harm is necessary. Accordingly, rather than reflect an anti-science position, precautionary science may generate more research.

Such a bias in the kind of error is not suitable for all scientific inquiry. However, where the public or the environment is endangered by the effects under question, precautionary science takes a precautionary approach and thereby helps to protect unwitting subjects of technologies. Such an approach seems preferable, even respectful,[22] from an ethical point of view because those who typically stand to gain most from the technologies (i.e., producers and proponents) also then bear the financial risk of demonstrating safety.[23]

Validity of Data

Characteristics of precautionary science described thus far (an inclusive peer community, expanded definitions of harm, broad frames of references, and minimizing errors that may harm the public) suggest a further need to redefine concepts of causality and valid data. For example, decisions about chemical use have been made on the basis of tests for carcinogenicity. Here, a standard of direct causality (i.e., chemicals causing cancer) was useful for testing single chemicals for a single harm.

However, at this time, we have no such protocol for endocrine or ecological disruption, and proving direct causality in complex systems over large geographies and time periods is an extremely onerous, if not impossible, standard. (In the United States, the Environmental Protection Agency is developing a testing protocol for endocrine disruption.) Recognizing such problems, the precautionary model effectively brackets questions of strict causality as the first order of business and places more emphasis on indirect and acausal relationships, such as correlation, pattern, and association. One example of this approach is the study on Agent Orange performed under the aegis of the U.S. Institute of Medicine (IOM). Scientists were asked to give the veterans the benefit of the doubt and to look for patterns and association rather than to determine whether Agent Orange certainly caused the veterans' illnesses. The IOM committee did in fact determine that Agent Orange was associated with some of the illnesses under question.[24]

This shift in emphasis relaxes the stringent requirement that all valid data are quantifiable and replicable and that all theories are predictive across wide-ranging circumstances. While not disposing of these principles altogether, precautionary science emphasizes, indeed requires, additional concepts of valid data. Specifically, a precautionary approach to environmental and health problems includes qualitative as well as quantitative data; places increased emphasis on inductive, context-sensitive inquiries, such as case studies; and values experiential as well as experimental information, including traditional, folk, and local knowledge. For example, native people in the Arctic have made extensive observations about changes in reindeer and whale health over the decades, if not centuries. Scientists seeking to understand these animals will perform a more rigorous analysis by incorporating the observations and knowledge of the native people into their research. Even better, scientists would do well to ask the native people what they should seek: The questions might be far more relevant to ecological well-being than those drawn up in the university setting.

Uncertainty

We suggest that precautionary science more effectively addresses the uncertainties inherent in complex environmental and health problems, because this model of science more explicitly confronts such uncertainties. Specifically, precautionary science directly addresses the complexity of issues (the number of, and relationships among, variables) and the uniqueness of biological, ecological, and social contexts, which render precise experimental replication problematic if not meaningless. In other words, precautionary science works on a general principle of indeterminacy rather than determinacy. The work of Brian Wynne and others is key in elucidating this concept.[25] Indeterminacy, as we interpret it, describes situations that are open-ended because of complex and dynamic interactions among physical, social, ethical, and cognitive elements.

Examples of the ubiquity and significance of indeterminate systems are eloquently narrated by Gary Nabhan in his book, *Enduring Seeds*.[26] Through vivid case studies of the conservation and extinction of Native American agriculture, Nabhan recounts how some seed varieties have "endured" many centuries of cultivation, while other varieties have been lost. The fate of such crops (e.g., the persistence of wild teosinte or the extinction of sump-weed) lies at the confluence of many open-ended and intertwined paths: biological and ecological processes, environmental patterns, beliefs and cultural traditions, knowledge systems, and political decisions. The outcome cannot be explained, much less predicted, by reductionist approaches to any one element. Such "uncertainty" cannot be categorized as either uniquely scientific or completely extrascientific. It is both an integral part of the scientific endeavor and inseparable from environmental and health issues. This type of uncertainty is also, however, profoundly uncontrollable, largely irreducible, and reveals our inherent ignorance of socially embedded, dynamic systems. In Wynne's terms, often we simply "don't know what we don't know." Such factors tend to be buried in mechanistic science and consequently are isolated from regulatory frameworks dependent on risk-cost-benefit calculations or with a "precautionary gloss."

However, ignorance need not be derogatory, and indeterminacy need not be debilitating. Acknowledging indeterminacy has profound repercussions for the methods and the role of science. It entails an appreciation for the naturally and socially constructed boundaries of our knowledge, and for our situation in, and our influence on, scientific research. Precautionary science invites us to make explicit the boundaries of our knowledge by unearthing complexity, ignorance, and values and thereby revealing how our concepts of certainty are defined.

The Role of Science in the Precautionary Principle

How, one might argue, can we be explicit about what we don't know? This is the strength of precautionary science and the vital role that it can play in implementing the Precautionary Principle. As discussed throughout this chapter, we can never reach absolute certainty, never obtain absolute proof, about cause and effect in complex systems. It is therefore more important and urgent to appreciate our limitations. The methods of precautionary science described previously are suited to the challenge. A common thread among these methods is a broader, more inclusive sense of community, a sense that encompasses both ecological and social levels. Recognizing the complexity of ecological communities and engaging with a broad community of peers will help to reveal questions that have previously been excluded, factors and perspectives that have not yet been considered. This insight can be, and should be, used as a resource for scientific inquiry, a guide for further study, and an impetus to responsible action.

The Precautionary Principle holds promise of a new direction for addressing complex environmental and health issues, a direction that requires alternative scientific methods, regulatory policies, and societal values. Already there are significant precedents for precautionary science in sustainable agriculture, preventive medicine, and public interest research, among others. In the interest of health and the environment, these initiatives ought to be nurtured and encouraged. If society chooses to implement the Precautionary Principle, the precautionary model can provide an ethically and scientifically sound footing.

NOTES

1. See Boehmer-Christiansen, S. 1994. "The Precautionary Principle in Germany—Enabling Government." In: T. O'Riordin and J. Cameron (eds.), *Interpreting the Precautionary Principle*. Earthscan, London.
2. For example, for Canada see *Regulation of Agricultural Products of Biotechnology*. 1993. Agriculture and Agri-Food Canada; for the United States see: *Exercise of Federal Oversight within Scope of Statutory Authority: Planned Introductions of Biotechnology Products into the Environment*. 1992. Federal Register 57, 6753-6762; for OECD, see *Safety Considerations for Biotechnology: Scale-Up of Crop Plants*. 1993. OECD, Paris.
3. Salter, L., with Levy, E., and Leiss, W. 1988. *Mandated Science: Science and Scientists in the Making of Standards*. Kluwer Academic Publishers, Boston.

4. The concepts of mechanistic and precautionary science are adapted from: Raffensperger, C., "Incentives and Barriers to Public Interest Research." Paper presented to Switzer Fellows, California. September, 1997. See also: Wynne, B. and Mayer, S., June 5, 1993. "How Science Fails the Environment." *New Scientist* 138(1876):33–35.

5. See Harrison, K. and Hoberg, G., 1994. "Risk, Science, and Politics." *Regulating Toxic Substances in Canada and the United States*. McGill-Queen's University Press. Montreal; Jasanoff, S., 1990. "The Fifth Branch." *Science Advisors as Policy-Makers*. Harvard University Press, Cambridge, MA.

6. For a discussion of boundaries in science, see Gieryn, T., 1983. "Boundary Work and the Demarcation of Science from Nonscience: Strains and Interests in Professional Ideologies of Scientists." *American Sociological Review* 48:781–795; Jasanoff, S., 1987. "Contested Boundaries in Policy-Relevant science." *Social Studies of Science* 17:195–230.

7. National Research Council, 1994. *Science and Judgment in Risk Assessment*. National Academy Press, Washington, D.C., pp. 140–141.

8. See Levidow, L., Carr, S., von Schomberg, R., and Wield, D., 1996. "Regulating Agricultural Biotechnology in Europe: Harmonization Difficulties, Opportunities, Dilemmas." *Science and Public Policy* 23(3):135–157.

9. Reviewed in: Altieri, M.A., 1998. The Environmental Risks of Transgenic Crops: An Agroecological Assessment. University of California, Berkeley. www.cnr.berkeley.edu/~agroeco3/risks.html.

10. For a review of GEO regulations among European Union members see: von Schomberg, R., 1998. "An Appraisal of the Working in Practice of Directive 90/22/EEC on the Deliberate Release of Genetically Modified Organisms." Final Draft Report for the Scientific and Technological Options Assessment of the European Parliament.

11. Shrader-Frechette, K.S., 1992. "Risk and Rationality." *Philosophical Foundations for Populist Reform*. University of California Press, Berkeley.

12. An influential and frequently cited document is the National Academy of Science report: "Introduction of Recombinant DNA-Engineered Organisms into the Environment: Key Issues." 1987. Washington, D.C.

13. See Wynne, B., 1992. "Uncertainty and Environmental Learning." *Global Environmental Change* 2:111–127.

14. Krohn, W. and Weyer, J., 1994. "Society as a Laboratory: The Social Risks of Experimental Research." *Science and Public Policy* 21(3): 173–183.

15. For example see: Wynne, B., 1992; Funtowicz, S.O., Ravetz, J.R. 1994. "Uncertainty, Complexity and Post-Normal Science." *Environmental Toxicology and Chemistry* 13:1881–1885.

16. The step-by-step approach has been strongly advocated by the OECD as early as 1986 and has been emulated by other national and international regulations. See OECD. 1986. *Recombinant DNA Safety Considerations*. OECD, Paris.

17. Shrader-Frechette, K.S. and McCoy, E.D., 1993. *Method in Ecology: Strategies for Conservation*. Cambridge University Press, Cambridge, England.

18. Beck, U. 1992. "Risk Society." *Towards a New Modernity.* Sage Publications, Thousand Oaks, CA.

19. Wilson, E.O. 1984. *Biophilia.* Harvard University Press. Cambridge, MA, pp. 39–45.

20. Colborn, T., Dumanoski, D., and Myers, J.P., 1996. *Our Stolen Future: Are We Threatening Our Fertility, Intelligence, and Survival?* Dutton, New York.

21. Penrose, R., 1995. "Must Mathematical Physics Be Reductionist?" In: Cornwell, J., *Nature's Imagination.* Oxford University Press, New York, p. 12.

22. Anita Bernstein, 1997. "Treating Sexual Harassment with Respect," *Harvard Law Review* Vol. 111.

23. Clearly this is a difficult problem. It is impossible to prove safety, beyond a shadow of a doubt. Indeed, that is the point of this paper. However, relations-based studies can, and should, be conducted and the financial cost of these studies should be borne by the proponents. These studies, should not, however, be conducted by proponents. Close ties between industry and government policy makers, and the relative ease of providing no data in support of no harm, may render such a "burden of proof" an advantage. Studies ought to be conducted by third parties.

24. Dr. David Kriebel, University of Massachusetts, Lowell, personal communication 1998.

25. Wynne, B. 1992.

26. Nabhan, G.P., 1989. *Enduring Seeds: Native American Plant Agriculture and Wild Plant Conservation.* North Point Press, San Francisco.

Chapter 7

THE POLITICAL ECONOMY OF PRECAUTION
R. Michael M'Gonigle

Outside the convention hall, the English media was filled with stories of combat on the high seas. Fishing boats from Scandinavia had surrounded a large incinerator ship (with the unfortunate name of the MV *Vulcan*) and had forced it to a halt. An anchor chain from a particularly aggressive fisher had been sideswiped around the *Vulcan*'s propeller and shut down its forward motion, terminating the toxic incineration process, and stranding the vessel in the middle of the ocean. As was usual with Greenpeace campaigns, this action in the North Sea was well calibrated to focus public attention on what was going on inside the meeting hall. There, at the London headquarters of the International Maritime Organization, the contracting parties to the ocean dumping convention[1] were gathered to discuss a proposed ban on ocean incineration. That meeting too was filled with conflict.

The convention (or London Dumping Convention (LDC), as it is generally called) is inherently controversial. Negotiated in 1973 after the conclusion of the Stockholm Conference on the Human Environment, the convention was primarily aimed to regulate the *disposal* of wastes, not their *prevention*.[2] But over the years, the debates had increasingly been pushed to consider broader prohibitions because, with many of the regulated substances such as low level radioactive waste and toxic metals, great long-term

123

uncertainty as to their effects has been attached to their release into the environment. These prohibitions, usually in the form of temporary moratoria, were always controversial. But nothing in my own background of international negotiations (including many years as Greenpeace's representative at the International Whaling Commission) prepared me for the vituperation of the exchanges that occurred that week in the fall of 1989.

The question was simple—on what basis should one make a decision whether to regulate or ban the incineration of wastes at sea? The answer was anything but simple. Scientific analyses underpin much of the decision making in the LDC, as in other regulatory bodies such as the stock assessments of the Scientific Committee at the Whaling Commission. The public posturing and negotiating around such analyses are usually left to the political representatives. But at this meeting, a major issue in the plenary session was the little-known Precautionary Principle, which, for the Greenpeace delegation and its scientific advisor, Dr. Peter Taylor, had become the foundation of its campaign to ban ocean incineration.[3]

Even now, ten years later, Dr. Taylor's patient explanations of the scientific basis of the principle, and why it necessitated a ban—and the angry denunciations he provoked from the scientific advisors of the major industrialized nations—ring in my ears. "This is not science, but politics" was the retort to Dr. Taylor, an exchange that was contemporaneously played out in a similarly vituperative running debate in the academic literature.[4] It continues to plague the issue today. But what was most startling about the exchanges was not the substantive differences as to the principle's scientific foundations, nor even the usual national concerns about the possible economic impacts of a ban on domestic industries. Instead, what caught everyone by surprise was the far more visceral, even personal, anger that advocacy of the principle elicited from its scientific opponents, for whom it seemed to challenge their scientific evidence, their academic credentials, and even their intellectual integrity.

Within a year, however, Greenpeace was successful and the LDC instituted a ban on ocean incineration. But the lessons of that process reveal a great deal more about the challenges of the precautionary approach than just its unique approach to scientific uncertainty. For one thing, reliance on the principle (at least in its essential form of *strong* precaution, as discussed here) can have serious negative consequences for the viability of established technologies and industries. For another, the decision to ban ocean incineration demonstrates the crucial role played by outside parties (in this case, nongovernmental organizations) in advancing new approaches to science and regulation over the recalcitrance of the major industrialized nations. In this regard, one of the lessons of that experience is that open decision-making institutions that grant public groups a participatory role are critical to institutional change.[5]

Overall, one can see in this single example that the substantive application of the Precautionary Principle and the process by which it is translated into action involve changing both the relations of economic and political power and the paradigms of analysis that are both embedded in and, in turn, underpin these relations. Moreover, as the examples discussed later reveal, the application of the principle extends well beyond toxic contaminants to encompass a range of resource management issues like forestry and fisheries, as well as larger concerns such as the maintenance of ecosystem integrity.

In short, inherent in the Precautionary Principle is a basic challenge to modern models of scientific regulation and, beyond that, to the way one does business and makes decisions about it. The existence of this challenge thrusts the debate into the realm of political economy. In the pursuit of a fuller understanding of the implications of the Precautionary Principle, this chapter addresses this political economy of precaution. More specifically, the chapter will develop a new perspective on this field through an explication of an *ecological* political economy (or "political ecology") of which the Precautionary Principle is an integral part. In turn, an appreciation of political ecology provides an invaluable guide for the principle's further development.

WHAT IS ECOLOGICAL POLITICAL ECONOMY?

To many people, academics and nonacademics alike, the phrase "political economy" causes the mind to cloud over. Politics we understand, even economics, but political economy is something else—a vague, nebulous concept. In summary, political economy can be understood as the study of society's way of organizing both economic production and the political processes that affect it and are affected by it. A good indication of the broad meaning of the concept can be discerned from the title of Adam Smith's classic early text on political economy, *An Enquiry into the Nature and Causes of the Wealth of Nations* (1776). Where, in other words, does wealth come from? How is it created? What is the role of the state in this process?

A century later, another classic text, Karl Marx's *Capital*, raised the stakes in the field by shifting the focus of political economy from the inherent workings of the self-regulating market (Smith's primary concern) to the broader dynamic of capital seen as a whole set of power relations. For example, scholars from all schools explain how market-based growth tends to concentrate capital in specific nodes of accumulation—hierarchical nodes (e.g., corporations), spatial nodes (cities), and class nodes (capital-owning individuals). In contrast, the more general process of market-based social development tends to be uneven. Thus, even resource-rich regions, such as farming and mining communities, that don't have economic capital or political power rely on the "trickle-down" effects of nodal accumulation.

Overall, therefore, the particular contribution of political economy as a field of inquiry has thus been its ability to uncover and explain what might be called the "system dynamics" of a society's processes of economic and political self-maintenance. In the process, political economy looks at the operation of a range of social processes such as economic mechanisms of production and exchange, legal rights including property and political rights, regulatory imperatives of bureaucracy, and so on. In so doing, it seeks to look at these institutions not just as isolated but as integrated aspects of a social whole, an important part of which are the cultural assumptions and values embedded in these structures. Referring to political economy as one of "the most sophisticated analytical techniques . . . applied to analyze social relations," one text describes political economy as "that part of economic theory which deals with the functioning of entire socioeconomic systems. In a somewhat looser sense it is also used to denote political–economic doctrines or comprehensive sets of economic policies such as liberal, conservative and radical."[6]

Our educational system ill prepares people to understand the concept of political economy, let alone its application by social scientists to the diverse associations of contemporary life. Put simply, most people are unaware of the dynamics embedded within our social institutions and how those dynamics drive the institutions and shape their evolution. Natural scientists have, in particular, been uncomfortable recognizing the existence (let alone the operation) of "social systems," preferring instead the precision of isolated variables, controlled environments, and quantitative calculation to which physical forces more easily lend themselves. Here again, the perception is that this study of political economy is not science (and, thus, not important). And yet, as postmodern theorists have demonstrated, it is our social structures that create and embody the forms of knowledge—including "science" itself—which we as members of these associations take as our real world, and on which in turn our institutions rely in their operations.[7] Truth, including the truths of science, is specific to the social context.

To speak of the political economy of precaution is thus to point to a more contextual and systematic way of understanding science, including its embodiment within the structures of economic production and political decision making and of visualizing reforms and alternatives. Indeed, an essential aspect of the Precautionary Principle is precisely its inherent tendency to prompt this sort of thinking. By revealing the scientific assumptions and biases built into existing patterns of production and regulation, the real relations of power inescapably emerge into focus. As a result, alternatives surface that would not otherwise reach the bargaining table.

Historically, the spectrum of political economy theorizing has been broadly delineated by the competing poles of "right" and "left," Smith and Marx, capitalist and communist. This polarity is, however, changing with the advent of ecological political economy (with which, as I have noted, the Precautionary Principle fits well) and its uncovering of aspects of economic production and political regulation not considered in the past. The huge historical importance of this development is still not widely appreciated by Western elites who, with the fall of the Soviet Union and the triumph of globalization, act as if economic growth has now left ideology behind. In contrast, political ecology identifies three major commonalities that underlie the visions of progress that have been held by both liberalism and marxism historically and that are still dominant today. These commonalities are, from an ecological viewpoint, heavily ideological.

First, both liberal and socialist approaches embrace the primary goal of attaining high levels of social production through economic growth, whether this growth is guided by the market or the state. This shared "productivism" translates into a shared embrace of another "ism," the one that has actually delivered the physical goods, namely, industrialism. Second, both approaches share a faith in the ability of large, central institutions to develop and guide this production, whether these institutions are private (corporations) or public (bureaucracies) or, as in the mixed economy of today, a melding of the two. This might be called the shared belief in managerialism. Third, these two approaches embrace the realm of science as the neutral or objective foundation on which both productivism and managerialism rest. Science drives increases in productivity, just as it provides governments with the tools to manage the impacts of this drive. This scientism is perhaps the most basic, and most subtle, of the shared assumptions on which the modern world has been built.

Political ecology challenges these beliefs[8] and does so in a manner that dovetails with the critique and program for reform offered by the Precautionary Principle. First, political ecology recognizes that the very fact of economic production entails physical damage, particularly through increasing entropy (or energy that cannot be used) that, in turn, is degrading the environment. Economic sustainability requires a move away from practices based on maximizing economic activity that requires high levels of resource throughput to practices that can achieve social goals by maintaining stocks of usable wealth with the least throughput. This new insight into the inherent costs of economic activity directly challenges the assumption of productivism that, ironically, equates the health of the economy with that very growth in the GNP which is the source of entropic decline.

Second, political ecology points to the self-interestedness and inefficiencies inherent in large centralized systems of power (whether these be corporations or bureaucracies) that depend on incoming flows of resources to generate the wealth that supports their operations. Given this dependence, such institutions are not easily able to adapt to new forms of production that are less resource intensive or more equitable. Conservation and demand management do not sit easily with Exxon and General Motors because they are sustained by linear resource flows and economic growth.

Similarly, the ability of government agencies to regulate these institutions is questionable where the agencies are themselves indirectly responsible for ensuring the continued economic well-being of the existing forms of production.

Third, political ecology recognizes that the scope of the decision-making processes utilized by these public and private institutions is narrowed by their prior acceptance of an undifferentiated growth ethic and by their predisposition to favor the sort of technical information/science that is compatible with it. In contrast, political ecology takes seriously the unwanted consequences of growth—the uncertainties, side effects, externalities—that all too often go unrecognized and unresolved by the dominant institutions.

The remainder of this chapter explores in greater detail these three areas of political ecology—the nature of knowledge, the nature of economic production, and the character of institutional decision making. In so doing, it will become clear that the large systemic critique offered by political ecology converges with that offered by the Precautionary Principle. It is, in fact, as if the Precautionary Principle is one face on the prism of political ecology, but that the light and vision through the prism is of common hue. Taking a larger view serves an important function, however, insofar as it reveals that the obstacles confronting the Precautionary Principle are *systemic*. These obstacles also impede attempts to act on conceptual innovations that have arisen in other areas contemporaneously with the Precautionary Principle (e.g., biodiversity and ecosystem health) that offer other perspectives into the ecological prism.

"SCIENCE" IN A WORLD OF DISPUTED KNOWLEDGE

As every student learns in high school, the Renaissance that marked the take-off of Western growth and expansion was also a time of scientific revolution. This revolution allowed the West to break free of inherited intellectual prejudices and material constraints and, at the same time, provided the new technologies to support this break. Underlying Western development since that time has been a belief in science as the neutral or objective process by which one can both know the world and harness its powers. From the development of the telescope and compass that was essential to the exploration and colonization of the New World, to the construction of the steam engine that powered the Industrial Revolution, to the proliferation of the computer that holds together today's globalized business networks, this science has demonstrated itself both technically and economically. Science has, literally, proven its worth.

In recent years, however, hitherto unseen attributes of this science have been a topic of growing debate. In this regard, the Precautionary Principle cannot be appreciated in isolation. On the contrary, its rapid rise to prominence over the past decade is really part of a larger challenge to our inherited notion of science as an exclusive, and reliable, form of knowledge. This challenge has been gaining strength for some thirty years. For example, until the 1960s, few understood the "paradigmatic" context for fields of scientific research.

When Thomas Kuhn published his now famous book, *The Structure of Scientific Revolutions* in 1962,[9] science had been generally understood to progress in a linear fashion, new knowledge piling on top of prior knowledge. Kuhn's great contribution was to demonstrate that, in diverse disciplines, scientific ideas developed through a series of conflicts involving fundamentally differing paradigms of thought. Before a conflict arises, Kuhn described how a dominant paradigm exists as an assumed way of thinking, its premises unconsciously accepted, with the explicit focus of research being on the individual experiments and studies existing within that paradigm (what he called "normal science"). When the paradigm is challenged, however, the conflict is often very acrimonious insofar as the whole way of thinking of a generation of scientists may be at stake, as may be the institutions of training and research in which the established paradigm of a discipline is embedded. New paradigms thus pose a threat to the established scientific infrastructure and, as often as not, to many industries that have developed around the dominant paradigm.

Kuhn's analysis was so illuminating that many of his concepts (such as the "paradigm shift") have become common parlance today. But his analysis also opened up Pandora's box. After all, if scientific thinking is bound up in precognitive paradigms, how is a scientist ever to find "the truth"?

In the 1970s, the scientific tradition came under an even broader attack from another direction as many began to explore the biases inherent in believing that one can know the world exclusively through rational processes. This challenge took the form of questioning the dualism of mind versus body, which dated from the 16th-century philosopher and scientist René Descartes. Thus, for example, it was noted that (as Kuhn had pointed out with regard to revolutions in specific disciplines), scientific progress comes first from a hunch, a feeling, an intuition. Indeed, as mathematicians had long ago discovered, no purely rational self-referencing system could be devised that did not depend ultimately on some term that took its meaning from outside the system. In contrast to a purely rationalist perspective, this larger approach implied that a scientist's science was affected by their larger lifeworld—their experiences, their values, how they worked, who they worked for, what they actually did. This approach also coincided with earlier demonstrations that no purely objective form of experimentation was possible either (the well-known Heisenburg uncertainty principle).

In this vein, the granddaddy of all challenges emerged with the so-called rise of "postmodernism" in the 1980s. This school of thinking—which is still new enough that its implications have yet to enter the broad cultural fabric—demonstrates that our supposedly rational ways of thinking are not neutral but are intimately embedded within a society's relations of power. Kuhn certainly demonstrated this with his historical depictions of the institutional context of paradigm conflicts, such as the prosecution of Galileo for his heresy that the Earth was not the center of the universe. But the postmodernists went further, challenging not just particular schools of thought but the whole ethos of modernism itself as somehow the embodiment of a progress that was ultimately rational and true. In contrast, to be "post"modern is to recognize that modernism is itself a phase, a historical and cultural creation, with no ultimate value of truth at all.

The cumulation of these critical analyses has seriously eroded one of the more specific tenets of faith that has underpinned scientific progress, the belief that the world is but a complex machine that can be understood given enough time and testing. This belief (which again dates from the scientific revolution, particularly from such towering figures as Francis Bacon and Isaac Newton) is closely associated with the scientific method of experimentation and inductive or deductive reasoning. This method has of course been enormously productive, allowing both for the discovery of many "natural laws" and for their manipulation into the creation of useful processes and products on which have been built modernity's sources of wealth. In recent years, however, its limitations have also become increasingly apparent, especially in escalating environmental and social problems.

In trying to understand the nature of these limitations, it has become commonplace to note that scientific procedures that seek to isolate variables can be dangerously reductionist. Such reductionism can easily overlook more complex processes that exist in the real world—systemic connections, synergistic effects, time lags, indirect patterns of causation, chaotic processes of change, and so on—processes that have led to concern for (in popular parlance) a more "holistic" approach. Because of these complexities, uncertainty is pervasive in human interactions with the natural world, say the new critics.

Not coincidentally, this broader understanding of the operation of natural systems goes hand-in-hand with an associated concern for the unforeseen consequences of continued undifferentiated economic growth on both natural and social systems. And, of course, all of this comes at a time when Western patterns of development have achieved both a microscopic reach and a macro-level scale never before seen in human history. Today, a single mistake can now provoke (in the word of the systems theorists) a "surprise" event or chain of events that can be globally catastrophic. Thus the identified limitations of traditional scientific procedures are of more than just incidental scientific interest. They are fundamental to social survival and social self-regulation.[10]

In this world, the inescapable presence of pervasive uncertainty in the scientific enterprise is (to use Thomas Kuhn's term) an "anomaly" not just for one discipline or paradigm, but for the scientific project as a whole. In this context, it was historically almost inevitable that the Precautionary Principle would arise, given its timely attention to the uncertainties associated with complex systems and to the limits of possible knowledge. The broad reevaluation that postmodernists call for in our claims to true knowledge is most assuredly applicable to science. And it is especially so in areas such as engineering, health, or resource management where science is *applied* and thus has implications for decisions that affect remote places in space and time, including future generations.[11] To say that a new chemical is safe, a risk acceptable, or a harvesting technique sustainable is thus to beg the question—for whom?

The nature of this quandary is demonstrated by considering briefly a recent example—the challenge posed to resource industries and agency managers over the past few years by those who base their claims on their "traditional ecological knowledge" or "local knowledge." Unlike institutionally based science, these forms of knowledge claim a veracity and reliability based not on the objective removal of the observer from the observed (the essence of the scientific method) but on precisely the opposite—on the observer's long practice, comprehensive exposure, and direct experience with the subject or object of study. Until recently, such claimants (who were often politically powerless aboriginals or rural peasants) were marginalized by powerful companies and government authorities, while their knowledge was dismissed as unreliable folkways by the scientific experts. Again, however, as part of the broader historical movement that is now challenging the larger political economy of science, these forms of knowledge are increasingly being recognized for the sophisticated understandings built into them and for the benefits to sustainability that would attend the operationalization of these understandings.

A dramatic example of the consequences of this conflict is the recent collapse of the east coast cod fishery, once one of the largest and most productive fisheries on the planet. Study after study has demonstrated how the rise of corporate-scale operations and technologies devastated local fish populations that were once sustainably managed on the basis of long-standing community traditions. As one rural sociologist has shown, part of the displacement process involved the substitution of a single variable, quantitative management technique (the "total allowable catch" or TAC) that could be monitored and controlled by remote experts for the more complex, and far more precautionary, management techniques associated with the preexisting "outport" fisheries. These latter techniques of management were dismissed as unscientific, at the same time as the displacement of the local techniques of production.[12]

Meanwhile, the large-scale production system was itself insulated from conservationist challenges by the refusal of the corporate or bureaucratic elites to consider small-scale techniques that had conservation practices embedded *within* them or to acknowledge deficiencies in a management system that simply could not control industrial producers that utilized practices based exclusively on profitability.[13] Not only was a precautionary science not embraced, critical science was actively suppressed. In its place, rhetoric drove decision making.[14] Critical commentators are, not coincidentally, calling for the use of both local knowledge and a precautionary approach in a dramatically reformed fisheries policy.

Overall, these recent historical developments come together in Barrett and Raffensperger's critique (see chapter 6) of "industrial science" and their call for the recognition of a "precautionary science" that better embodies a precautionary approach. The dichotomous character of these two forms of science is set out by the authors where, again, we see the clear outlines of a paradigm conflict. This conflict is now so commonplace that decision makers increasingly confront situations that are not informed by clear evidence based on "normal" science within an accepted paradigm. Instead, their choices are located in situations of paradigm conflict—between toxic dumpers and precautionary recyclers, between high-tech doctors and acupuncturists, between industrial foresters and ecoforesters, between corporate factory trawlers and artisanal small-boat handliners.

Given the argument of this chapter—what is deemed to be good science or true knowledge is contextually determined—the matter does not end with a new body of such knowledge. Instead, the key to resolving paradigm conflicts is to open up the institutional setting within which any new substantive knowledge can be generated. The failure to do so has instead characterized governmental responses to the communal assertion of the importance of traditional or local knowledge. Often, officials have ignored the demand to create new institutions of cooperative management suitable to the local cultural context and instead have attempted to incorporate the knowledge itself as bits and pieces of a new information base into otherwise unchanged structures of production and control.[15] On the contrary, the pursuit of good science can only be met by new institutions that can mediate the differences of a "postnormal science."[16]

In summary, how we practice science determines what that science is, and for whom. We must act on this recognition. Knowledge *is* contextual; it is embedded within economic and political institutions that have their own imperatives and their own dynamics. Beyond the specific focus on the conflict associated with the Precautionary Principle, we are now able to situate this conflict within the longer historical pattern by which old established sciences (and institutions) have always been challenged by new ones from the margins. Ultimately, this is a matter of authority, voice, and power.

TWO MODES OF ECONOMIC PRODUCTION

Seen in its historical context, the Precautionary Principle can be understood as part of a larger movement to create a new approach to science and knowledge. Its singular contribution is that it mandates us to take uncertainty seriously: Uncertainty is pervasive; it is cumulative; and, contrary to the long-held assumption of the scientific tradition, it cannot be conquered. Unlike the systems that we have built on the basis of achievable certainty, this new recognition points to the need to develop new systems, ones that do not incautiously push at the boundaries of unforeseen consequence. Not coincidentally, this science-based movement is paralleled in the efforts of those seeking to create a new, sustainable, foundation for the growth economy.

As discussed briefly previously, one of the achievements of the field of ecological economics has been to expose the long-held assumption of productivism that underpins our inherited notions of economic progress, whether on the "left" or "right." Productivism means continued high levels of economic throughput. Indeed, with the added characteristic of free competition, the contemporary market economy not only seeks to maintain, but also to continuously increase, its level of throughput. Ecological economists explain that this process of productivity-driven economic growth runs counter to the second law of thermodynamics (the law of entropy). As a cost associated with creating specific order somewhere (building a car, heating a house), more disorder is created generally in the world (through the effects of mining ores and smelting, drilling oil and refining it, transporting the raw materials, producing the finished products, discharging the associated pollution, disposal of the wastes, etc.).[17] This understanding has serious social implications for the growth economy.

The problem of energy/entropy can be put in another way by recognizing that the throughput economy is fundamentally a *linear* economy, and it is so in several ways. Most obviously, according to the second law of thermodynamics, it operates by taking order in from the world around it and sending disorder out. From a spatial perspective, economic growth has also meant the taking of resources (fish, trees, coal, oil) from one place, using them for productive growth somewhere else, and then discharging the wastes still somewhere else. Unlike previous schools of political economy (such as those spawned by Smith and Marx), almost by definition the signal characteristic of an ecological political economy is its attention to these *physical* and *spatial* relations. In this regard, the throughput economy is marked not only by its excessive flows of energy and resources, but also by the development of particular institutional forms that accompany these flows.[18]

As noted earlier, the focus of political economy is on understanding the dynamics of these social systems. In the historical evolution of the Western growth economy, one can identify two structural trends that have both driven, and in turn been supported by, the rise of the flow economy: (1) the growing dominance of *hierarchical institutions* that (2) are themselves increasingly reliant on ever more distant (spatially and temporally) *nonlocal resources*. At the same time, these structural characteristics are not specific just to the Western growth economy. Indeed, organizing the world to maintain the continuous flow of resources toward, and up into, these centralist hierarchies—whether the papal institutions in Rome, to the Anglo-Saxon Crown in Westminster, or the Exxon corporate office in New York—is what "progress" has always been about. Such patterns of hierarchical social growth mark a variety of civilizations over many millennia where nonmarket forms of production, exchange, and hierarchical power long predated modern forms of growth.

The problem of high throughput/entropic institutions is thus not new. The market-based growth economy is but a recent, and particularly effective, mechanism building and sustaining such institutions. In this repetitive historical pattern, however, one can also discern a basic contradiction that besets all forms of centralist growth. In short, the rise of central power is, and always has been, sustained by the territorial structures that precede that rise, and it cannot survive without them. Yet, driven to grow, centralist institutions consume the very territorial processes on which they depend and, in doing so, await their own demise. This is the story of countless civilizations past that have risen, only to fall.

If this analysis might seem overblown to some readers, one might bear in mind the fact that, for the first time ever, civilization has achieved a global scale and, at the same time, so has the nature of related environmental problems (from ozone depletion to PCB contamination to deforestation). Ecological linearity is fundamental to the entropy-creating globalization spreading into every corner of the globe, a linearity that seeks unconstrained access to all remote sources of sustenance, and new repositories for garbage, while it systematically discounts the impacts of present flows on future generations.

In summary, the precautionary or ecological economic critique can be distilled into three basic points. First, the linear growth economy *inherently* generates externalities in all aspects of its operation, from the depletion of resources that are secured as raw materials for processing, to the massive energy costs associated with the processes of production, to the pollution costs associated with the disposal of wastes and spent goods. Second, the institutions that have been constructed on the throughput model inherently resist changes that might constrain this throughput. For example, the economic concept of "externality" (i.e., a cost or benefit not included in the market price of a good) implies that something should be "internalized" into the economic accounting in order for the market to reflect the full costs of production. In practice, however, negative environmental and social externalities continue, and on a massive scale, because their creation is inherent to the throughput economy. As a result, techniques such as benefit-cost analysis and risk assessment are of limited utility in addressing issues of environmental or economic sustainability because they are essentially self-referencing, taking market values as given while not addressing the operation of the market as a whole social or institutional mechanism.

Third, the combination of necessary externalities and resistant/self-referencing institutions creates a systemic contradiction, one that increasingly besets the resolution of environmental problems. This contradiction can be simply expressed: the greater the crises caused by growth, the greater is the institutional imperative *not* to change. The fewer resources that are left, the more intense the competition to get at them in order to continue to secure the structural subsidy (the "externality") to throughput production that access to those resources represents.

Because both the crisis-generating character and the modus operandi of the growth economy are based on the maintenance of high levels of linear economic flows, the shift from linear to *circular* systems is *the* issue of sustainability. As explained more fully later, this is exactly what the Precautionary Principle is about, which is why it is also so controversial. Restraining established *linear* processes while fostering countervailing *circular* alternatives is what both political ecology and the Precautionary Principle are all about.

In looking at the implementation of the Precautionary Principle, many critics have decried the principle as vague and lacking in technical rigor.[19] However, as others (such as Nicholas Ashford) have pointed out, the principle is, in the first instance, a directional guide that needs broad interpretations and applications to the design of decision-making institutions.[20] As Dovers and Handmer put it, "[t]he primary use of the Precautionary Principle is to fly the flag of ignorance and uncertainty and hold the policy fort until a framework is developed to improve our overall understanding of the problem, guide the application of available techniques, and identify areas where appropriate techniques are lacking."[21]

Contrary to the charge of imprecision, the direction envisioned by precaution is both clear and identifiably new. Because it embodies a reorientation away from linear to circular economic processes, precaution directs economic activity toward *processes of low throughput that minimize external costs.* This reorientation dictates an approach to all forms of economic activity that is similar to Ken Geiser's goal (see chapter 20) for industrial production to seek "optimal efficiency and zero emissions." Overall, by taking energy limits, externalities, and the needs of resource renewability seriously, the Precautionary Principle directs actions toward the development of a renewables economy.

This model reverses the dominant approach that takes the economy as the starting point and the environment as a secondary concern that needs to be developed and used, managed, priced, and made more productive to fit into the needs of economic growth. Instead, precaution seeks to situate economic activity *within* ecological bounds and to develop those forms of knowledge, those institutions of production, and those forms of decision making that will allow this reversal to occur. A critical aspect of this re-embedding economic processes within ecological processes is the need to avoid the great mistake of the throughput economy—the creation of institutional overextensions and dependencies that resist adaptation, even when the information is telling us to change.

"Closing the loop" in production is the most common way of expressing the Precautionary Principle's approach to circularity. But this approach actually embodies a wide range of very precise approaches to *economic* planning that have been emerging (often independently of the Precautionary Principle), including, for example, the advent of "demand management" planning in the 1980s and "industrial ecology" in the 1990s. Similarly, the popular social support since the 1970s for a strategy of "reduce, reuse, recycle" mirrors precautionary thinking. This thinking has found its ultimate expression in pollution prevention and "clean production," an approach that is an explicit outgrowth of the Precautionary Principle. In all of these methodologies and design strategies, a common element is the removal of the historic Western separation between profit-driven production and government-enforced environmental protection by redesigning production processes so as to embed them within the limits of ecological function.

Precaution is, however, evident in a range of emerging renewable resource strategies that are not often addressed by the usual focus on industrial production. These strategies ultimately take us beyond simply "closing the loop" but even more explicitly lead us to address the contextual (i.e., spatial) character of economic production in a manner that fits economic processes within limited ecological spaces.

A good example of how this works can be found in the recently concluded United Nations Convention on Straddling Stocks and Highly Migratory Species (1995).[22] This fisheries treaty includes an extensive provision in the text, and a special annex, on the application of the Precautionary Principle to the development of new and existing fish populations.[23] As in many other international treaties, these and other conservation-oriented provisions were included as a result of the concerted diplomatic efforts of a broad range of activist environmental, social justice, labor and community groups.[24]

At the same time, these groups pressed for a range of specific provisions that would have mandated parallel *production-oriented* standards. Although the enormous opposition of the industrialized fishers precluded these from being included in a substantial way, a number of more hortatory principles were accepted that demonstrate how precaution would translate into techniques of fisheries production. These include the need to protect artisanal fisheries, minimize by-catches and waste, promote selective gear types, and eliminate overcapacity.[25]

In addition, the convention recognized the ecological limitations imposed by existing jurisdictional divides, particularly that between domestically controlled waters with the 200-mile Exclusive Economic Zone, and the internationally high seas outside. To overcome these unnatural divisions, the convention mandated the application of a "consistency principle" that ensures common controls are applied throughout the ecosystem regardless of legal boundaries.

Pressure for similar changes—that is, changes that apply precaution to the redesign of production practices within the limits of ecological space—also exist in other areas of renewable resource use. In forestry, for example, a global debate is ongoing concerning the paradigmatic challenge to industrial forestry by "ecoforestry," a debate that is precisely parallel to the scientific and economic debates discussed previously.[26] As Frederick Kirschenmann demonstrates (see chapter 18), this is mirrored in the debate between industrial agriculture and "agroecology." Indeed, his call for developing practices at a watershed level parallels yet another concept that has arisen contemporaneously with the Precautionary Principle, that is "ecosystem-based management."

In the late 1980s, forest managers in the United States articulated the need to replace the reductionist management of trees and timber with a more holistic "ecosystem management."[27] A division quickly developed around this new concept and the role uncertainty and precaution play in defining it (although this was not as explicit in the debate as it has been with the Precautionary Principle). Under the incautious industrial model of forest industry development, levels of cut are set economically, and negative (i.e., external) effects on other forest values are contained to some degree by legal regulation. Ecosystem management was first articulated by the U.S. Forest Service as an incremental improvement in its management strategies by allowing it to broaden the focus of control. Many scientists noted, however, that it was in fact impossible to "manage" an ecosystem because of its huge complexities, a task that was made all the more difficult where the predominant mode of forest exploitation was clear-cutting.

Instead, many scientists advocated an ecosystem-*based* approach, the starting point of which is to design the whole regime of economic exploitation to whatever shape was required to allow for the maintenance of ecological "integrity" by retaining the "composition, structure, and function" of the forest ecosystem. The level of cut, and cutting practices, would be set within that regime.[28] Thus, the attention turns from managing forests (or fish, or watersheds) to managing forestry (or fisheries or farming). Only this latter model embodies precaution, and it does so by redesigning the very nature of the economic activity being undertaken to make it inherently compatible with its context, rather than overrunning it. This model for renewable resources is clearly similar to the regulatory shift in pollution control where one moves from managing specific ambient environments by setting "end of pipe" standards to managing whole industries by devising and mandating clean production techniques.

The list of such sectoral overlaps goes on.[29] Indeed, how all these approaches converge can perhaps best be seen in capsule form when they are applied to that most centrist of institutions—the modern city. Historically, the rise of the city has gone hand-in-hand with the rise of the linear economy, with urban areas drawing resources from afar, processing them for the accumulation of city wealth, and disposing of the effluent into the larger commons. Today, however, the cutting edge of urban planning involves redesigns that would make these complex systems more responsive to the functioning of both the local ecosystems within which cities are embedded and the distant ones on which they have an impact.

And this is, again, to be achieved by closing the loop so as to move cities from their preeminent places as spatial nodes of linear accumulation to becoming leading laboratories of circular invention and redevelopment. This is, for example, the overall thrust of the section of the Rio Declaration (from the 1992 United Nations Conference on Environment and Development) dealing with urban areas, the so-called Local Agenda 21.[30] This is also the thrust of another contemporary concept that reflects precautionary thinking, "industrial ecology."[31]

In summary, the Precautionary Principle is part of a larger movement articulating both the need and potential for a *systemic* transformation in the processes and institutions of production.[32] Based on an ecological political economy, the Precautionary Principle entails a shift in the unit of analysis from that of the individual chemical or timber stand or fish stock to that of the high throughput economy itself. And it does so by challenging the fundamentals of linear centralism by proposing instead to resituate the processes of production within the circular confines of territorial sustainability. And this change entails a political, as well as an economic, transformation.

PRINCIPLED POLITICS

Just as a precautionary approach envisages a broader science and a reconstituted framework of economic production, so too it will require a lot more than just one or two pieces of new legislation. Quite the contrary, achieving a precautionary transition inevitably involves a reconsideration of the very nature and function of the state.

Public policy practitioners have long noted the mismatch between government's dual obligations of supporting industry's efforts at competitive success and of protecting the public's safety and environmental health. Political economists explain this mismatch by demonstrating how the primary function of the state historically has been to facilitate the accumulation of wealth by ensuring continued economic grow. Legal regulation (including environmental protection) has taken second place within that primary function, and in a manner designed not to upset it. This is the source of much conflict as the state is caught between supporting accumulation and seeking public legitimacy.[33]

This situation has direct applicability to many of the concerns for the design of decision-making institutions addressed in this book. The allocation of the burden of proof onto the victim of pollution, while allowing industrial polluters the luxury of a regime of "permissive regulation," is one manifestation of the economically driven design of legal rules.[34] More broadly, the general character and degree of "public participation" is shaped by these concerns such that final decisions at the end of participation processes inevitably rest with government authorities, such decisions being reviewable by disgruntled activists on largely procedural, rather than substantive, grounds.[35] In the process, many of the dominant assessment techniques—such as benefit/cost analysis, environmental and social impact assessment, and risk assessment—are oriented to evaluating how a project is to be undertaken, rather than whether the project should be undertaken.

Internalizing a precautionary approach into public decision-making implies, not surprisingly, a very different conception of both the function and roles of the state. Functionally, rather than supporting continued undifferentiated economic growth, the critical function of government would be both to constrain incautious economic behaviors and to facilitate precautionary behaviors. It should be noted that this joint role is radically different than traditional liberal conceptions that, taking the existing economic trajectory as given, have largely positioned government environmental interventions as negative constraints that impose new burdens on industry. But because the throughput economy inherently generates externalities in its operations, these interventions are self-defeating.

In contrast, where the regulatory process is designed so that the constraints on one type of activity are balanced with supports for alternatives, then the pressure to undo regulations (which is omnipresent in the current approach) becomes rechanneled into innovation and facilitation of a transition to precautionary economic activities.

In this light, as other authors in this volume point out (see especially

Nicholas Ashford, chapter 11, and Mary O'Brien, chapter 12), the critical regulatory innovation is the "alternatives assessment" procedure. Such a process is primarily directed at developing and mandating new economic approaches that lead to risk *reduction*, rather than just working with existing technologies and undertaking a quantified risk *assessment* of their impacts. Risk assessment would have a place in this regulatory process, but only at a subsidiary level where it was used to compare the candidates remaining for regulatory approval after the more risk-creating options were screened out. In addition to its role in technology forcing, such a refashioned regulatory process would stimulate new information gathering and analytical techniques to help diffuse innovation on a sector-wide basis.[36]

Alternatives assessment is the core technical/institutional innovation of the Precautionary Principle. But this technique will not achieve its potential unless designed in a manner that ensures that a full, democratic debate occurs and, as O'Brien points out, unless real accountability occurs in the decision-making process. Where it is well designed, alternatives assessment will become an invaluable political mechanism for the general transition to an ecological economy. The sort of inquiry and debate, which the process will generate, is exactly what is proposed for the resolution of "postnormal" scientific conflicts.

But here, too, the prerequisites for success—public access to information, institutional transparency, political accountability, long-term planning—are the types of democratic proposals that progressive proponents of state reform have been demanding for decades.[37] The difference today is that a good process of alternatives assessment, by revealing workable new directions for innovation and sustainability, offers a way out, an escape valve against the constraints imposed by environmental regulations on economic growth. The unwillingness of democratic governments to contemplate such a transformative tack underlies much of the frustration about the lack of implementation of the Precautionary Principle. Nevertheless, support for these democratic innovations is the hallmark of what I would call "ecological liberalism," that is, an attitude to state action that will allow the necessarily radical, but gradual, transformation to be achieved over time.

There is, as well, another level of political reform that follows from the recognition of uncertainty and the role of precaution, and that is the invigoration of ecosystem-based community management. In the American context, with the antigovernmental posturing of many resource-hungry industries and states (one thinks especially of the "sagebrush" rebels of the western rangelands), the espousal of community governance will be met with justified skepticism. Yet, there is great potential here. Some authors, for example, argue that the extent of environmental uncertainty is so great that centralized management is simply not workable and that new forms of community regulation *based on truly sustainable community economies* are scientifically inevitable. As one interdisciplinary scholar noted:

> The complex nature and the relevant scale of the biological problems is the underlying reason why self-governance is likely to be a superior form of management. As a consequence, we believe it is not possible or even desirable to separate the biological and socio-economic aspects of management. . . . Decentralization will be necessary because the information problem created by the spatial and temporal diversity within these systems demands an attention to detail that cannot be achieved by a centralized authority. But higher level social restraints are undoubtedly also necessary to maintain the system within the boundaries set by the higher level biological parameters and to prevent mutually self-defeating policies among decentralized authorities.[38]

With the central government setting minimum standards and ensuring democratic forms of decision making, "ecosystem-based management" offers huge opportunities to offset the traditionally centralist powers of industrial fishers, corporate foresters, and so on. In this way, the political, indeed, constitutional, objectives of global social movements from community forestry to aboriginal self-government dovetail with those of the advocates of the Precautionary Principle.[39]

PRECAUTION AND SYSTEM REFORMATION

Implementing precaution is not a discreet task but is intimately enmeshed in the development of an ecological society. It calls for dramatically new forms of knowledge and cutting-edge technological innovation. Economically, the transformation will involve incremental steps but ones that favor wholly new forms of development, such as the use of closed-loop systems in industrial processes (clean production). In the process, it necessitates larger changes that favor whole new industries (organic agriculture, ecoforestry) that inherently tend to incorporate the Precautionary Principle within their basic operating premises. More generally still, these industries are part of a larger commitment to an ecologically based economy that moves from a dependence on ever-growing linear processes of resource and energy flows to one that minimizes such flows in favor of self-regenerating, circular processes.

Such changes will not happen spontaneously but imply a parallel reconstruction in the processes of governance that will shed new light onto decision making. In the end, it is critical to open the state to the social and community movements that understand the potential of the Precautionary Principle and that will create the political momentum for a new trajectory of economic and political evolution.

NOTES

1. *International Convention on the Dumping of Wastes at Sea.* 11 I.L.M. 1291 (1972).
2. Under Article IV (a), provisions exist for prohibition of a limited range of substances when they are listed in Annex I, a provision that applied to a very limited range of substances.
3. For a broader account of this, see Kevin Stairs and Peter Taylor, "Non-Governmental Organizations and Legal Protection of the Oceans: A Case Study" in *The International Politics of the Environment: Actors, Interests, and Institutions,* eds. A. Hurrel and B. Kingsbury (Oxford: Oxford University Press, 1992), 110–141.
4. The debate in the journal *Marine Pollution Bulletin* has run on now for over a decade. See V. Pravdic, "Environmental Capacity—Is a New Scientific Concept Acceptable as a Strategy to Combat Marine Pollution?," *Marine Pollution Bulletin.* 16 (1985): 295–296; K.K. Sperling, "Protection of the North Sea: Balance and Prospects," *Marine Pollution Bulletin* 17 (1986): 241–246; J.S. Gray, "Statistics and the Precautionary Principle," *Marine Pollution Bulletin* 21 (1990): 174–176; J.S. Gray, D. Calamari, R. Duce, J.E. Portmann, P.G. Wells, and H.L. Windom, "Scientifically Based Strategies for Marine Environmental Protection and Management," *Marine Pollution Bulletin* 22 (1990): 432–440; R.C. Earll, "Common Sense and the Precautionary Principle—An Environmentalist's Perspective," *Marine Pollution Bulletin* 24 (1992): 182–185; R.M. Peterman and M. M'Gonigle, "Statistical Power Analysis and the Precautionary Principle," *Marine Pollution Bulletin* 24 (1992): 231–234; A.R.D. Stebbing, "Environmental Capacity and the Precautionary Principle," *Marine Pollution Bulletin* 24 (1993) 287–295; C.C. ten Hallers-Tjabbes, J.F. Kemp, and J.P. Boon, "Imposex in Whelks (Buccinum Undatum) from the Open North Sea: Relationship to Shipping Traffic Intensities," *Marine Pollution Bulletin* 28 (1994) 311–313; L. Buhl-Mortensen, "Type II Statistical Errors in Environmental Science and the Precautionary Principle," *Marine Pollution Bulletin* 32 (1996) 528–531.
5. For a detailed discussion of the London Dumping Convention, see R.M. M'Gonigle, "'Developing Sustainability' and the Emerging Norms of International Environmental Law: The Case of Land-Based Marine Pollution Control," in *The Canadian Yearbook of International Law* 28 (Vancouver: UBC Press, 1990), 169–225.
6. Branko Horvat, *The Social Science Encyclopedia* eds. A. Kuper and J. Kuper, (London: Routledge and Kegan Paul, 1985) at 612.

7. As the philosopher of science, Mary Hesse, noted:

    [T]alk of the "truth" of science, and of the ontology of objects which
    it presupposes, becomes wholly internal to scientific theory itself.
    Truth and existence-claims are determined, not by the world, but by
    the postulates of theory: for our physics there are fundamental parti-
    cles and fields, a space-time continuum, forces, and persisting physi-
    cal objects; for other cultures there are spirits, witches, telepathic
    communications, persons not uniquely and continuously space-time
    locatable, and so on." *In Defense of Objectivity* (London: Oxford Uni-
    versity Press, 1973), 10.

8. For an excellent short exposition of this critique, see Robyn Eckersley, "The
    Ecocentric Challenge to Marxism." In *Environmentalism and Political Theory*
    (Albany, NY: SUNY Press, 1992), 75.

9. Kuhn, Thomas. *The Structure of Scientific Revolutions* (1962). Chicago: Univer-
    sity of Chicago Press.

10. See R. M. M'Gonigle, T. L. Jamieson, M. K. McAllister, and R. M. Peterman,
    "Taking Uncertainty Seriously: From Permissive Regulation to Preventative
    Design in Environmental Decision Making," *Osgoode Hall Law Journal*, 32,
    no. 1 (1994): 99–169.

11. For a good discussion of the implications of pure vs. applied science, see Lene
    Buhl-Mortensen, "Type-II Statistical Errors in Environmental Science and the
    Precautionary Principle," *Marine Pollution Bulletin*, 32 (1996): 528–531. The
    author notes that when "one moves from pure science to applied science affect-
    ing policy, the question of what is rational moves from epistemological consid-
    erations to both ethical and epistemological ones" (at 530). See also
    K.S.Shrader-Frechette, *Risk and Rationality: Philosophical Foundations for Pop-
    ulist Reforms* (Berkeley: University of California Press, 1991), especially at
    134–135.

12. See Barbara Neis, "Fishers' Ecological Knowledge and Stock Assessment in
    Newfoundland," *Newfoundland Studies*, 8 (1992): 155–178.

13. A. Charles, "The Atlantic Groundfishery: Roots of a Collapse," *Dalhousie Law
    Journal* 18 (1995): 65.

14. For a brief discussion of the role of rhetoric, see D. Fleming, "The Economics
    of Taking Care: An Evaluation of the Precautionary Principle." In *The Precau-
    tionary Principle in International Law: The Challenge of Implementation*, eds. D.
    Freestone and E. Hey (The Hague, The Netherlands: Kluwer Law Interna-
    tional, 1996), 146–167, especially 151–152.

15. For a good discussion of some related issues in this area, see the special issue of
    Current Anthropology, 39, No. 2 (April 1998).

16. For a good summary of this perspective, see Silvio O. Funtowicz and Jerome R.
    Ravetz, 1993, "Science for a Post-Normal Age," *Futures*, 7(1993): 739–744.

17. For the leading texts on ecological economics, see Nicholas Georgescu-Roegen,
    The Entropy Law and the Economic Process (Cambridge, MA: Harvard Univer-
    sity Press, 1971); Herman E. Daly, *Steady-State Economics: The Economics of*

Biophysical Equilibrium and Neutral Growth (San Francisco: W.H. Freeman, 1977); *Ecological Economics: The Science and Management of Sustainability*, ed. R. Costanza, (New York: Columbia University Press, 1991); R. Costanza, J. Cumberland, H. Daly, R. Goodland, and R. Norgaard, *An Introduction to Ecological Economics* (Boca Raton, Fla: St. Lucie Press, 1997).

18. For a fuller theoretical discussion of this point, see R. M. M'Gonigle, "Ecological Economics and Political Ecology: Towards a Necessary Synthesis," *Ecological Economics* 28(1998): 11–26.

19. For example, D. Bodansky argues that the Precautionary Principle is too vague, does not indicate when it should be applied, what factors to consider, how costs should be incorporated, and how one should choose between risks. In: "Scientific Uncertainty and the Precautionary Principle," *Environment* 33 (1993):4–5, 43–44. Similarly, A. Nollkaemper argues that the Precautionary Principle has nothing new to offer but is simply a way to label concepts already present in the international community, "The Precautionary Principle in International Environmental Law," *Marine Pollution Bulletin*, 22 (1991): 287. See also, A.R.D. Stebbing, "Environmental Capacity and the Precautionary Principle," *Marine Pollution Bulletin*, 24(1994): 287–295. F.B. Cross, "Paradoxical Perils of the Precautionary Principle," *Washington and Lee Law Review*, 53 (1996): 851, and J.S. Gray, "Integrating Precautionary Scientific Methods into Decision-Making." In D. Freestone and E. Hey (eds.), op. cit. at 133–146, especially 135–136.

20. In addition to his article in this volume, see another article that points to both a new direction for decision making that reflects precaution but also to a range of technical components of implementation, see N.A. Ashford, "An Innovation-Based Strategy for the Environment." In: *Worst Things First? The Debate Over Risk-Based National Environmental Priorities*, eds. A. Finkel and D. Golding (Washington, D.C: Resources for the Future, 1994), 275–314.

21. Stephen R. Dovers and John W. Handmer, "Ignorance, the Precautionary Principle, and Sustainability," *Ambio* 24 (1995): 97.

22. *Final Acto of the United States Conference on Straddling Fish Stocks and High Migratory Fish Stocks*, 34 I.L.M. (1995).

23. Article 6, and Annex II.

24. For a discussion of this, see M. M'Gonigle and D. Babicki, "The Jurist and the Ecologist: Shifting Paradigms in the International Law of Conservation and Exploitation: The Case of the Straddling Stocks." In: *Globalism and Regionalism: Options for the 21st Century* (Proceedings of the Twenty-Fourth Annual Conference Canadian Council on International Law, Ottawa, 1995), reprinted in Y. Le Bouthillier, D. McRae, and D. Pharand (eds.) *Selected Papers in International Law: Contributions of the Canadian Council on International Law, the First 25 Years* (London, U.K.: Kluwer Law International, 1998), 48–56; R.M. M'Gonigle, "A New Naturalism in International Fisheries Law?" in *Fisheries* 22 (March 1997), 24–25.

25. These provisions are contained in Articles 5 and 24.

26. See, for example, *Ecoforestry: The Art and Science of Sustainable Forest Use*

(eds.) A.R. Drengson and D.M. Taylor (Gabriola, BC: New Society Publishers 1997), and R. M. M'Gonigle, "Structural Instruments for Sustainable Forestry: A Political Ecology Approach" in Chris Tollefson (ed.) *The Wealth of Forests: Markets, Regulations, and Sustainable Forestry* (Vancouver: University of British Columbia Press, 1998).

27. For a good summary of this, see R.E. Grumbine, "What Is Eco-System Management?" in *Conservation Biology*, 8, No. 27 (1994): 41–47, and, "Reflections on 'What is Ecosystem Management?'" *Conservation Biology*, 11, No .1 (1997): 41.

28. See, for example, the innovative report by The Scientific Panel for Sustainable Forest Practices in Clayoquot Sound. Report 5: Sustainable Ecosystem Management in Clayoquot Sound: Planning and Practices, 1995, Victoria, B.C.: Cortex Consulting. For a more general treatment, and proposal for legislative reform, see also C. Burda, D. Curran, F. Gale, and R.M. M'Gonigle, *Forests in Trust: Reforming British Columbia's Forest Tenure System for Ecosystem and Community Health*, July 1997, Report 97-2,2 (Victoria, B.C.: Eco-Research Chair of Environmental Law and Policy, 1997).

29. One could cite other examples endlessly. A particularly pertinent one here, however, involves an escalating reaction against "fish farms" (aquaculture), especially net-cage fish farms that are situated in open water areas with a huge range of associated environmental problems (from pollution to disease transmission and genetic mutations to wild stocks through escapements). The proffered solution is, not surprisingly, closed-loop tank systems. For an excellent discussion of this, see David Ellis, *Net Loss: The Salmon Netcage Industry in British Columbia* (Vancouver: David Suzuki Foundation, 1996). (Copy available from the foundation at 219-2211 W. 4th Avenue, Vancouver, B.C., Canada V6K 4S2.)

30. See Chapter 7 of *Agenda 21: Programme of Action for Sustainable Development* (United Nations Publication, 1993); see also *The Local Agenda 21 Planning Guide: An Introduction to Sustainable Development Planning* (Toronto: International Council for Local Environmental Initiatives, 1996).

31. To examine the concrete implications of this concept, see the new journal, *Industrial Ecology*.

32. See B. Commoner, "Pollution Prevention: Putting Comparative Risk Assessment in Its Place," in *Worst Things First*, op. cit., 203–228.

33. On the operation of the dialectic of accumulation and legitimacy, see J. O'Connor, *The Fiscal Crisis of the State* (New York: St. Martin's Press, 1973); J. Habermas, *Legitimation Crisis* (Boston: Beacon Press, 1973). As Habermas notes, a "legitimization crisis" occurs when "the legitimizing system does not succeed in maintaining the requisite level of mass loyalty while the steering imperatives taken over from the economic system are carried through" (at 46). Habermas notes the state's role in necessarily compensating for the "politically intolerable consequences" of market functions (at 55).

34. See M'Gonigle et al., "Taking Uncertainty Seriously," op. cit.

35. This problem of procedural, rather than substantive, accountability is funda-

mental to the regulatory process in countries such as Canada and Great Britain where the legislative and executive branches overlap through the parliamentary process. In republican systems, such as the United States, the legislative branch has more incentive to rely on fixed standards in legislation, and thus can more easily be held accountable on substantive criteria. The general problem is evident in all jurisdictions, however. For a comparison of U.S. and Canada's treatment of environmental policy, including a comparison of institutional structures, see George Hoberg. "Governing the Environment: Comparing Canada and the United States," in *Degrees of Freedom: Canada and the United States in a Changing World*, eds. K. Banting, G. Hoberg, and R. Simeon (Montreal: McGill-Queen's University Press, 1997), 310–340.

36. For a good discussion of this, see Nicholas Ashford, "An Innovation-Based Strategy for the Environment," in *Worst Things First? The Debate over Risk-Based National Environmental Priorities*, eds. Adam M. Finkel and Dominic Golding (Washington, D.C.: Resources for the Future, 1994), 275.

37. For a good discussion of how this might be translated into international agencies, see B. Weinraub. 1992. "Science, International Environmental Regulation, and The Precautionary Principle: Setting Standards and Defining Terms," *NYU Environmental Law Journal* 1: 173, especially at 197ff.

38. J.A. Wilson, J.M. Acheson, M. Metcalfe, and P. Kleban, "Chaos, Complexity and Community Management of Fisheries." In: *Marine Policy* 18 (1994), 291–305 at 292 and 299.

39. For a detailed review by the author and others of this issue of community management applied to sustainable forestry, and a proposal that utilizes an extensive trust arrangement between local communities and the central government, see Burda et al., *Forests in Trust*, op. cit.

Chapter 8

Anita Bernstein

PRECAUTION AND RESPECT

Anita Bernstein

Sexual harassment brought me here: I'm uncredentialed and unlettered in environmental policy. Sexual harassment? Having spent some years writing about this area of the law, I mean to suggest that ideas and experiences from this area raise possibilities for implementing the Precautionary Principle.

In order to connect these two subjects, envision sexual harassment the way researchers and specialist lawyers do, somewhat contrary to popular and artistic views of the phenomenon. Literary works portray sexual harassment as blackmail and extortion—Isabella imperiled in *The Merchant of Venice*, Scarpia making threats in the second act of *Tosca*, or Felix menacing a housemaid in Trollope's *The Way We Live Now*. The bestseller *Disclosure* continues the tradition. But this behavior, called "quid pro quo" harassment in American case law and scholarship, describes only a small fraction, probably less than 5 percent, of litigated sexual harassment cases. The great majority of statutory actions accuse employers of having maintained a "hostile environment" or an "abusive working environment."[1] And so after a few years of writing on sexual harassment law—with nary a word about ecology or pollution—I now wonder whether I haven't been working all along in the area of environmental law.

If the environment of hostile-environment sexual harassment shares common characteristics with the environment of environmental law and policy, then these two growing domains can assist each other in identifying concerns and strategies.[2] In writing about sexual harassment, I've pleaded for both enhanced regulatory effort[3] and attention to particular environmental-law insights,[4] but it was only an invitation to participate in these sessions that provoked me to pull the two points together and to think about how environmental law can help sexual harassment reform. Here I offer a vector in the opposite direction: Sexual harassment law reform can aid environmental law and policy.

THE RESPECTFUL PERSON

Among sexual harassment specialists, "environment" generally refers to the workplace. This environment is regulated by many state and federal statutes, notably Title VII of the 1964 Civil Rights Act. Although Title VII was enacted with race discrimination in mind, it also proscribes discrimination on the basis of sex. About twenty years ago federal judges began to agree that hostile-environment sexual harassment is a form of unequal treatment that is "based on sex"—just like the more familiar unequal pay, failure to hire, refusal to promote, and so forth—and therefore actionable under the statute.[5]

Soon the definitional question—or, as one critic wrote recently, "the what" of sexual harassment—emerged.[6] Quid pro quo harassment, if explicit, is easy to perceive and denounce. And nonharassment sex discrimination claims can be simple: Unequal salaries paid to equally qualified workers speak for themselves; the ratio of men to women at a worksite can be counted. But what does it mean to say that an employer has committed, or ought to be liable for, hostile-environment sexual harassment? Courts identify an objective referent. In addition to proving that the individual felt aggrieved at the time, the plaintiff must also prove that his or her work environment was one that would seem hostile and abusive to a reasonable person.[7]

Over time, this "reasonable person standard" provoked some dissent. Judges, lawyers, and scholars reminded readers that until recently the reasonable person was called the reasonable man.[8] Even today, looking through the current edition of the leading law dictionary after the entry for "reasonable man," you won't find "reasonable person," let alone "reasonable woman."[9] Thus the reasonable person as an objective referent in sexual harassment cases was deemed falsely pluralistic, and biased against women, in an area where women suffer most of the harm and bring nine-tenths of the cases. To replace the reasonable man, courts and commentors offered the "reasonable woman," the "reasonable person of the same gender and race or color as the plaintiff," and so on, nearly *ad infinitum*.[10] Defenders of the reasonable person standard retorted that these particularistic measures did not fit the problem of attempting to state a universal standard.[11] The Supreme Court hinted that it prefers "reasonable person."[12] Late in the debate, I weighed in. I believe the problem is the adjective rather than the noun, and courts ought to ask whether the defendant employer conformed to the standard of *the respectful person.*

This proposal identified reason and reasonableness as inapposite to hostile-environment sexual harassment, for reasons that pertain not only to theory but also to the pragmatics of sexual harassment.[13] As it turns out, these pragmatics pertain also to environmental harm. Reason comes with a tradition of elite privilege, whereas women, racial minorities, and the less educated—groups that are hurt by sexual harassment beyond proportion to their numbers—have all helped to advance understandings of respect.[14] Jurors, co-workers, supervisors, managers, and other laypersons could make better use of "Is X behaving as a respectful person?" than "Are the conditions (for which X is responsible) conditions that a reasonable person would find hostile or abusive?" when called on to evaluate a work environment. The obligation of respect centers on an actor, or agent; whereas the "reasonable person" of sexual harassment focuses on reaction, the respectful person looks at the actor accused of misconduct.[15]

What then is misconduct, according to the respectful person standard? We may now move beyond sex in the workplace. To guide both actors and those who judge them, this inquiry needs precepts that are widely accepted, consonant with moral instructions, and pervasively taught. Kantian ethics, aided by the near-universal Golden Rule, provide a good beginning. Though occasionally criticized, these ethics have enjoyed a great following over many centuries and through hundreds of societies. Conscientious individuals have found these tenets broad enough to cover many situations yet specific enough to avoid inanity, and they tend to point clearly in one direction rather than another. The respectful person starts here.

This person has a duty to refrain. "What you do not want done to yourself, do not do to others," goes the Confucian version of the Golden Rule.[16] Confucian ethics posit a respectful person who knows his place in the social order, rather than one trying to fulfill a religious or spiritual ideal. This social-reformer's approach to the Rule may be better suited to law reform than are religious models that mandate active benevolence.[17] For purposes of legal doctrine, then, the duties of a respectful person ought to be seen as a negative rather than an affirmative.

The restraint of a respectful person may be described in various ways. Here I focus on two aspects of this restraint. The first pertains to the Kantian injunction against treating another person simply as the means to an end. The second type of restraint is a duty not to reject or deny the personhood of another.

Treating another person simply as the means to an end is *at best* (i.e., only if the end deserved approval) a misapplication of utilitarian or consequentialist ethics. Recognition respect cannot be rendered or withheld for utilitarian reasons. Whereas instrumentalist designs look forward to a future time of greater utility, recognition respect is a function of past or present characteristics, as the etymology of "respect" implies."[18]

The negative duty to refrain from violating the personhood of another reflects a concern about boundaries. Another person is a bearer of wishes, stances, and commitments that cannot be buried under the designs of the agent. The agent may not proceed heedlessly as though the object had no personhood, no distinct identity. This duty implies an obligation to receive communication from the object. In order to accept communication, the agent must hesitate, await guidance, and think about consequences to other persons. This receptiveness may demand, on occasion, that the agent take "no" for an answer.

Although sexual harassment law provides an especially good forum to explore the construct of the respectful person—sexual harassment is familiar, accessible, open to commentary from ordinary people and an array of experts alike, and frequently litigated—the legal duty to respect covers numerous other legal domains. The value of respect is implicit all over the law.[19] It is implicit also in the Precautionary Principle.

COMPETING VALUES OF THE PRECAUTIONARY PRINCIPLE

"In the face of scientific uncertainty and the possibility of societal or environmental harm, act with precaution."[20] This phrasing of the Precautionary Principle appears true. Who would challenge it? And what exactly does the principle require? I think the two questions are related. Hostility to the Precautionary Principle forms a kind of negative photographic image around it, revealing the contours of the principle. Opposition has emphasized three competing values: the "assimilative" powers or "dilution" capacity of the environment,[21] scientific criteria of certainty based on a demonstrated causal relationship,[22] and cost-benefit analysis.[23]

References to the assimilative powers or dilution capacity of the environment declare that because pollution can cleanse itself, or be diluted or assimilated into oblivion, the stringencies of a robust Precautionary Principle cannot be justified by the magnitude of environmental risks.[24] Pollution, it turns out, does not entirely cleanse itself, and dilution-capacity rationales against the Precautionary Principle sound especially wrongheaded today: "There is a growing certainty that the planet (and human beings) can no longer contend with the waste we create and put into the environment."[25] My sexual harassment analogy to the dilution argument recalls the time before 1976, prior to the coining of "sexual harassment" as a legal and cultural phrase, when courts stumbled in the dark. "It was just something that happened to you," explains one architect of current sexual harassment doctrine.[26] Dilution arguments, like the posture of sexual harassment law that endured for centuries, insist that no cure is needed, because no problem exists. A harassed worker must have known perfectly well that she was entering a rough environment: *tant pis* for her, say the assimilationists and diluters.[27] And *tant pis* for the silent oceans, forced to absorb discharges until policy makers are overwhelmed by evidence that they can no longer absorb. In sexual harassment as well as environmental law, such claims are now seldom taken seriously.

A stronger stance against the Precautionary Principle juxtaposes the knowledge and confidence of science against the fretting of worriers who cannot articulate what they fear. According to one critic, the principle reduces to a fatuous slogan: Better safe than sorry.[28] Another critic takes a more flowery turn: "Scientific uncertainty, rather than the normal verified hypotheses of cause and effect, becomes the basis for policy. The subjunctive becomes governmental imperative."[29] These dissenters find the Precautionary Principle confounding and even incoherent. If one may apply the principle against chemicals and industry, they contend, then why not apply it to every possible technology and regulation—even to the Precautionary Principle itself?[30] Policy makers cannot follow this questionable teaching, it is argued. Trust instead the methods and teachings of science. Demand proof of risks before regulating them.[31]

The self-confidence of this message, its craving to submit to elites, and its preference for arcane rather than accessible information all remind me of how sexual harassment doctrine has exalted reason and reasonableness at the expense of respect.[32] I do not mean to denigrate scientific knowledge, any more than I want to attack reason. My life depends on them both. But what is at stake in both environmental policy and sexual harassment law are the contributions that get cast arrogantly aside when policy makers act without caution. As many have pointed out, the Precautionary Principle is not adverse to knowledge, and it gives science-grounded information its full measure of attention and deference.[33] Three Canadian observers underscore the rationality of the Precautionary Principle when they write that the principle "implies that we should judge performance in the light of what was known and not known to the institutional actors at the time, rather than what we know now or may know in the future about costs and benefits."[34] Similarly, as I have elaborated elsewhere, a respectful person will render respect to the ideal of reason, and reason is important in adjudicating sexual harassment claims.[35]

Some who claim the mantle of scientific certainty distort the Precautionary Principle. One comprehensive attack declares that the principle has a perverse bias in favor of old risks rather than safer new technologies;[36] that it neglects risks from alternatives, risks from remediation, and foregone benefits;[37] and that it ignores the costs of implementing the principle when the public needs the same money for guns or butter:[38] "Richer is safer," after all.[39] This attack has a veneer of logic, but it criticizes using a caricature, much as some commentators have accused those who dissent from the reasonable person standard for sexual harassment of being hysterical.[40] Thoughtful advocates of the Precautionary Principle do not reflexively prefer the old, the familiar, and the expensive. Sited realistically in the present—just as the respectful person identifies duties based on past or present characteristics of an object, rather than by looking toward a utilitarian future[41]—the Precautionary Principle has as good a claim to rationality and intelligence as any science-credentialed rationale.

The last reference to costs and benefits brings us to the third antithesis: The Precautionary Principle is contrary to microeconomic cost-benefit analysis.[42] It does not condition precautions on their cost-effectiveness.[43] True, some versions do phrase the Precautionary Principle as something like "lack of full scientific certainty shall not be used for a reason for postponing cost-effective [precautionary] measures,"[44] but this caveat about cost-effectiveness is foreign to the principle and originates from sources that are hostile to it, notably the U.S. government.[45] Precautions usually bear direct costs. The German scholar Harald Hohmann, a passionate defender of the Precautionary Principle, has filled a book detailing these costs.[46] Hohmann provides a list of some minimums that the principle demands: "extensive technical and regional planning, environmental impact assessments, limitation of discharges through emission standards and treatment using the best available technology" as well as "product substitution" and "reduction or recycling of wastes."[47] Expensive measures indeed.[48] Because precautionary duties are triggered where the risk-neutral decision maker would forego precautions, the principle violates the microeconomic tenet of risk neutrality.[49]

Here a sexual harassment lawyer recalls fondly the apologists for harassment who argue—I have never actually heard them myself—that "the market" cures harassment by paying harassed workers a wage premium or that the utility of harassment to harassers offsets the disutility that workers suffer.[50] This false equilibrium borrows from the other two antitheses: Dilution capacity and scientific certainty similarly depend on an argument that precautionary approaches to environmental protection are not justified, at least not yet (a question-begging statement, to be sure) and therefore unjustifiably encroach on polluters' prerogatives. As with sexual harassment, "conservatives" condone the infliction of irremediable pain and damage by arguing that proponents of change have not proved the economic value of the remedies they propose. I place quotations around "conservatives" because it is odd for persons of that predilection to tolerate the violent disruption of environments without procedural safeguards or a showing of need from the would-be violator. The more authentic conservatism of the Precautionary Principle combines beautifully with progressive efforts.

CONNECTING TRADITION WITH PROGRESS

Like any other law reform proposal, the respectful person standard for sexual harassment claims held conservative as well as progressive inclinations. Reforms that are not eccentric or utopian comport with the past, not just the ideal future. And so the respectful person is fashioned out of ancient ingredients: old-fashioned human dignity, 18th-century Enlightenment views of personhood, statutes demanding respect for inanimate objects, and established constitutional values. The conservative *bona fides* of the Precautionary Principle are equally beyond challenge. To the extent that the Precautionary Principle seeks to replace or amend existing paradigms for environmental regulation, connection to the past and present will buttress its legitimacy, and ought to be emphasized. Precaution and respect both connect tradition with progress.

Another common trait of precaution and respect is their resonance with nonexperts and individuals who have been excluded from elite policy making. Confident predictions about assimilative capacity, assertions that environmental risks are small and insignificant, quantitative data applied to activists like a bludgeon, and elaborate defenses of corporate prerogative have a tendency to offend some nonexpert listeners. Their resistance is often countered with a condescending sneer: What do you know? It turns out that these dissenters know the Precautionary Principle, which gives content to concerns about pollution that have been proved, at least occasionally, to be far from wrong or irrational. In turn, popular discontent with authoritarian assurances lends strength to the Precautionary Principle.

Applied by nonexperts such as jurors, mediators, and arbitrators, a respectful person standard would allow variety in legal disputes to influence the development of one close analogue to the Precautionary Principle. For this reason, I am looking forward to encountering the respectful person as part of a standard sexual-harassment jury instruction.[51] Environmental law and policy have explored the Precautionary Principle as a regulatory device and a part of international law built away from adjudication, thereby keeping closed an avenue of public comment. The effort to define and implement a Precautionary Principle will benefit from lay articulations of precaution and respect.

NOTES

1. By "statutory actions," I refer to the principal federal employment discrimination statute, Title VII of the Civil Rights Act of 1964, as significantly amended in 1991. This statute is used in many litigated cases. A host of other legal remedies exist for sexual harassment, among them the common law of torts, state, and local anti-discrimination laws and other statutes that address housing, education, and consumer transactions. The thoughts expressed in this chapter apply as well to these alternative legal remedies.

2. An effort along these lines is Angela P. Harris, *Criminal Justice as Environmental Justice*, 1 J. GENDER, RACE & JUSTICE 1 (1997).

3. See Anita Bernstein, *Law, Culture, and Harassment*, 142 University of Pennsylvania Law Review 1227, 1292–1295 (1994). American sexual harassment law has a weak record in prevention and regulation. The only significant regulation is a set of Equal Employment Opportunity guidelines, now nearly twenty years old. Sexual harassment lawyers have much to learn from the regulatory successes of environmental law.

4. See Anita Bernstein, *Treating Sexual Harassment with Respect*, 111 HARV. L. REV. ____ (1997) (hereinafter *Treating*).

5. See Williams v. Saxbe, 413 F. Supp. 654, 657 (D.D.C. 1976), *rev'd on other grounds sub nom. Williams v. Bell*, 587 F.2d 1240 (D.C. Cir. 1978); *Barnes v. Costle*, 561 F.2d 983 (D.C. Cir. 1977); *Tompkins v. Public Serv. Elec. & Gas Co.*, 568 F.2d 1044 (3d Cir. 1977).

6. See Kathryn Abrams, *The New Jurisprudence of Sexual Harassment*, 83 CORNELL L. REV. (1998).

7. See *Harris v. Forklift Systems, Inc.*, 513 U.S. 17 (1993); *Meritor Savings Bank, FSB v. Vinson*, 477 U.S. 57 (1986); *Henson v. City of Dundee*, 682 F.2d 897 (5th Cir. 1982).

8. See Leslie Bender, A Lawyer's Primer on Feminist Thoery and Tort, 38 J. LEGAL EDUC. 3 (1988); Ronald K.L. Collins, Language, History and the Legal Process: A Profile of the "Reasonable Man," 8 RUT.-CAM. L.J. 311 (1997).

9. See BLACK'S LAW DICTIONARY 1266 (6th ed. 1990).

10. See *Ellison v. Brady*, 924 F.2d 872 (9th Cir. 1991) ("reasonable woman" and

"reasonable victim"); *Stingley v. Arizona*, 796 F. Supp. 424, 428 (D. Ariz. 1992) ("reasonable person of the same gender and race or color"); for variations in the law review literature, *see* Jane L. Dolkart, *Hostile Environment Harassment: Equality, Objectivity, and the Shaping of Legal Standards*, 43 EMORY L.L. 151 (1994); Caroline Forell, *Essentialism, Empathy, and the Reasonable Woman*, 1994 U. ILL. L. REV. 769.

11. *See* Kathleen A. Kenealy, *Sexual Harassment and the Reasonable Woman Standard*, 8 LAB. LAW. 203 (1992); Robert Unikel, Comment, *"Reasonable Doubts": A Critique of the Reasonable Woman Standard in American Jurisprudence*, 87 Nw. U. L. REV. 326 (1992).

12. *See Harris v. Forklift Systems, Inc.*, 513 U.S. 17, 22 (1993).

13. "Pragmatics" would include how to recognize it around you, how to know when you might be doing it, how to stop it, and when to recognize that an employer should not be blamed for it. *Cf.* [analogy from conference].

14. *See Treating, supra* note 4.

15. This focus on the actor is another virtue of a respect-oriented standard of behavior. Civil liability imposes blame, and it is fairer to ask whether an individual has done wrong than to judge behavior only in terms of the reaction it provokes.

16. *See* H.D.T. ROST, THE GOLDEN RULE: A UNIVERSAL ETHIC 49 (1986).

17. *See Treating supra* note 4, at n.256.

18. Respect derives from the Latin "respicere," to look back, or to take a second look. Robin S. Dillon, *Respect and Care: Toward Moral Integration*, 22 CAN. J. PHIL. 105, 108 (1992).

19. *See Treating, supra* note 4 (discussing intellectual property rules, legal protections for certain nonhuman entities such as corpses and animals, and judicial interpretations of the First Amendment freedoms of speech, religion, and association).

20. Carolyn Raffensperger and Joel Tickner, Conference Synopsis, Sept. 12, 1997.

21. *Id.*, "Conference Rationale."

22. *See* Daniel Bodansky, *Scientific Uncertainty and the Precautionary Principle*, ENVIRONMENT, Sept. 1991, at 4.

23. *See* Harald Hohmann, Precautionary Legal Duties and Principles or Modern International Environmental Law 341 (1994).

24. *See* Naomi Roht-Arriaza, *Precaution, Participation, and the "Greening" of International Trade Law*, 7 J. ENVTL. L. & LITIG. 57, 60–61 (1992).

25. *See* Conference Rationale, *supra* note 21.

26. CATHARINE A. MACKINNON, FEMINISM UNMODIFIED: DISCOURSES ON LIFE AND LAW 106 (1987).

27. *See Rabidue v. Osceola Refining Co.*, 584 F. Supp. 419, 430 (E.D. Mich. 1984) (stating that sexual harassment law was not designed "to bring about a magical transformation in the social mores of American workers"), aff'd, 805 F.2d 611 (6th Cir. 1986), cert. denied, 481 U.S. 1041 (1987).

28. *See* Frank B. Cross, *Paradoxical Perils of the Precautionary Principle*, 53 WASH. & LEE L. REV. 851 (1996).

29. Patrick Michaels, Senior Fellow, Cato Institute, quoted in Gregory D. Fullem, Comment, *The Precautionary Principle: Environmental Protection in the Face of Scientific Uncertainty*, 31 WILLIAMETTE L. REV. 495, XXX (1995).

30. *See* Cross, *supra* note 28, at 861.

31. Several years ago two Bush administration officials published an account of their opposition to the Precautionary Principle, endorsing in its place a "no regrets" or "wait and see" approach. Though unclear, this perspective seems to demand a high degree of scientific certainty before making environmental policy. C. Boyden Gray & David B. Rivkin, Jr., A *"No Regrets" Environmental Policy*, FOREIGN POLICY, Summer 1991, at 47.

32. Carolyn Raffensperger has put it well: "Usually we contrast precaution with certainty. Certainty is a concept of reason. Precaution is a concept of respect." E-mail of April 1997.

33. *See* Tim O'Riordan and James Cameron, INTERPRETING THE PRECAUTIONARY PRINCIPLE 62–68 (1994) (describing scientific responses to uncertainty).

34. Michael Trebilcock, Robert Howse, and Ron Daniels, *Do Institutions Matter? A Comparative Pathology of the HIV-Infected Blood Tragedy*, 82 VA. L. REV. 1407, 1412 (1996).

35. *See Treating, supra* note 4.

36. *See* Cross, *supra* note 28; *see also* Peter Huber, *Safety and the Second Best: The Hazards of Public Risk Management in the Courts*, 85 COLUM. L. REV. 277 (1985).

37. *See* Cross, *supra* note 28; *see also* Louis W. Sullivan, *Chemical Villains: A Case Unproved*, L.A. TIMES, Apr. 1, 1996, at B5.

38. *See* Cross, *supra* note 28, at 915–16.

39. *Id.* at 917.

40. *See* Tama Starr, A Reasonable Woman: The *"Reasonable Woman" Standard for Assessing Sexual Harassment Cases*, REASON, Feb. 1994, at 48; *cf. Michigan Rejects "Reasonable Woman" Standard in Sexual Harassment Cases*, Mich. Empl. L. Letter, July 1993. ("Courts utilizing the reasonable woman standard pour into the standard stereotypic images of women [as] sensitive, fragile, and in need of a more protective standard.")

41. *Ibid.*

42. *See* David Pearce, *The Precautionary Principle and Economic Analysis*, in O'Riordan and Cameron.

43. *See* Konrad von Moltke, *The* Vorsorgeprinzip *in West German Environmental Policy*, in ROYAL COMMISSION ON ENVIRONMENTAL POLLUTION, TWELFTH REPORT 57, 58 (1988). (Contrasting the *Vorsorgeprinzip*, the German antecedent of the Precautionary Principle, with international programs based on "the prevention principle"; the prevention principle "is carefully linked to some criterion of economic feasibility, whereas the Vorsorgeprinzip initially has no such qualification attached to it.")

44. *See* Principle 15, Rio Declaration on Environment and Development, U.N. Doc A/CONF. 151/5/Rev.1 (1992), *reprinted in* 31 I.L.M. 874 (1992); *see also* United Nations Framework Convention on Climate Change.

45. *See* Daniel Bodansky, *The United Nations Framework Convention on Climate*

Change: A Commentary, 18 YALE J. INT'L L. 451, 504 (1993) (noting that "cost-effective" caveat was injected into the Convention by the United States and Saudi Arabia to blunt the effect of the Precautionary Principle).

46. *See* Hohmann, *supra* note 23.
47. *Id.* at 190.
48. As the critic Frank Cross elaborates, the Precautionary Principle is also undeterred by indirect costs. Cross provides examples of those costs, including the water contamination that would result from a ban on chlorine, harms caused by asbestos removal, and the impoverishing effect of spending too much on the environment. See Cross, supra note 28.
49. *See* Pearce, *supra* note 42.
50. *See* Mark G. Kelman, *Trashing*, 36 STAN. L. REV. 293 (1984).
51. A proposed model jury instruction appears in *Treating*, *supra* note 4.252.

Part III

~~~

INTEGRATING PRECAUTION
INTO POLICY

This part is the heart of the volume. Parts I and II examined some of the history, rationale, and foundations of the Precautionary Principle. While the principle appears as a foundation of almost every contemporary international environmental treaty and national statements of environmental policy, few treaty secretariats or nations have established structures for determining when precaution should be triggered and what action should be taken once such a trigger is established. As a result, policy makers in the United States (and to some extent abroad) have not been enthusiastic about integrating precaution into decision making without a substantive, analytical framework that they can defend to industry, the public, and the courts. These chapters offer some of the answers.

Chapters in this part outline *how* we implement the principle. They address questions such as: "How could the Precautionary Principle be integrated into current and future decision-making structures?" and "What policy tools would most effectively implement precautionary action?"

Joel Tickner develops a structure for operationalizing the Precautionary Principle in law and policy in chapter 9. He argues that such a structure switches the questions asked when making decisions under scientific uncertainty, as well as presumptions of harm and safety. Components of that structure are duties and goals for environmental health, shifting burdens of proof, focusing on alternatives to potentially harmful activities, increasing knowledge of the impacts of human activities, and democratic decision making.

Scottish Natural Heritage lays out a practical procedure for applying the Precautionary Principle in development and other activities that might harm natural resources in chapter 10. This chapter offers step-by-step guidance for how to determine if precautionary action should be taken to protect the natural world, depending on the availability of options and uncertainties about future harm.

In chapter 11, Nicholas Ashford discusses the barriers posed by cost-benefit analysis to the implementation of the Precautionary Principle. He proposes an alternative analytical framework, called "trade-off analysis," in which trade-offs between health and economic gains of an activity to workers, companies, and the public are expressed in their natural units, instead of being monetarily quantified. Who bears the costs and who reaps the benefits from a policy are thus not hidden in the aggregation of costs and benefits. Ashford goes on to suggest the implementation of technology options analyses as a mechanism to facilitate the shift toward safer and clearer technologies developed and chosen by firms.

Mary O'Brien presents the concept of alternatives assessment as a critical way to focus environmental and health policy on seeking solutions to potential hazards, rather than endlessly analyzing and determining acceptable levels of risk. Alternatives assessment consists of publicly examining a full range of alternatives to a potentially damaging activity, as well as its underlying need. The chapter includes examples of alternatives assessment in practice as well as an outline of essential elements of such a mechanism to carry out precautionary action and its benefits.

In chapter 13, Robert Costanza and Laura Cornwell propose an economic mechanism for implementing precaution, called "precautionary assurance bonding." This mechanism would require initiators of a potentially dangerous activity to post a bond equal to the cost of the worst possible impacts of that activity prior to starting. They argue that this type of economic mechanism would encourage those undertaking a potentially dangerous action to reduce impacts or find suitable alternatives upfront. They provide examples of how this type of mechanism has been used in the past and how it could be used in implementing the Precautionary Principle.

Sanford Lewis describes mechanisms for decreasing artificial or "smoke-screen" uncertainty in decision-making processes through corporate disclosure mechanisms. He argues that better information and greater access provide an economic and public accountability mechanism for those undertaking potentially harmful activities to take greater precautions. Such disclosure mechanisms can be implemented through public policy, private actions, or "good neighbor agreements" between local communities and companies.

Dick Sclove and Madeleine Scammel offer tools for increasing public participation in decision making regarding science and technology, as the Precautionary Principle requires an open, informed, and democratic process. They argue that while in a democracy decisions affecting all citizens should be made democractically, decisions and policies regarding harmful industrial activities and technologies are made to the systematic exclusion of affected publics. They present two mechanisms for providing better citizen access and participation in science and decision making: community research networks (or "science shops") and technology assessment by citizens (consensus or laypersons conferences), which have a history of successes in northern Europe and are growing in importance in the United States.

Finally, Gordon Durnil describes his first encounter, as a commissioner on the U.S.–Canadian body monitoring water quality in the Great Lakes, with scientific uncertainty in the face of evident environmental harm. The commission on which Durnil served became one of the first bodies in which U.S. representatives advocated and implemented the Precautionary Principle. A conservative politician, Durnil urges environmentalists to develop a broad public base of support for precaution.

Taken together, the chapters in this part provide shape and substance to the Precautionary Principle, as well as stepping stones toward its implementation. They widen the perspective of environmental and public health decision-making processes, which currently focuses on quantitative scientific and cost analyses of activities after those activities have begun. All of the chapters in this part seek to avoid problems before they occur and take action before proof of harm has accrued. Integrating these concepts into law and policy will lead from a permissive, acceptable risk approach, which portrays the risks of modern society as a "given," to one based on preventing harm and protection of human and ecosystem health. They will also aid in making the decision-making process more transparent, accountable, and open to the participation of the potentially affected public, who need a say in the scientific and technological decisions that affect their lives.

Chapter 9

❧

A MAP TOWARD PRECAUTIONARY DECISION MAKING

Joel A. Tickner

While the Precautionary Principle has been explicitly or implicitly endorsed by many nations through treaties and national law, its components have yet to be clearly defined, and no comprehensive framework to guide decision making under the Precautionary Principle currently exists. This lack of a structure for precautionary decision making has limited the widespread use of the principle and in some instances has led to strong resistance to its implementation. As such, there is a critical need to develop a map or structure and tools to operationalize the Precautionary Principle.

This chapter presents an overview of the essential components of a structure for implementing the Precautionary Principle in environmental and public health decision making. After a discussion of the basis of such a framework, I will describe its fundamental structure, components, and criteria for decision making. Finally, I will provide some examples of how components of the Precautionary Principle have been implemented in practice. Detailed examples of specific components of a precautionary decision-making structure are described elsewhere in this volume. The examples and basic framework structure presented in this chapter are focused mainly on toxic chemical concerns but are applicable to any environmental or public health hazard.

Basis of a Framework for Operationalizing the Precautionary Principle

The basis for operationalizing the Precautionary Principle is twofold: (1) to change questions asked when making decisions under scientific uncertainty and (2) to change the presumptions about the harm of a particular activity, action, or substance. First, the Precautionary Principle forces scientists and policy decision makers to ask a different set of questions about activities and potential hazards. Current decision-making approaches ask questions such as: "How safe is safe"; "What level of risk is acceptable"; and "How much contamination can a human (usually a healthy adult male) or ecosystem assimilate without showing any obvious adverse effects?" The Precautionary Principle asks a different set of questions such as "How much contamination can be avoided while still maintaining necessary values?"; "What are the alternatives to this activity that achieve a desired goal (a service, product, etc.)?"; and "Do we need this activity in the first place?"

Changing the questions we ask about a problem (the problem frame) engenders a totally different set of public policies. Policies based on the principle are preventive, whereas those based on current decision-making approaches tend to focus on pollution control and remediation. Precautionary approaches are goal and alternatives oriented, lending themselves to technology innovation, pollution prevention, and impact assessment. Policy responses based current decision-making approaches generally lead to add-on, end-of-pipe technologies, personal protective equipment, and medical treatment for those negatively affected. In essence, the Precautionary Principle moves the focus of decision making (and hence the questions asked by decision makers) from one of risks, which are highly uncertain and difficult to measure, to one about solutions to problems, for which we can often have a greater level of certainty.

In addition to changing the questions decision makers ask about issues, the Precautionary Principle shifts the presumptions used in decision making. Rather than presume that a specific substance or activity is safe until proven dangerous, a process that takes substantial time and resources, the principle places a presumption in favor of protecting the environment and public health. This shift of presumption places the responsibility for demonstrating safety and preventing harm on those undertaking potentially harmful activities. Accordingly, humans and the environment receive the benefit of the doubt under terms of scientific uncertainty and ignorance, rather than a particular substance or action.

The Precautionary Principle establishes a type of "speed bump," which creates bottlenecks in the development process but does not stop flows. It establishes a process of seeking the least hazardous alternative to achieve a specific purpose, continuously updating knowledge to avoid harm. The goal of instituting a speed bump is to create options in any given situation that are the most "error friendly," those that would be least prone to environmental or health damage or for which harm would be most reversible (von Weizacker, 1996). Operationalizing the Precautionary Principle does not mean that current decision-making tools, such as risk assessment and cost-benefit analysis, are discarded. It does mean, however, that these tools are used simply to inform decision making in order to protect health and the environment, rather than to make decisions themselves. In this respect, these tools are relegated to a second tier in the decision-making process. Instead of using these techniques to quantify an "acceptable" risk, they are used to compare alternatives to an activity (or to establish priorities) that is a much less complex and often more clear-cut activity, requiring less rigorous quantitative analysis and less uncertainty.

The precautionary approach to decision-making consists of the following elements that are explained in the following sections: (1) A general duty to take precautionary action in the face of uncertainty; (2) goal setting for environmental and public health protection; (3) shifting burdens of proof to initiators of potentially harmful activities; (4) tools to aid decision making under uncertainty (e.g., decision-making criteria); (5) prevention-oriented methods to carry out precaution-based decisions (such as clean production); (6) economic incentives to promote precaution; (7) means to continuously measure potential adverse effects of both current and alternative activities; and (8) democratic decision-making structures.

We can think of a framework for operationalizing the Precautionary Principle as having two separate applications: for decision making regarding new activities and for decision making based on potential hazards that already exist. Being precautionary about new activities may be easier (politically, economically, and scientifically) than acting precautiously on an existing hazard.

THE OVERALL STRUCTURE FOR PRECAUTIONARY DECISION MAKING

Figure 9.1 presents a schematic diagram of the structure for operationalizing the Precautionary Principle. The components of such a framework are described in detail in the next section.

Under traditional linear decision-making structures (boxes in bold), evidence of harm is collected, its probability is examined through a risk assessment process (looking at both hazard and exposure), and then a risk man-

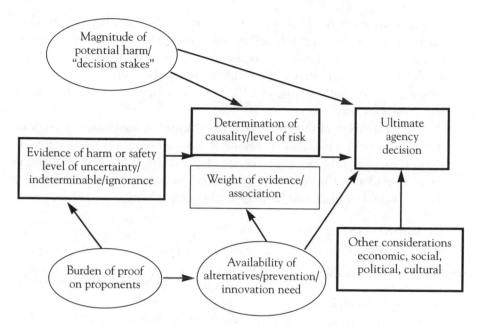

FIGURE 9.1. A Precautionary Decision-Making Model

agement decision is made to take or not take action, considering the costs and benefits of regulation as well as other factors. Causality and level of risk are the central elements of this structure.

Under a precautionary decision-making structure, evidence of harm is considered, as well as evidence of alternatives and the magnitude of possible harm from an activity. The latter two are considered just as important in the decision-making process as the determination of causality. In this regard, if there is information about safer alternatives or if the magnitude of potential harm from an activity is great, it may be possible to partially or entirely bypass the costly and often contentious determination of causality that is central to current decision-making structures. For example, if an activity could cause widespread, irreversible harm or if it could harm sensitive members of a population (i.e., children), it might be prudent to take action, even though there is not clear-cut evidence of harm. It is also necessary under a precautionary decision-making structure to consider uncertainty, indeterminacy (large-scale uncertainty), and ignorance, which are rarely evaluated under current structures. Indeterminacy about cause–effect relationships would favor action to prevent harm while the problem is studied further. That is action taken in advance of certainty.

Decisions about the likelihood of harm are made under this structure on the basis of a "weight of evidence" approach, taking into consideration information from various kinds of sources, the magnitude of impacts, and the availability of alternatives. This differs from the current quantitative approach to decision making, which considers singular types of information and quantifies risk.

A central aspect of this structure is the shifting of the burden of proof to the proponent of a potentially harmful activity who must provide information on its safety, the need for the activity, and the availability of alternatives. If evidence (in contrast to certainty) of harm is present, then the activity would be presumed harmful until conclusively proven otherwise.

COMPONENTS OF THE FRAMEWORK

This section presents the components of a policy framework for operationalizing the Precautionary Principle. While their order does not reflect their relative importance, it does reflect a logical order to their incorporation into legislation and regulation. Depending on the case, some components of the framework may be more important than others and some may not be applicable.

The elements of a policy framework for operationalizing the Precautionary Principle would consist of the following.

Definition and General Duty to Take Precautionary Action

An important first step in developing a precautionary approach framework is the creation of a working definition for the Precautionary Principle. The definition will establish a broad-based goal of precaution in the face of scientific uncertainty, which will set the stage for subsequent decision-making strategies. The Wingspread Conference on Implementing the Precautionary Principle defined the principle as follows: "When an activity raises threats of harm to the environment or human health, precautionary measures should be taken even if some cause and effect relationships are not fully established scientifically." Several international treaties contain definitions of the Precautionary Principle, but the Wingspread Statement is one of the first to include specific elements of precaution.

Through a clear definition, the principle can be incorporated as a general duty in environmental and other public health legislation. This is how the principle has been used to date on a national and international level. General duty statements provide important powers to government agencies to take action on perceived threats that may or may not be specifically regulated.[1] The U.S. Occupational Safety and Health Act General Duty Clause

(OSHA section 5) has been used in numerous instances to protect workers from chemical or ergonomic hazards. This clause states that "each employer shall furnish to each of his employees employment and a place of employment which are free from recognized hazards that are causing or are likely to cause death or serious physical harm to his employees."[2] The U.S. Clean Air Act, Section 112r, Risk Management Plan Rule, contains a general duty on companies to prevent chemical accidents, which regulators could use to require options analyses for safer processes. The ill-defined, uncertain outcome of this type of general duty created by the Precautionary Principle may be critical to its success by creating a certain air of uncertainty in regulation that may lead companies to take action that would normally not be taken if regulatory obligations were clearly defined.

Aggressive Goals for Reductions in Hazardous Substances, Processes, Products, and Practices

The Precautionary Principle requires a jurisdiction to establish goals of where it would like to be with regards to preventing, reducing, or eliminating a specific hazard. For example, the Swedish government has set ambitious goals for the reduction of some heavy metals and persistent organic chemicals (Ministry of Environment, 1997). These goals, with interim targets, are periodically measured and reexamined. They are established in cooperation with a wide cross-section of society so as to minimize economic displacement and to allow development of alternatives. Many countries, such as the United States, fail to establish goals for environmental health other than to cure specific diseases like cancer. This results in a rudderless policy that fails to prevent those illnesses in the first place.

Rather than trying to forecast potential impacts and uncertain futures as is often done with tools such as risk assessment, under precautionary decision making, government agencies, industry, workers, and the public would develop a vision of where society (or a country, state, etc.) should be and then work backward to determine steps toward that goal, a process called "backcasting." Backcasting is the opposite of forecasting. Forecasters determine where we are and try to predict where we will be in the future. Backcasters decide where we should be and then figure out how to get there. For example, the Dutch government establishes five-year environmental plans with clear goals and then works with municipalities, industry associations, and specific companies to establish "covenants." These covenants are voluntary agreements between the government and the industry that establish interim and final goals but that place responsibility on the company to achieve those goals in the most efficient way possible (without creating new risks). The covenants are backed by strong enforcement and regulation if goals are not met (van Dunne, 1993).

Such a backcasting exercise could be combined with some sort of scenario development to promote understanding of the effects of different means to achieve goals or different levels of goals. This would also allow for flexibility in both goals and the methods undertaken to achieve them.

Shifting the Burden of Proof to Those Undertaking Hazardous Activities

A key function of the Precautionary Principle is to require proponents of a potentially hazardous activity to demonstrate safety or prove that no safer alternative exists before engaging in that activity. This reverses the presumption of environmental and public health policy away from "safe until proven harmful." As government authorities never have sufficient resources to study every chemical, factory, or ecosystem, it becomes critical for those undertaking a potentially dangerous activity (and who will ultimately benefit most from that activity) to have to prove that their activities will not harm humans or the environment. Also, since wildlife, workers, and the public rarely have the power to control development, technologies, the production of products, or production processes, those undertaking activities have a responsibility to prevent harm.

Manufacturers currently have the burden of proof of safety for pharmaceutical products (and sometimes pesticides) in many countries throughout the world, though these burdens are often couched to some degree on economic benefits or effectiveness of the drug or pesticide. Nonetheless, these laws could serve as a model for future environmental and occupational health and safety regulations.

The burden-of-proof requirement has its most important role with regard to new activities, but it can also be applied to existing hazards. Before a potentially dangerous activity can start, the proponent should demonstrate that no harm would occur and that there were no safer alternatives to an activity.[3] The government agency would then have the duty of permitting, restricting, or prohibiting such an activity and/or offering alternatives. In the case of existing activities, a suspicion of harm shifts the burden of proof from the government agency or the public to the company or actor who has undertaken an activity (e.g., when it is found that an emission is harmful, it would be up to the emitting company to prove that there is no safer way to carry out that activity). Finally, as those undertaking a potentially dangerous activity have a vested interest in proving its safety and control scientific information, it has been suggested (Raffensperger, 1997) that any claim of safety undergo an independent peer review funded by the proponent of the activity who would provide all information (demonstrating harm or safety) used in the analysis.

Criteria and Structure for Decision Making About Harm Under Uncertainty

A structure for operationalizing the Precautionary Principle needs to provide clear instructions to inform decision makers on how to weigh scientific and other evidence about the likelihood of harm. Deciding whether there is enough evidence of potential or actual harm to take action is perhaps the most contentious aspect of the decision-making process. There are two important questions that must be asked in developing criteria and structure for precautionary decision making: First, does a legal regime for the Precautionary Principle establish some standard of proof of harm at which level precautionary action would be taken?[4]; and second, what information should be included in decision making?

A method for analyzing evidence and associations under a Precautionary Principle structure is described later. Decision making may differ depending on whether a current or proposed activity is being examined. For a proposed activity where evidence of potential harm has surfaced, focusing on alternatives can allow decision makers to bypass the complex determination of causality and proof of harm.

Decision making about associations or likelihood of harm under the Precautionary Principle should be based on a "weight-of-evidence" approach, rather than on some quantitative probability of harm (as is the case with risk assessment approaches). The weight-of-evidence approach to decision-making takes into account the cumulative weight of information from numerous sources that address the question of injury or the likelihood of injury to living organisms (IJC, 1995).[5] Types of information that might be considered include observational studies, worker case histories, toxicological studies, exposure assessments, epidemiologic studies, and monitoring results. Based on the weight of evidence, a determination is made as to whether an activity has caused or is likely to cause harm and the magnitude of that harm.[6]

Lists of criteria for evaluating information on causal associations (cause-and-effect relationships) and potential harm have been proposed by numerous authors. These criteria guide the collection and analysis of information, as well as the questions asked by decision makers. Some of these lists are included in lists A–E in the chapter appendix. Some of the criteria address causal inference (such as the Hill criteria and those of the Massachusetts Weight-of-Evidence Committee) while others address magnitude of harm and considerations for weighing evidence of potential harm. The Massachu-setts criteria, which were developed within an ecological risk assessment framework, provide some important criteria for assessing cause and effect relationships, but many of those criteria would require substantial quantita-tive evidence before such a relationship can be established, and this could undermine precautionary action. The Dovers and Ludwig criteria indicate that a determination of causal association may not be necessary when an activity could potentially have irreversible, widely distributed, or multigen-erational impacts. In other words, they address the "decision-stakes" of a par-ticular decision under uncertainty (Funtowicz and Ravetz, 1991).

The weight-of-evidence determination (or the determination of whether to allow an activity to continue or restrict it) would vary depending on the range and scale of impacts and the availability of alternatives (or other means) to prevent the hazard. List F in the chapter appendix presents some proposed criteria for precautionary decision making, divided by causal infer-ence and decision-stakes (magnitude) criteria. Taken together, these guide weight-of-evidence analyses and the amount of information needed before precautionary action is taken.

The decision-making criteria and weight-of-evidence determinations can be incorporated into a decision tree/process type format. The analysis would consist of two parts. The first branch would deal with existing hazards. In this case if the weight of the evidence indicated actual or possible harm, pre-ventive action would be taken. Preventive action could consist of stopping the activity, requiring an analysis of alternatives to the proposed activity, or undertaking mitigating measures. If insufficient evidence of harm was avail-able to arrive at a weight-of-evidence determination, the proponent of the activity would have the burden of providing evidence of no harm (subject to independent evaluation). If this evidence was simply not available and uncertainty remained, precautionary measures, such as those mentioned ear-lier, would be taken.

A second part of the analysis targets new chemicals, products, or work activities. The initiator of an activity would conduct an initial impact state-ment identifying potential impacts of the activity, potential alternatives, and the proposed action. Precaution would serve as a default presumption until the weight-of-evidence determination demonstrated that there was no safer alternative for the activity that would fulfill the needs of the initiator, and that there is a necessity for such activity—or that the activity posed no real risk. The weight-of-evidence analysis would also identify potential adverse impacts of that activity and monitoring/investigation requirements for the initiator.

Depending on the level of uncertainty about cause-effect and the potential magnitude of the impacts of an activity, different levels of precautionary action might be warranted. Different levels of evidence of harm could lead to different types of responses ranging from weak to strong precaution (e.g., study requirements or substantive requirements, such as mitigation or alternatives development). For example, an activity for which we have only minimal evidence of harm and for which harm, if it were to occur, is minimal, would possibly lead to increased monitoring;[7] an activity for which we have some evidence of harm would require preventive or remedial action and monitoring (depending on the magnitude of the problem); and an activity for which we are fairly certain of harm or for which damage, should it occur, would be large or irreversible would be limited or prohibited.

Prevention-Based Tools for Implementing Precautionary Action

The Principle of Precautionary Action calls for preventive actions before proof of harm has been established. These measures should be taken in advance of scientific evidence and, when possible, at the design stage of a potentially hazardous activity. The Precautionary Principle does not fulfill its purpose unless preventive methods for carrying out precaution are implemented.[8] There are numerous tools for carrying out precautionary policies that have been used throughout the world. Many of these tools are discussed in accompanying chapters in this volume. They include the following:

- *Bans and phaseouts.* At least eighty countries ban the production or use of a small number of highly toxic substances. The Nordic countries have particularly advanced the use of bans as a public health strategy. Bans and phaseouts are the only way in which to reduce to zero the risk of injury or disease from a very toxic chemical or hazardous activity. Several chemicals are now being phased out in Sweden, including cadmium and mercury, and a phaseout of persistent organic chemicals has been proposed. The U.S. International Joint Commission recommended a phaseout of industrial chlorine in the Great Lakes region as the only way to effectively reduce the health and environmental risks posed for organochlorine contamination in the region.
- *Clean production/pollution prevention.* Clean production and pollution prevention include changes to production systems or products to reduce pollution at the source (in the production process or product-development stage). By focusing impacts within the firm, these methods can cost-effec-

tively reduce or eliminate pollution from production processes and products. A majority of U.S. states have some form of pollution prevention program, which have documented success stories in reducing industrial pollution. Some, such as New Jersey and Massachusetts, require companies to identify alternatives to reduce their emissions and use of toxic chemicals and waste. On an international level, cleaner production activities are beginning to address the dangers of products themselves and the raw material and energy inputs necessary for their creation. This is leading to a focus on product redesign on the basis of dematerialization and detoxification, in addition to a fundamental questioning of need.

- *Alternatives assessment.* Rather than asking the risk-assessment/cost-benefit analysis question of what level of contamination is safe or economically optimal, alternatives assessment asks the question of what activities can be undertaken to reduce or eliminate the hazard. For example, the U.S. National Environmental Policy Act (NEPA) calls on the federal government to investigate alternatives (in an Environmental Impact Statement), including a no-action alternative, for all of its activities (or activities it funds) determined to have potential environmental impacts. Citizens have the right to appeal decisions if a full range of options (or potential impacts) is not considered. Several European countries have initiated such programs for all private industrial activities.

    Nicholas Ashford at the Massachusetts Institute of Technology has developed a structure for chemical accident prevention called "Technology Options Assessment" (TOA) (Ashford et al., 1993). Under this scheme, companies would be required to undertake comprehensive assessments of alternative primary prevention technologies and justify their decision if safer alternatives were not chosen.

- *Health-based occupational exposure limits.* Over a period of several years, a group of occupational health experts in the United States has developed health-based occupational exposure limits that are based on the lowest exposure level at which health effects have been seen (Ziem et. al., 1990). Under a precautionary exposure limit scheme proposed by Stijkel and Reijnders (Stijkel and Reijnders, 1995), any substance that is a known, probable, or suspected carcinogen or reproductive toxicant would have a maximum occupational exposure limit of 0.1 mg/m^3. This low limit takes into account the fact that for most carcinogens there is no known threshold level. Additional testing could lead to lower limits for individual substances. In its 1996 annual meeting, the American Public Health Association passed a resolution calling for the rapid adoption by occupational safety and health authorities of these "precautionary exposure limits," given the limited information and testing on chemical effects.

Use of the "Polluter Pays" Principle

Under a framework based on the Precautionary Principle, those engaging in potentially hazardous activities would be responsible for the costs of the damage they caused. This concept, called the "polluter pays principle," is a fundamental principle of environmental law in several European countries and in some international treaties. The difficulties of such a principle lie in assessing and quantifying damage and attributing harm to those who have caused it. There are two specific economic mechanisms that could be used to implement the polluter pays principle and encourage more precautionary decision making: assurance bonding and strict and joint and several liability.[9]

In chapter 13, Robert Costanza and Laura Cornwell examine the precautionary polluter pays principle as an incentive for firms to make preventive decisions (see also Costanza and Cornwell, 1992). This principle encompasses precautionary assurance bonding to ensure that companies undertaking potentially hazardous activities have sufficient resources available to pay for any damages created. Companies would be required to pay a premium before commencing a potentially dangerous activity that is based on the worst potential damage that the facility could possibly cause. This assurance bond could be returned if the potential damage does not occur. This type of system would move costs of potential damage to the present, where they will have the greatest impact on decision making, thus leading the firm to invest in safer alternatives.

A Precautionary Principle decision-making structure could also impose both strict liability and joint and several liability on those undertaking a potentially dangerous activity. Strict liability, or liability without fault, means that a company or initiator of an activity need not be found negligent to be assessed blame for harm (e.g., ignorance or a poor understanding of potential impacts does not serve to eliminate responsibility for harm). Joint and several liability means that harm caused by several actors can be attributed to a single actor or to all of them.

The polluter pays principle provides an economic/market incentive for investing in cleaner and safer technologies, products, and activities. However, governments will need to remove subsidies from dangerous practices, technologies, and products and redistribute them to cleaner and safer ones in order to reduce the penalties for undertaking more environmentally and healthier actions.

Evaluating Alternative Activities, Technologies, and Chemicals

The Precautionary Principle will not serve its role if it prevents one hazard while creating another. The potential creation of new, unintended risks by application of the principle is one of the most common criticisms of the principle (Cross, 1996). A systematic and comprehensive scheme must exist to examine the impacts of alternatives to potentially hazardous activities to ensure that one hazard is not replaced by an unknown, yet potentially greater one. This scheme would examine the entire life cycle of the product or activity, including raw material extraction, production, use of product, and disposal. For example, the decision to phase out a particular chemical would be inextricably linked to the development and analysis of alternatives to fulfill the services provided by that chemical. This type of scheme would encourage those undertaking an activity to systematically and comprehensively study (and thus understand) the range and types of potential impacts before initiating an activity so as to avoid or make plans to monitor or mitigate such impacts.

Ongoing Monitoring, Investigation, and Information Dissemination

The Precautionary Principle calls for more, not less, science to better understand the complexity of ecosystems and the impacts of different stressors on them. Harm cannot be prevented if it is unknown or poorly characterized.[10] However, the need to further study a problem should not forestall preventive actions.

Under the Precautionary Principle, those undertaking potentially dangerous activities have a responsibility for understanding their impacts. A lack of knowledge cannot be considered the same as a lack of impacts. For example, recent research has found that for the vast majority of high production volume chemicals, we do not have the most basic toxicological information (EDF, 1997). Yet, these chemicals are still widely used and released into the environment, and decisions are still based on an analysis of risks (EDF, 1997). To obtain critical chemical information, the Danish government has proposed a chemicals categorization scheme that considers a chemical as toxic as the most toxic one in its class if there is no evidence to the contrary (Bro-Rasmussen et al., 1996).

While potential harm should be studied (and avoided) before an activity commences, the promoter of an activity should be required to conduct ongoing analyses of potential impacts, to inform regulators and the public of the results of these analyses (allowing the public to independently review these analyses and conduct its own analyses as is currently done under some "good neighbor" agreements), and to take action when potential impacts are identified. This duty could consist of periodic assessment/audit requirements, long-term monitoring, and premanufacture/start-up impact statements any time a change is made to a product, process, or activity. A suspicion of harm would require notification of the public and government authorities so that subsequent actions to restrict or further study the potential impact could be initiated.

Methods for Participative and Democratic Decision Making

Democratic decision-making processes are an important component of the Precautionary Principle. Decisions made under great uncertainty are policy decisions. Science only provides information about the kind of harm that is likely; it cannot tell us what to do so that we avoid the harm. Additionally, decisions regarding whether to undertake or stop an activity are public decisions because of their potential to impact ecosystems, public health, and the commons. Consequently, these decisions require holistic analysis, qualitative measures, values, and the weighing of various types of evidence. This differs considerably from decision making using current methods, such as risk assessment, which is based on closed, assumption-laden, quantitative models and the supremacy of objective science that virtually excludes the public. It is almost impossible for citizens to participate in "expert" driven risk science because of the complexities of mathematical models and heavy reliance on assumptions, all of which obscure the inherent values in those assumptions. Nevertheless, research has shown that assistants of experts are often more accurate in predicting uncertain future events than the experts themselves, as a result of differences in the way they assimilate and analyze information (IJC, 1995).

Structures must exist for citizens (workers, communities, teachers, etc.) to participate both in the collection of information on which to base decisions and in the decisions (scientific, technological, and political) themselves. A simple right to know is insufficient to ensure democratic decision making, as is the right to participate without adequate information. Some of the most interesting work on democratic decision-making structures is currently being undertaken in Denmark, Norway, Sweden, and The Netherlands. In the United States, the Loka Institute is trying to develop, refine, and disseminate these models. Three particular models for democratic participation in decision making have been discussed, two of which are described in further detail in chapter 15.

- *Consensus conferences.* Consensus or "layperson" conferences are processes to involve citizens in decision making regarding complex issues. These are not multistakeholder type negotiations where often those with greater power have a greater say in the outcome (or have greater access to information and resources). In these conferences, a randomly selected group of citizens receives background on a specific issue (e.g., genetic engineering), participates in a hearing with experts, and then deliberates the problem, issuing a detailed report and recommendations. These panels have taken place on numerous occasions in Denmark. In Norway, a consensus conference on genetic engineering led that country's government to ban genetically modified crops, because of the potential risks involved and the lack of a need for additional food (Sclove, 1997). While these types of panels would appear to work best for big issues with large societal repercussions, a hybrid of the consensus conference, such as a citizen grand jury invoked for six-month periods, might be useful.

- *Scenario workshops.* Similar to backcasting, scenario workshops provide a participatory method for addressing broad societal *how* questions such as how to reach sustainability. In these workshops, different stakeholders come together to analyze different scenarios to answer a particular question or problem. In doing this, they address barriers to solving a problem and an action plan for solving it.

- *Science shops.* A key element to citizen participation in decision making is the ability of citizens to independently and critically examine evidence and conduct their own studies. This type of "citizen science" leads to new types of qualitative and quantitative input into decision-making processes. Science shops are independent university (or nonprofit institution)-based centers that provide sophisticated referral services for communities and community groups needing to undertake environmental research. The science shop responds directly to the citizen questions and engages in participatory research. The citizens or citizen's group must be both willing and capable of undertaking social action as a result of the research (Sclove, 1997).

These democratic decision-making methods lead to a greater transparency in the science and assumptions used to reach decisions and the decisions themselves. While they may be difficult to implement on a wide scale and for each individual decision, they provide a model of how to involve citizens in complex decisions under uncertainty. On a smaller scale, citizens (with appropriate financial and technical assistance) can be involved on government-sponsored panels to make decisions regarding permitting, siting, impact assessments, and whether to take action on a potentially harmful activity.

Strong Enforcement
Strong enforcement is necessary to ensure that precautionary, preventive actions are actually carried out (and that ongoing monitoring is occurring). Strong regulations alone will not achieve precautionary results. Enforce-

ment, combined with strong regulation, would encourage the development of safer, cleaner technologies and practices. Even the threat of enforcement can create the necessary uncertain conditions for those undertaking an activity to act with precaution and to think about possible consequences. Enforcement should be coupled with financial and criminal repercussions for those failing to act in a precautionary manner. Strong enforcement should also be coupled with outreach to firms, farmers, developers, and others undertaking potentially harmful activities (as well as the public itself) to assist them in understanding their impacts and in developing and implementing safer and cleaner technologies and activities that fulfill their needs and serve the public good.

EXAMPLES OF THE PRECAUTIONARY PRINCIPLE IN USE

This section describes several examples of where components of the Precautionary Principle have been incorporated into environmental or public health decision making. While some of these cases represent proposed legislation or policy decisions that have yet to be fully implemented, they provide examples of how the principle can be used. Although these are examples regarding toxic substances, there are cases, particularly in countries like Australia, Scotland, and Norway, where the principle has been invoked to prevent harm to ecosystems from development, farming, fishing, deforestation, and genetic engineering.

Toxics Use Reduction and Precaution in Massachusetts

The Massachusetts Toxics Use Reduction Act (TURA) is an example of how the Precautionary Principle can be applied to industrial processes and toxic chemicals. Passed in 1989, the act requires that manufacturing firms using specific quantities of some 900 industrial chemicals undergo a bi-yearly process to identify alternatives to reduce the use of those chemicals. The process involves understanding what you are trying to do, how you are doing it, measuring impacts and progress, and systematically searching for and analyzing alternatives on a regular basis.

There are several aspects of the Toxics Use Reduction Act that make it a good example of precautionary action: (1) The Commonwealth established a goal of a 50 percent reduction of toxic by-products (waste) through toxics use-reduction techniques; (2) the act does not instruct industrial facilities to identify the "safe" level of use, emissions, or exposure to chemicals. Rather, the act instructs firms to identify ways to redesign production processes and products to reduce waste and, subsequently, use of those chemicals—any amount of use is considered too much; (3) the act instructs companies to go through an alternatives assessment process whereby they understand why they use a specific chemical (what "service" it provides) and how it is used in the production process. They also conduct a comprehensive financial, technical, environmental, and occupational health and safety analysis of viable alternatives. The firm is not required to undertake any particular option, but in many cases the economic and environmental/health and safety benefits provide enough justification for action; and (4) companies are required to measure their progress yearly at reducing their use of toxic chemicals. This information is publicly available.

In 1997, the Commonwealth conducted an analysis of the act. From 1990–1995, companies in Massachusetts reduced their toxic chemical emissions by more than two-thirds, their total chemical waste by 30 percent, and their total use by 20 percent. On the cost side, the act saved Massachusetts industry some 15 million dollars. This figure does not include the public health and environmental benefits gained through the program (TURI, 1997).

To supplement the successes in toxics use reduction, in early 1997 State Representative Pamela Resor introduced a bill to the Massachusetts legislature entitled "An Act to Establish the Precautionary Principle as the Guideline for Developing Environmental Policy and Quality Standards for the Commonwealth." This bill was introduced in response to industry-supported bills that would require risk assessments and cost-benefit analyses for any major environmental regulation and that would dismantle the successful Toxics Use Reduction Act, in favor of a voluntary pollution prevention scheme. The proposed legislation calls for state agencies to apply the Precautionary Principle where there "are reasonable grounds for concern that a procedure or development may contribute to the degradation of the air, land and water of the commonwealth." The act also calls on agencies to implement the Precautionary Principle through pollution prevention, evaluation of alternatives, and research into the effects of potentially environmentally harmful activities (The Commonwealth of Massachusetts, 1997).

London Dumping Convention

The London Dumping Convention (LDC) came into effect in 1972 to protect the marine environment from the dumping of wastes and other matter. Similar to other international agreements, the Precautionary Principle was introduced into later revisions of the LDC. In 1996, parties to the convention, noting the need for more stringent measures to protect the marine environment, thus passed a Protocol to the Convention. The protocol first establishes a list of materials that can be dumped at sea (a reverse list)—all other materials are prohibited from dumping. Second, those applying to dump-allowed wastes at sea must first undergo a waste assessment audit and demonstrate that there is no other way to reduce, reuse, or recycle those wastes. If a country determines that there are other alternatives for the waste, a permit can be denied. Permits are regularly reviewed to ensure that there are no alternatives to the dumping (IMO, 1996).

International Joint Commission

In its *Sixth Biennial Report on Great Lakes Water Quality*, the International Joint Commission (IJC) addressed progress in achieving the goals of the Great Lakes Water Quality Agreement (including virtual elimination of the input of persistent toxic substances into the Great Lakes). The IJC noted the damage caused by persistent and bioaccumulative substances in the Great Lakes Basin and the critical need to address those. They also noted that attempts to manage such chemicals on the basis of the notion of assimilative capacity in the environment (that the environment and humans can render a certain amount of contamination harmless) had failed miserably. As a result, the commission proposed a strategy calling for the sunseting of all persistent toxic substances in the Great Lakes ecosystem, because they cannot be managed safely. "Such a strategy should recognize that all persistent toxic substances are dangerous to the environment, deleterious to the human condition, and can no longer be tolerated in the ecosystem, whether or not unassailable scientific proof of acute or chronic damage is universally accepted." In devising its policy, the IJC has specifically rejected risk assessment and cost-benefit analysis, noting the difficulties and complexities involved in such assessments and the fact that they are not relevant to the virtual elimination commitment.

To identify candidate chemicals for action, the IJC has developed a policy framework for addressing areas of scientific uncertainty, based on the weight-of-evidence approach, so that unequivocal evidence of cause-effect is not needed before taking action. The commission notes the following:

> If taken together, the amount and consistency of evidence across a wide range of circumstances and/or toxic substances are judged sufficient to indicate the reality or strong probability of a linkage between certain substances or class of substances, a conclusion of causal relationship can be made. This conclusion is made on the basis of common sense, logic and experience as well as formal science. Once this point is reached, and taking a precautionary approach, there can be no defensible alternative to recommending that the input of those substances to the Great Lakes be stopped. The burden of proof must shift to the proponent of the substance to show that is does not or will not cause the sus-

pected harm, nor meet the definition of a persistent toxic substance.

Using this rationale, given the wide range of adverse known and possible human health and environmental impacts associated with organochlorine compounds, the IJC has called for a phaseout of the manufacture and use of chlorine and chlorine-containing compounds as industrial feedstocks in the basin (IJC, 1992, 1994a, 1994b, 1995).

CONCLUSIONS

In this chapter I have attempted to develop a framework to guide decision making using the Precautionary Principle. The components of such a structure attempt to shift both the questions asked by decision makers and the underlying presumptions regarding public health and environmental protection. While it would be unrealistic to believe that all of these components could be implemented into law and policy in the short term, there are parts of this framework that should receive immediate attention. For example, the first step toward bringing the Precautionary Principle to the forefront is its incorporation as a general duty into local, state, and national laws. This would increase both awareness of its role in decision making under uncertainty and the public accountability of agencies and those undertaking potentially harmful activities. Simultaneously, efforts could be undertaken to refine existing laws to incorporate a greater emphasis on prevention, alternatives assessment, burden shifting, performance bonds, and ongoing monitoring of impacts. Finally, all parties will need to work to increase public participation in decision making and to revamp the scientific method so that it supports precautionary decision making.

In the short term, concrete examples of where the Precautionary Principle has been implemented help to provide evidence of the benefits of a precautionary decision-making structure and greater acceptance of the principle as a guide to environmental and public health policy. Operationalization of various components of the principle will be slow and contentious at times, as precaution requires a new way of addressing and solving environmental problems. But our growing public health crisis necessitates a fundamentally new vision, new tools, and swift action if we are to reach sustainability.

Appendix to Chapter 9

LIST A: CRITERIA FOR SCALING AND FRAMING POLICY PROBLEMS IN SUSTAINABILITY (DOVERS, 1995)

Problem-Framing Attributes:

1. Spatial scale of cause of effect: local-national-regional-international-global
2. Magnitude of possible impacts (on both humans and natural systems)
3. Temporal scale of potential impacts
    timing (near, medium, long term)
    –longevity (short, medium, long)
4. Reversibility (easily/quickly reversed or expensive/irreversible)
5. Measurability of factors and processes (well-known, ignorance)
6. Degree of complexity and connectivity

Response-Framing Attributes:

7. Nature of cause(s) (simple, systemic)
8. Tractibility (availability of means, acceptability of means)
9. Public concern

LIST B: ATTRIBUTES RELATED TO STRENGTH OF ASSOCIATION BETWEEN ASSESSMENT AND MEASUREMENT ENDPOINTS

Massachusetts Weight-of-Evidence Workgroup (MWEW, 1995)

1. Biological linkage between measurement endpoint and assessment endpoint
2. Correlation of stressor to response

3. Utility of measure for judging environmental harm
4. Extent to which data quality objectives are met
5. Site specificity
6. Sensitivity of the measurement endpoint for detecting changes
7. Spatial representativeness
8. Temporal representativeness
9. Quantitativeness
10. Use of a standard method

LIST C: PRINCIPLES OF EFFECTIVE ENVIRONMENTAL MANAGEMENT (LUDWIG ET AL., 1993)

1. Consider a variety of plausible hypotheses
2. Consider a variety of possible strategies
3. Favor actions that are robust to uncertainties
4. Hedge
5. Favor actions that are informative
6. Probe and experiment
7. Monitor results
8. Update assessments and modify policy accordingly
9. Favor actions that are reversible
10. Solicit all available information from all stakeholders (taken from Raffensperger, 1997)

LIST D: HILL CRITERIA FOR DISTINGUISHING BETWEEN ASSOCIATION AND CAUSATION IN EPIDEMIOLOGIC STUDIES (HILL, 1965)

1. Strength of evidence
2. Consistency of evidence
3. Specificity of effect
4. Temporality of effect
5. Dose response of effect
6. Plausibility of effect
7. Coherence with existing knowledge
8. Experimental evidence
9. Analogy (structure activity)

LIST E: CRITERIA FOR MANAGING UNCERTAINTY AND REGULATION IN PUBLIC POLICY (GEE, 1997)

1. Who/what gets the benefit of scientific doubt?
2. Who takes the burden of proof?

3. What level of proof is appropriate?
4. Which are the toxicologically dominant and strategically dominant causes (i.e., is it of public health significance but not of statistical significance)?
5. What are the multiple benefits of risk reduction?
6. What is the optimal balance between time spent establishing causation and time spent reducing risk?
7. What is the likely size and distributions of false negatives and false positives?
8. What is the optimum mix of policy instruments, targets, and timetables that will maximize overall cost-effective public policy?
9. What mechanisms are needed for establishing need in public policy on new products and services so as so optimize the balance between innovation and risk?

LIST F: PROPOSED CRITERIA TO GUIDE PRECAUTIONARY DECISION MAKING

Association/Uncertainty Criteria

1. Strength of evidence (experimental and observational)
2. Amount and consistency of evidence across a wide range of circumstances
3. Temporality of effect
4. Coherence with existing knowledge
5. Plausibility of effect
6. Have all evidence and all plausible hypotheses been considered
7. Power of study(ies) to detect an effect
8. Have false negatives (Type II errors) been minimized?
9. Is the evidence statistically significant or of public health significance?
10. Is there some presumption of causal relatedness based on previous experience which would lower the evidentiary standard (i.e., is there evidence from any other similar case that would lead one to believe that a similar impact could be considered in the present case)?
11. What is the adverse effect being studied and is it the correct one?

Decision-Stakes Criteria

1. Spatial scale of cause of effect: local-national-regional-international-global
2. Magnitude of possible impacts (on both humans and natural systems)
3. Temporal scale of potential impacts
   –timing (near, medium, long term)
   –longevity (short, medium, long)

4. Reversibility (easily/quickly reversed or expensive/irreversible)
5. Measurability of factors and processes (well-known, ignorance)
6. Degree of complexity and connectivity
7. Is the action robust to uncertainties (error friendly)?
8. Do alternatives or measures exist to reduce or eliminate potential harm (ease of prevention)?
9. What is the trade-off between further study and potential impacts?

NOTES

1. One example of how to include the Precautionary Principle in environmental legislation is described in the Preface by Ken Geiser.
2. The basic presumption of the OSH Act, found in its preamble and legislative history, is the protection of workers from industrial hazards when there is suspicion of harm. However, in practice, and as a result of some critical court decisions, this initial focus of the act has been lost to some degree.
3. Obviously, it is impossible to prove complete safety. As such, the burden-of-proof requirement could be constrained by the lack of a definition of what is safe or acceptable. So it may be most important that the burden be that no safer way to conduct an activity is possible (or to prove that an activity is necessary).
4. I would argue that that specific level of proof required for action is not as important as the weight of the evidence of harm. The determination of weight of evidence (or level of proof needed) for action will depend on criteria of causal inference in addition to criteria of "decision-stakes" or magnitude of potential impacts. Arguably, the amount of proof needed before taking precautionary action is much lower than that required under current decision-making schemes.
5. The use of the weight of evidence approach by the U.S.–Canada International Joint Commission is described in the next section and in chapter 16.
6. The weight of evidence determination could be made by a government agency acting alone, independent review panel, or some other participatory process.
7. If the magnitude of potential harm were great (e.g., global warming) and widespread, this would lead to more preventive or remedial action.
8. Several analysts (Bodansky, 1994) have noted that the Precautionary Principle is integrated into some environmental laws that require best available technology, such as the Clean Water Act and the Clean Air Act. However, implementation of these laws has generally been based on single media end-of-pipe control technology, failing to integrate a pollution prevention approach. While establishing technology or science-based emissions limits in the face of uncertainty about the range of effects of an emission could be considered precautionary, if pollutants are shifted between media or new risks are created, preventive action has not occurred.
9. There are other economic mechanisms that could stimulate precautionary action. For example, product take-back legislation, where producers are

responsible for the end-of-life disposal costs of a product, can encourage the design of safer, longer lasting products with decreased life-cycle impacts.

10. It is possible that, because of poor understanding of a problem, scientists fail to identify specific impacts or fail to recognize a more wide-scale problem. For example, previously, scientists rarely considered the impact of toxic chemicals on hormonal systems, focusing mainly on cancer as an adverse outcome.

REFERENCES

Ashford, N., et al. *The Encouragement of Technological Change for Preventing Chemical Accidents: Moving Firms from Secondary Prevention and Mitigation to Primary Prevention* (Cambridge, MA: Center for Technology, Policy and Industrial Development, Massachusetts Institute of Technology, 1993).

Bodansky, D. "The Precautionary Principle in U.S. Environmental Law." In: T. O'Riordan and J. Cameron (eds.) *Interpreting the Precautionary Principle* (London: Earthscan, 1994).

Bro-Rasmussen, F., et al. *The Non-Assessed Chemicals in the EU: Report and Recommendations from an Interdisciplinary Group of Danish Experts* (Copenhagen: Danish Board of Technology, 1996/5).

The Commonwealth of Massachusetts. *An Act to Establish the Principle of Precautionary Action as the Guideline for Developing Environmental Policy and Quality Standards for the Commonwealth*, H. 3140 (1997).

Costanza, R. and Cornwell, L. "The 4P Approach to Dealing with Scientific Uncertainty," *Environment*, 34:9 (1992).

Cross, F. "Paradoxical Perils of the Precautionary Principle," *Washington & Lee Law Review*, 51: 851–925 (1996).

Department of Environment of the United Kingdom. *Ministerial Declaration* (London: Second International Conference on Protection of the North Sea, 1992).

Dovers, S. "A Framework for Scaling and Framing Policy Problems in Sustainability," *Ecological Economics*, 12: 93–106 (1995).

Environmental Defense Fund (EDF). *Toxic Ignorance: The Continuing Absence of Basic Health Testing for Top-Selling Chemicals in the United States* (New York: Environmental Defense Fund, 1997).

Funtowicz, S. and Ravetz, J. "A New Scientific Methodology for Global Environmental Issues." In: R. Costanza (ed.) *Ecological Economics: Science and Management of Sustainability* (New York: Columbia University Press, 1991).

Gee, D. Criteria for Managing Uncertainty and Regulation in Public Policy. Draft manuscript (1997).

Hill, A.B. "The Environment and Disease: Association or Causation?" *Proceedings of the Royal Society of Medicine*, 58: 295–300 (1965).

International Joint Commission (IJC). *Sixth Biennial Report on Great Lakes Water Quality* (Windsor, Ontario: IJC, 1992).

———. *Applying the Weight of Evidence: Issues and Practice* (Windsor, Ontario: IJC, 1994a).

———. *Seventh Biennial Report on Great Lakes Water Quality* (Windsor, Ontario: IJC, 1994b)

————. *1993–95 Priorities and Progress Under the Great Lakes Water Quality Agreement* (Windsor, Ontario: IJC, 1995).

International Maritime Organization (IMO). *1996 Protocol to the Convention on the Prevention of Marine Pollution by Dumping of Wastes and Other Matter, 1972 and Resolutions Adopted by the Special Meeting*, LC/SM 1/6, 14 November 1996.

Ludwig, D., et al. "Uncertainty, Resource Exploitation, and Conservation: Lessons from History," *Science*, 260: 17, 36 (April 2, 1993).

Massachusetts Weight-of-Evidence Workgroup. *Draft Report: A Weight-Of-Evidence Approach for Evaluating Ecological Risks* (Boston: Massachusetts Department of Environmental Protection, 1995).

Ministry of the Environment (Sweden). *Towards a Sustainable Chemicals Policy* (Stockholm: Government Official Reports 1997:84).

Raffensperger, C. "Regulatory Agency Accountability Under the Precautionary Principle," draft paper prepared for the Wingspread Conference on Implementing the Precautionary Principle, January 23–25, 1998.

————. "Incentives and Barriers to Public Interest Research," Switzer Fellows Speech, the Headlands, CA, September 1997.

Sclove, R. "Democratizing Science and Technology," lecture given as part of the University of Massachusetts Lowell Center for Competitive Enterprise Lecture Series, December 5, 1997.

Stijkel, A. and Reijnders, L. "Implementation of the Precautionary Principle in Standards for the Workplace," *Occupational and Environmental Medicine*, 52: 304–312 (1995).

TURI (Massachusetts Toxics Use Reduction Institute). *Massachusetts Is Cleaner and Safer: Report on the Toxics Use Reduction Program* (Lowell: Massachusetts Toxics Use Reduction Institute, 1997).

van Dunne, J.M. *Environmental Contracts and Covenants: New Instruments for a Realistic Environmental Policy?* (Vermanda Lelystad, 1993).

von Weizacker, C. "Lacking Scientific Knowledge or Lacking the Wisdom and Culture of Not-Knowing." In: A. Van Dommelen (ed.) *Coping with Deliberate Release: the Limits of Risk Assessment* (Tilburg: International Centre for Human and Public Affairs, 1996).

Ziem, G., et al. "Health-Based Exposure Limits and Lowest National Occupational Exposure Limits, Health-Based Exposure Limits Sub-Committee, American Public Health Association." Draft 9/1/1990. Appendix: An Overview of Decision-Making Criteria.

Chapter 10

～♘～

APPLYING THE PRECAUTIONARY PRINCIPLE IN PRACTICE: NATURAL HERITAGE CONSERVATION IN SCOTLAND

Scottish Natural Heritage

Scottish Natural Heritage (SNH) is a government agency whose general aims and purposes in law are to secure the conservation and enhancement of, and foster understanding and facilitate the enjoyment of, the natural heritage of Scotland and to have regard to the desirability of securing that anything done, whether by SNH or any other person, in relation to the natural heritage of Scotland is undertaken in a manner that is *sustainable*. The "natural heritage" is defined as "the flora and fauna of Scotland, its geological and physiographical features, its natural beauty and amenity."

SNH was the first public body in the United Kingdom (U.K.) to be given a statutory remit that mentioned sustainability. While responsibility for sustainability was initially commonly seen to reside primarily with environmental interests, latterly the role played by a whole range of social and economic actors in sustainable development is becoming more widely appreciated.

SNH has endeavored to interpret broad sustainability principles in terms of its statutory responsibilities. The following policy guidance note on the application of the Precautionary Principle was one of the fruits of this labor. It considers more general concepts of uncertainty and significance in terms of impacts on the natural heritage and creates a framework for deciding whether the Precautionary Principle should apply and what to do if it does apply. It reflects SNH policy on what the principle means and how it should be used. The guidance begins with a step-by-step procedure, to be followed when agency staff believe the principle might be relevant.

Although produced by SNH in 1996, the procedure described here could benefit from being more widely tested than has been possible internally. It draws attention to the importance of reversibility/adaptability in development, which may have more general applicability than the relatively limited circumstances under which SNH has tried to apply it. Staff would be interested to receive any feedback or suggested modifications or simply to hear the experiences of those who try to use it. SNH is planning to encourage further discussion on the principle in coming years.

EXPLANATORY NOTES ON APPLYING THE PRECAUTIONARY PRINCIPLE

Applying a step-by-step approach requires some explanation and interpretation. In this section, the rationale for the procedure outlined in the schematic overview is described.

Has a Range of Development Options Been Considered?

This question is designed to draw attention to how the Precautionary Principle should ideally be applied at the strategic level of decision making. Strategic policies (in structure plans, corporate plans, policy statements) and project proposals should promote consideration of a range of possible development and locational options so that those with the least uncertainty and risk are favored. They should also encourage *adaptability* in development design where developments may put the natural heritage at risk. The distribution of the main protected areas gives an indication of the location of important elements of the natural heritage, and hence an initial "sieve" in this options appraisal. In this way, the need for further detailed application of the principle can be minimized by avoiding protected areas. This requirement can be taken forward in the strategic environmental assessment (SEA) process that should promote the inclusion of policies that seek to develop a range of adaptable options to meet the development needs for an area.

In practice it may be necessary to move straight to step 2 as the SEA process is not yet established. The general presumption in favor of development means that it is largely developer led: regulatory systems (such as the

Below is provided a schematic overview of the procedure that the Scottish Natural Heritage has proposed for decision makers to apply the Precautionary Principle to decisions that might adversely affect the natural heritage. This decision-tree is then described in detail in the following section.

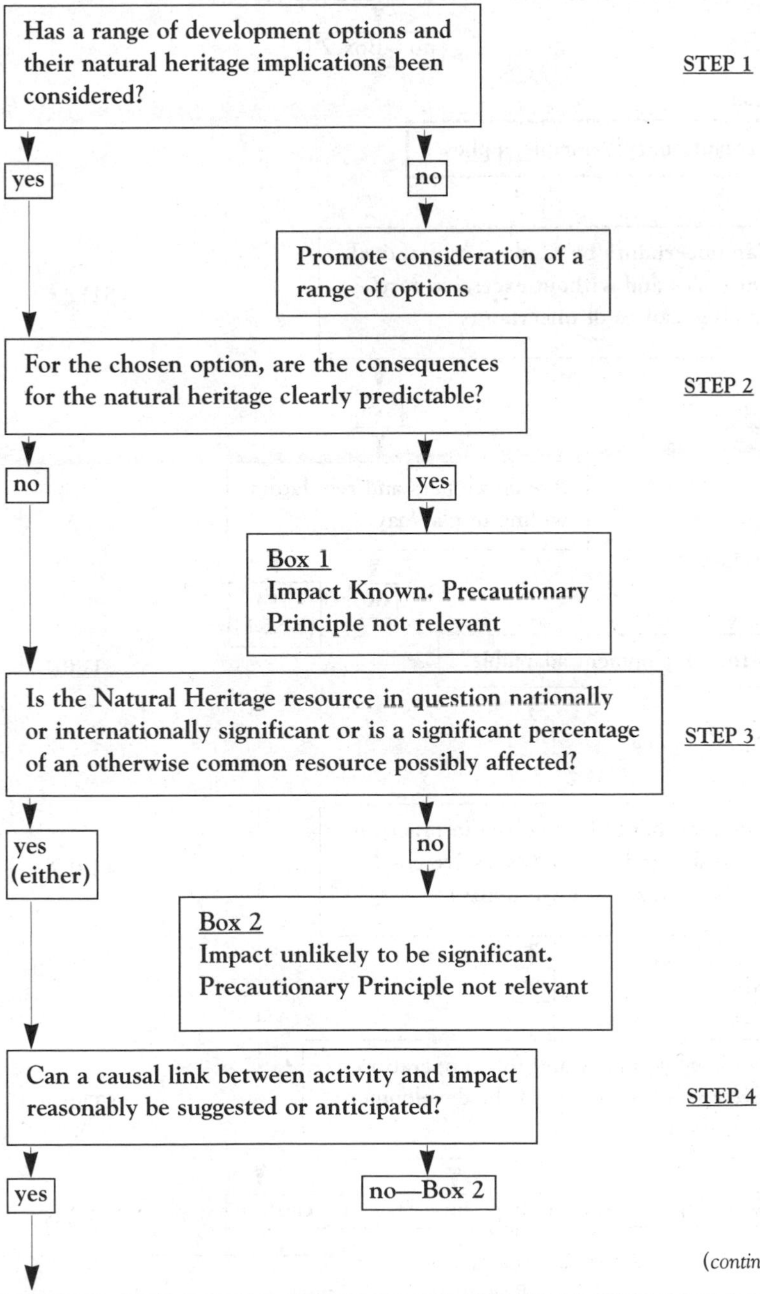

Has a range of development options and their natural heritage implications been considered? — STEP 1

↓ yes   ↓ no

Promote consideration of a range of options

For the chosen option, are the consequences for the natural heritage clearly predictable? — STEP 2

↓ no   ↓ yes

Box 1
Impact Known. Precautionary Principle not relevant

Is the Natural Heritage resource in question nationally or internationally significant or is a significant percentage of an otherwise common resource possibly affected? — STEP 3

↓ yes (either)   ↓ no

Box 2
Impact unlikely to be significant. Precautionary Principle not relevant

Can a causal link between activity and impact reasonably be suggested or anticipated? — STEP 4

↓ yes   ↓ no—Box 2

↓

(continues)

FIGURE 10.1. A Procedure for Applying the Precautionary Principle

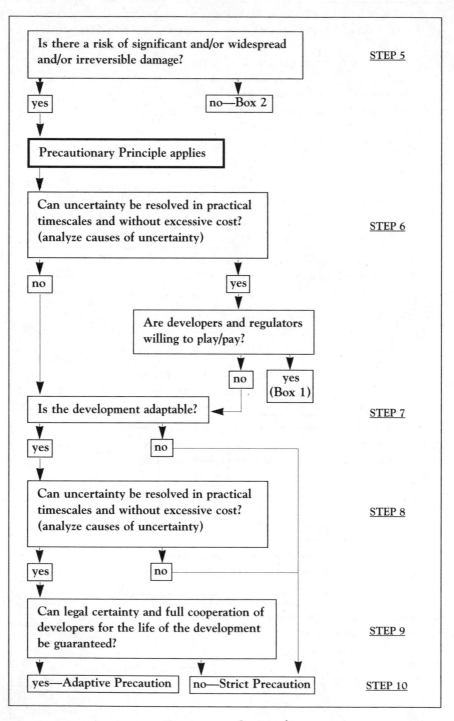

FIGURE 10.1. *Continued*

planning system) only have a certain leverage in ensuring option appraisal. Nevertheless, there is still scope to encourage policies promoting the consideration of development options and adaptability. Also, it is important to promote the strategic approach to businesses themselves. Applying the principle in the absence of such policies (i.e., "late" in the development process) and where developers are "cold" to the ideas may mean additional development costs and a greater chance of conflict.

Step 2: Are the Consequences for the Natural Heritage Clearly Predictable?

Many activities have a combination of both certain and uncertain (but possible) impacts. The aim here is to list the areas of certainty and uncertainty. The wider the range of the kind of people involved in this process the better to facilitate the exchange of information and opinions. If there are no uncertainties (i.e., either that there is certain to be an impact or that there is certain to be no impact) the procedure ends here, as the Precautionary Principle is not relevant. A preventive approach, however, might be appropriate. If there are *any* uncertainties move onto the next step, where their significance will be assessed.

Step 3: Is the Natural Heritage Resource of Significant Importance?

It is unrealistic to expect to apply the Precautionary Principle in all cases where there is a risk of environmental damage. Indeed, to do so would discredit the principle and SNH. Official SNH policy is that it should only be applied if the balance of likely costs and benefits justifies it (the so-called "proportionality" principle). This balance is not measured but is rather a matter of judgment. In SNH's view, an important aspect of this cost-benefit assessment is judging whether the natural heritage resource in question falls within any of the following special protection categories:[1]

- International/national designation system: For example, Natura 2000 sites, Ramsar Sites, National Scenic Areas, National Nature Reserves, Sites of Special Scientific Interest
- International/national conservation status: For example, Berne/Bonn Convention species, specially protected species on schedules or annexes to the Wildlife and Countryside Act and the Habitats and Species Directive
- Significant proportion of an otherwise common resource: For example, over 20 percent of a common species' population

If an activity does fall within one of these categories, then an initial cost-benefit test is met, in that there may well be a significant cost of not taking precautionary action.

Step 4: Can a Causal Link Be Reasonably Suggested?

The general presumption in favor of development means that in most cases in which SNH is involved, it will be necessary for SNH to be able to justify a linkage between cause and effect in order to warrant precautionary action. This presumption is switched only in rare cases (see below). According to the Department of the Environment (DoE), the principle "is not a license to invent hypothetical consequences." Environmental assessments should be helpful here, provided they are given due attention to uncertainties.

The definition of "reasonableness" (can a causal link *reasonably* be suggested?) is a key point: There is no easy answer to this other than it is what a reasonable person might think.

It is not necessary to demonstrate quantitatively a causal link (i.e., through statistically valid experimentation), although if experimentation is at all feasible it should be considered (see below). The principal way of answering this question is to review information from any other similar situation elsewhere and consider how relevant it is to the case in question. If a causal link can be justified, move on to step 5.

There is one area of relevance to SNH where the general presumption in favor of development, and thus the burden on SNH to prove that an impact will happen, has been altered; that is, developments likely to have a significant effect on special areas of conservation and special protection areas. Article 6(3) of the European Union (EU) Directive on the Conservation of Natural Habitats and of Wild Flora and Fauna states that in the case of a project likely to have a significant effect on such a site "competent national authorities shall agree to the plan or project only after having ascertained that it *will not* adversely affect the integrity of the site concerned . . ." (emphasis added).

Step 5: Is There a Risk of Significant and/or Widespread and/or Irreversible Damage?

Assessment of the "significance" of a development on the natural heritage is a judgment SNH has always had to make; it is assessed for a site in relation to the implications of a development on the objectives of the designation and for a species in relation to impacts on its conservation status.

The risk of "widespread damage" applies in the case of natural heritage resources that may not be rare or nationally significant but that may be subject to an adverse impact across the whole of their range or distribution, rather than in relation to a specific site.

The risk of "irreversible damage," including damage that is reversible but only at high cost and over a long time period, is an additional consideration in either situation that reinforces the need for precaution.

If any of these considerations apply and especially if there is a risk of irreversible damage, the Precautionary Principle applies and can be cited, and move on to step 6. SNH interprets the principle as full scientific proof of a possible environmental impact is not required before action is taken to prevent that impact.

It does not automatically mean that SNH will seek to prevent a development proposal (see below). The emphasis of the procedure now is to see whether uncertainties can be resolved beforehand and whether sufficient safeguards can be applied to a development to allow it to proceed in a cautionary manner.

Step 6: Can Uncertainty Be Resolved in Practical Timescales Without Excessive Cost?

Here it is necessary to pin down the causes of environmental uncertainty. There are three main causes: complexity, lack of data, and lack of understanding.

Complexity

Natural systems can be highly complex owing to their interconnected and dynamic nature. This natural complexity is often increased by the existing effects of human activities and where the system in question is extensive. As a general guide, the sea, the atmosphere, and many freshwater environments are likely to rank as highly complex because of their high connectivity, extent, and the complicating effects of existing management activities. The need here is to itemize the main reasons for complexity and decide whether they are a cause of the relevant uncertainty. If so, research is likely to be needed to try and identify the links between cause and effect.

Data

Lack of survey data is a common reason for uncertainty. This simply may be due to the cost of obtaining data or may be due to technical and geographical problems with collection. In practice, the two have the same effect. Specific impacts on known sites of extraction are predictable if survey data for the sites have been collected.

Understanding

Research information or survey data may be available but may be difficult to interpret or there may be different interpretations of what they mean—both leading to uncertainty. In this situation, it is important to ensure that all interested parties have access to and can reach a common understanding of available information and remaining uncertainties.

Classifying the causes of uncertainty in this way will help in the assessment of whether they can be resolved through, for example, further research, experimentation, or survey or through further debate on the interpretation of existing information. If so, can it be done in practical timescales and without excessive cost? If there is agreement that uncertainties can be resolved and they are, the procedure ends here, as the Precautionary Principle is not relevant.

Step 7: Is the Development Adaptable?

If uncertainty cannot practically be resolved before a development proceeds (i.e., there remains a risk of adverse impact but we cannot be sure), the question of whether the development can proceed but in a modified or strictly regulated way inevitably will arise.

The key question here is the following: If the development proceeds, can it physically be adapted in the light of new information on its impacts, effectively to remove any harmful impacts?

For a development to proceed in circumstances that would by now apply at this stage in the procedure, the initial focus must be on the development itself; that is the underlying cause of the risk, *rather* than any "secondary" risks arising from use or operation of the development.

Given the importance of the resource in question, it is necessary to be sure that the risk, if it translates into a real impact, can be removed and that it is not going to re-occur at some future date. This requires careful consideration of whether flexibility can be designed into the deveopment itself. This is a central message of the precautionary approach and reinforces the requirement in step 1 for strategic policies which promote both the consideration of a range of development options per se as well as a range of design options for particular developments that have associated risks of natural heritage impact. If the risk of impact arises not from the development itself but from its operation or use, the viability of mitigating measures should be considered in relation to steps 8 and 9.

Step 8: Is It Technically Feasible to Establish a "Feedback" Monitoring Regime?

By now the possibility of an adaptable development has been established. In practice, the adaptive approach will only work if it is technically possible to monitor the impacts of the development in both its establishment and operating phases. There may be problems with establishing a monitoring regime because of remoteness, durability of equipment, and/or the nature of the environment (e.g., underwater). Standards of proof to trigger action will need to be agreed by all parties, and these will need to be high if they are to result in changes in the development.

Step 9: Can Full Cooperation and Legal Certainty for the Life of the Development Be Guaranteed?

By this stage it will have been established that, in theory, the development is both adaptable and its impacts can accurately be monitored. The key remaining questions are the following:

- Are the developers willing to cooperate fully in redesigning the development, monitoring for potential impacts, and linking monitoring results with action? Cooperation is an obvious requirement. In practice, the absence of anything other than full cooperation would prejudice this course of action. For it to work, a developer would need to be willing to consider adaptations in the first place, would be willing to build in flexibility, would need to fully cooperate in the monitoring regime, and be prepared to implement required action. A key requirement is the need for the developer to demonstrate that resources are available to carry out such action and will continue to be available.
- Can an effective legal framework be established? It is insufficient solely to rely on goodwill. Agreements need to be legally enforceable and underpinned by planning conditions and agreements or other appropriate regulations. Conversely, goodwill is essential for such agreements to work in practice—the parties must enter an agreement willingly.

There are always ways in which agreements can be subverted or only partially implemented. The important point here is to recognize that absolute certainty is not possible. If, after a period of time, SNH's experience is that agreements do not work in practice, we would need to strengthen our policy presumption in favor of strict precaution in implementing this procedure. For this reason, agreements should be carefully monitored.

Step 10: *Adaptive or Strict Precaution?*

If the answer in step 6 is "no" and in steps 7, 8, and 9 it is "yes," the Precautionary Principle can be applied in an adaptive way. Each of these steps then becomes a focus for attention. For example, in step 7 SNH would engage in discussion on alternative "fall-back" development designs that would be taken forward if monitoring information reveals adverse impacts. In step 9, the attention would be on the detailed terms of agreements and controls.

If the answers to any of the questions in steps 7, 8, and 9 are "no," SNH will seek to prevent the development through a strict application of the Precautionary Principle—on the grounds of unacceptable risk to an important natural heritage resource. This will involve objecting at the appropriate stage in the development control process. Clear justification, in the form of a full documentation of the step-by-step procedure, will be required for this position.

CONCLUSION

Scottish Natural Heritage's step-by-step guidance represents an attempt on the part of a government agency to codify the Precautionary Principle and define what this principle means in practice. It recognizes the need for good science in decision making but also that environmental complexity and consequent uncertainty limit the ability of obtaining scientific proof of impact within decision-making timescales.

The above-stated guidance reflects several important aspects of the Precautionary Principle:

- A precautionary approach can be operationalized through a series of iterative steps with feedback loops, with each step being adequately justified. Central to this procedure are an analysis of alternative development options as well as analyses of the causes of uncertainty and potential impacts.
- Postconstruction adaptability is at the core of this approach. What is needed in conditions of inevitable ecological uncertainty is to identify the most ecologically adaptive course of action, one that can most easily be changed in relation to subsequent, new information. Development should not be a once-and-for-all process but a flexible learning process, where assessment is continuous and a range of alternative development options is always identified.
- Reasonable pragmatic standards of evidence are required to activate the principle. In practice, this means gathering together whatever information is available on possible impacts from, for example, other similar situations. It may be necessary to treat developments that proceed in circum-

stances of sufficient uncertainty as "real-life experiments" and develop a database so that information can be used subsequently to better understand actual and potential impacts. Such an approach to precaution requires a different *modus operandi* for largely science-based agencies and will only be improved upon through ongoing trials and learning.

Note

1. The United Kingdom has conservation obligations deriving from national legislation and policy addressing the need to protect species and habitats that are nationally scarce or declining. The U.K. also has international obligations deriving from the European Union or international conventions to conserve species or habitats that might be scarce or declining on an international scale—even if national populations appear secure. Not all features of importance are legally protected; nonetheless, SNH is determining here that a straightforward and convenient test of "significant" importance would be whether a proposal affects a protected area or species. Hence, the presence of these nationally/internationally important sites or species is a criterion of "significance."

Chapter 11

❧

A CONCEPTUAL FRAMEWORK FOR THE USE OF THE PRECAUTIONARY PRINCIPLE IN LAW

Nicholas A. Ashford

The Precautionary Principle has been applied in various ways in decisions about health, safety, and the environment for about twenty-five years, much longer than recent commentaries would have us believe. In the United States, for example, in interpreting congressional legislation, the courts have argued that federal regulatory agencies are required to err on the side of caution in protecting workers and to protect public health from emissions to air with an ample or adequate margin of safety. States place strict liability on certain dangerous industrial and manufacturing operations. In this decade, these precautionary inclinations of the American and Anglo-Saxon jurisprudential systems have found their way into multilateral environmental agreements and international law.

What brings the Precautionary Principle into sharp political focus today is the increasing pressure to base governmental action on more "rational" schemes, such as cost-benefit analysis and quantitative risk assessment. The Precautionary Principle has been criticized as being both too vague and too arbitrary to form a basis for rational decision making. The assumption underlying this criticism is that any scheme not based on cost-benefit analysis and risk assessment is both irrational and without secure foundation in either science or economics. This chapter makes more explicit the rational tenets of the Precautionary Principle within an analytical framework as rigorous as uncertainties permit and one that mirrors democratic values embodied in regulatory, compensatory, and common law. Unlike other formulations that reject risk assessment, this chapter argues that risk assessment can be used within the formalism of trade-off analysis—a more appropriate alternative to traditional cost-benefit analysis and one that satisfies the need for well-grounded public policy decision making.

The recent crescendo of commentary on the legal application of the Precautionary Principle, following its increased incorporation into national and multilateral environmental agreements, has focused on situations in which there are significant uncertainties about the health and environmental effects of products, technologies, and other human activities. Where those certainties do not exist, it is often conceded that cost-benefit analysis is an appropriate approach to designing policies. This chapter will argue that the Precautionary Principle is a more fitting basis for policy even when large uncertainties do not exist. Furthermore, it will offer an approach to making decisions within an analytic framework, based on equity and justice, to replace the economic paradigm of cost-benefit analysis.

ELEMENTS OF THE PRECAUTIONARY PRINCIPLE

The application and discussion of the Precautionary Principle have focused on action to prevent, or refrain from contributing to, possible serious irreversible harm to health and the environment—whether on an individual basis or in terms of widespread environmental or health consequences. In particular, the Precautionary Principle has become embodied in regulations directed toward persistent and bioaccumulative toxic substances.

However, the Precautionary Principle need not be restricted to cases of irreversibility or large uncertainty of effect. It might also be applied to mitigate a harm that is ultimately reversible—if reversing the damage could be more costly than preventing it. And what of the cases in which there are no uncertainties—for example, when we know that future generations will be harmed? Cost-benefit analysis is biased against investing heavily in the present to prevent such future harm, because of the use of discounting of cost and benefit streams over time. And there are many situations in which we are aware of our ignorance: For example, we know that only a very small percentage of all chemicals in commerce have been tested for toxic effects. In these cases, too, precaution is appropriate.

However, the Precautionary Principle is not amenable to replacing cost-benefit analysis as a "decision rule" for action. Attempts to establish a threshold of harm above which the Precautionary Principle is triggered, for example, have been less than satisfactory. Rather, the Precautionary Principle is most useful in guiding the selection of policies and aiding in the establishment of priorities in an attempt to deliver justice and fairness.

Precaution rightly focuses on uncertainty and irreversibility as two important factors, but others must be considered as well. A complete list of the important elements must include:

- Seriousness and irreversibility of the harm addressed;
- Societal distribution of possible costs and benefits of policies and technologies;
- Technological options for preventing, arresting, reversing, or mitigating possible harm and the opportunity costs of selecting a given policy option; and
- Society's inclinations regarding erring on the side of caution and erring on the side of laxity.

Uncertainties in all of these elements are relevant to the Precautionary Principle. Since most attention has been focused on the first element, this chapter will give special attention to the other three.

The Limits of Cost-Benefit Analysis in Addressing Distributional Concerns

During the past two decades, cost-benefit analysis has become the dominant method used by policy makers to evaluate government intervention in the areas of health, safety, and the environment. In theory, cost-benefit analysis of a policy option enumerates all possible consequences, both positive and negative; estimates the probability of each; estimates the benefit or loss to society should each occur, expressed in monetary terms; computes the expected social benefit or loss from each consequence by multiplying the amount of the associated benefit or loss by its probability of occurrence; and computes the net expected social benefit or loss associated with the government policy by summing over the various possible consequences (Ashford, 1981).

The analysis examines economic effects, health and safety effects, and environmental effects of a particular policy initiative for various groups: producers, workers, consumers, and others. Initially, the consequences are represented in their natural units: Economic effects are expressed in monetary units; health and safety effects are expressed in mortality and morbidity terms; and environmental effects are expressed in appropriate descriptive terms. Then, traditional cost-benefit analysis translates all consequences into their current monetary value and calculates a net benefit (or cost).

These representations pose two problems. The first problem is the difficulty, even arbitrariness, of placing a monetary value on human life, health, and safety and a healthy environment. The second problem is that by translating all of these consequences into equivalent monetary units, discounting each to present value (since a dollar invested now is expected to earn interest over time), and aggregating them into a single dollar value, the effects on the economy from investing now in future health, safety, and environmental benefits are weighted far more heavily than those benefits that occur in the future, including those to future generations.

As a decision-making tool, cost-benefit analysis offers several compelling advantages. It clarifies choices among alternatives by evaluating consequences systematically. It professes to foster an open and fair policy-making process by making explicit the estimates of costs and benefits and the assumptions upon which those estimates are based. And by expressing all gains and losses in monetary terms, cost-benefit analysis permits the total impact of a policy to be summarized in a single dollar figure.

This final step, however, may be stretching analytic techniques too far. An alternative approach, called trade-off analysis, begins in the same way as cost-benefit analysis but stops short of assigning monetary values to nonmonetary consequences (Ashford, 1981). Instead, all effects are described in their natural units. The time period in which each effect is experienced is fully revealed, but future effects are not discounted to present value. Uncertainties are fully described. Trade-offs between worker health or environmental improvements and costs to producers and consumers are made apparent, because the different cost and benefit elements are not aggregated.

Using trade-off analysis, politically accountable decision makers could make policy choices in a transparent manner. Who bears the costs and who reaps the benefits from a policy option would not be hidden in a single, aggregate dollar figure. Decisions would be based on accountability rather than accounting.

Promoting Rational Technology Choices

One important element often left out of the traditional cost-benefit matrix has been the consideration of technological alternatives (Ashford, 1998). Regulatory agencies have a mixed history disseminating information about cleaner and safer technologies and promoting their adoption. Agencies could help prevent pollution and accidents by helping firms to think about their technological options in a more formal and systematic fashion.

Options for technological change must be considered according to a variety of criteria—economic, environmental, and health and safety factors.

Identifying these options and comparing them against the technology in use is called "Technology Options Analysis" (Ashford, 1994, 1998). Unlike traditional technology assessment, Technology Options Analysis does not require absolute quantification of all the variables: One has only to demonstrate, in a comparative manner, that a particular technology is better or worse than another in performance, health, safety, ecological effects, and so forth. It is likely to be less sensitive to initial assumptions than, for example, cost-benefit analysis and would enable industry and government to identify more creative cost-effective solutions. Government might require industries to undertake Technology Options Analysis, instead of traditional technology assessment focusing on technologies already existing within, or easily accessible to, the firm or industry. The latter would likely address only the technologies industry puts forward; it may thus miss the opportunity to identify and subsequently influence the adoption of superior technological options.

Once superior existing technologies—or technologies within easy reach—are identified, industries may be motivated to change their technology out of economic self-interest or in order to avoid future liability. On the other hand, government might either force the adoption or development of new technology or provide technical or financial assistance. Requiring firms to change technology can itself be a risky venture. Developing a new technology or adopting a technology new to a firm or industry introduces new uncertainties and financial risks. If this is done, policy should allow for error and accommodate industry for failures in bona fide attempts to develop new technologies (e.g., by allowing more time or sharing the financial risk).

Whichever route is taken by government, the Precautionary Principle requires the investigation of technology options for the development and adoption of cleaner and inherently safer (i.e., sustainable) technologies.

Which Errors Are Worse?

Policy makers must address both uncertainty about (1) the nature and extent of health, safety, or environmental risks and about (2) the performance of an alternative technology. First, they must choose whether to err on the side of caution or risk. With regard to the first type of uncertainty, two mistakes can be made. A Type I error is committed if society regulates an activity that turns out later to be harmless and resources are needlessly expended. Another error, a Type II error, is committed if society fails to regulate an activity that finally turns out to be harmful (Ashford, 1998).

Similarly, where uncertainty exists on the technology side, Type I errors can be said to be committed when society mandates the development or adoption of a technology that turns out to be much more expensive or not as effective at reducing risks as anticipated and resources are needlessly or foolishly expended. Type II errors might be said to be committed when, because of insufficient commitment of resources or political will, a significant missed opportunity is created by which society fails to force or stimulate significant risk-reducing technology.

Value judgments clearly attend decisions whether to lean toward tolerating Type I or Type II errors with regard to both risk and technology choices. This is because the cost of being wrong in one instance may be vastly different from the cost of being wrong in another. For example, banning a chemical essential to a beneficial activity such as the use of radionuclides in medicine has potentially more drastic consequences than banning a nonessential chemical for which there is a close, cost-comparable substitute. It may be perfectly appropriate to rely on "most likely" estimates of risk in the first case and on "worst-case" analysis in the second. A Type II error on the technology choice side was committed in the case of the Montreal Protocol banning CFCs by creating a scheme by which DuPont and ICI, the producers of CFCs, were allowed to promote the use of their own substitute, HCFCs, rather than adopt a more stringent protocol that would have stimulated still better substitutes.

Evaluating errors and deciding which way to favor is not a precise science. However, making those evaluations and valuations explicit within a trade-off analysis will reveal the preferences on which policies are based and may suggest priorities.

FURTHER GROUNDS FOR INVOKING
THE PRECAUTIONARY PRINCIPLE

Two other considerations must be examined when developing a conceptual framework for the Precautionary Principle. They are the following:

- *Democratic decision making.* The extent to which affected parties participate in identifying, evaluating, and selecting a protective policy may influence the acceptability of the policy. In the case of a possible, but highly uncertain harm, an equitable outcome may depend more on an equitable decision-making *process* than on a defensible argument about the technical correctness of an outcome based on existing information. The Precautionary Principle may be invoked to ensure a fair decision-making process, as much as to prevent harm.
- *Burdens of persuasion and proof.* Part of the perceived fairness of the process involves the burden of persuasion—that is, the designation of which party

has the burden of demonstrating or refuting a presumed fact. This is distinct from the burden of proof—a term referring to the strength of the evidence (data and information) needed to justify taking action. Both terms are relevant in the formulation of the Precautionary Principle.

Much discussion has focused on cause-and-effect relationships between exposure to other events and harmful effects for which a high statistical confidence level or strength of association is traditionally required. To escape the rigors of these requirements, some proponents of the Precautionary Principle argue that the burden of persuasion should be shifted to the proponents of a potentially harmful technology. Opponents argue against such a radical shift, pointing out that negatives are harder to prove.

Of course, uncertainties of cause-and-effect relationships are by no means the only determinations to which the Precautionary Principle should be, or is applied. Others are the following:

- The complex sets of rights and duties embodied in so-called right-to-know situations, including
    - the duty of potential wrong-doers to *generate* information
    - the duty to *retain* information
    - the duty to *provide access to* information to the potential victims of possible harm
    - the duty *to warn* the potential victims of possible harm;
- Providing funds to mitigate actual future harm to health or the environment;
- Compensating victims of unmitigated harm; and
- The duty to prevent harm.

The strength of the evidence required for these other, equally important, factually informed determinations may be much less than the traditional standard of proof in usual cause-and-effect determinations.[1]

Other standards (burdens) of proof commonly invoked in public policy determinations include, in decreasing order of stringency: "strict liability for harm" (in the area of compensation, the "polluter pays principle" is sometimes invoked in statutory language or by the courts in fashioning equitable relief to victims); "clear and convincing evidence"; "more probable than not" or "preponderance of the evidence"; "substantial cause or factor"; and "contributing factor." This "sliding scale" of evidentiary strength can be thought of as invoking the Precautionary Principle by expanding the "allowable possible error" in factual determinations. An alternative to shifting the burden of persuasion to another party is to lessen the burden of proof required to trigger an intervention to prevent or mitigate harm to health, safety, or the environment.

Also ignored by many commentators is the fact that burdens of persuasion often *shift in the course of fact finding*. Thus, depending on the nature of the intervention (notification, control, prevention, compensation), even if it is necessary for the regulator or potential victim initially to prove a [potential] harm, that proof is often not a high burden. A presumed fact (through a rebuttable presumption) might even be established by statute on the showing of certain other factual elements, such as the very existence of harm. Then, the burden of persuasion shifts to the intended regulated industry or alleged (potential) wrong-doer to refute the presumed or initially established fact, often with a higher burden of proof. Legal injunctions against potentially harmful action are granted by the courts as *equitable remedies*. The commentators on the Precautionary Principle have often ignored a rich and important set of policy interventions or actions that are informed, but not dictated, by factual determinations. Regulatory agencies themselves—depending on their statutory mandates—are not bound by traditional burdens of proof. Further, reviewing courts usually give *deference* to factual findings by the agencies, as long as they stay within the "zone of reasonableness" defined by those mandates.

PRECAUTION IN HINDSIGHT

It would be instructive to see how well we have fared with the implementation of the Precautionary Principle over the past twenty-five years. Scientific knowledge related to emerging health, safety, environmental, or public health problems began with a suggestion—sometimes a mere whisper—that trouble was brewing. Those suggestions and whispers ultimately ripened into full-fledged confirmations that our worst fears were not only true, reality often exceeded those fears. Examples that come to mind include asbestos-related cancer and the toxic effects of benzene, lead, and Agent Orange, to name just a few.

The frightening, but enlightening, reality is that with few memorable exceptions, the early warnings warranted heeding and the early predictions were certainly in the right direction—even understated. In retrospect, not only were all precautionary actions justified, we waited far too long to take those actions.

Barry Commoner, in *The Closing Circle*, warned us to avoid exposures "not consonant with our evolutionary soup." Theo Colborn has assembled in *Our Stolen Future* striking examples of why this is so. Endocrine-disrupting chemicals present an opportunity to act earlier, although some damage has already been done. Similarly, intervening now to prevent the next generation of developmentally or immunologically compromised, chemically intolerant persons (Josephson and Josephson, 1996), or otherwise chemically damaged individuals, many of them children, is both possible and necessary.

NOTE

1. Much of the discussion of the Precautionary Principle focuses on *cause-and-effect relationships* for which a high statistical confidence level ($p = 0.05$) or strength of association is traditionally required in scientific publications. It should be remembered that the convention of requiring a p-value no higher than 0.05 was an arbitrary historical choice. Critics of those wishing to invoke the Precautionary Principle by reducing the strength of causal proof would do well to remember this. In addition, other ways of knowing besides statistical correlations might be pursued (Ashford and Miller, 1998).

REFERENCES

Ashford, Nicholas A., 1981, "Alternatives to Cost-Benefit Analysis in Regulatory Decisions," *Annals of the New York Academy of Sciences*, Volume 363, April 30, 1981, pp. 129–137. See also Ashford, Nicholas A., and Caldart, Charles C. "Economic Issues in Occupational Health and Safety," in *Technology, Law and the Working Environment*, Revised Edition, Island Press, Washington, D.C., 1996.

———. 1988, "Science and Values in the Regulatory Process," in *Statistical Science*, Institute of Mathematical Statistics, Volume 3, Number 3, August, pp. 377–383.

———. 1994, "An Innovation-Based Strategy for the Environment and the Workplace" in *Worst Things First? The Debate over Risk-Based National Environmental Priorities*, A. M. Finkel and D. Golding (eds.), Resources for the Future, Washington, D.C., 275–314.

———. 1998, "The Importance of Taking Technological Innovation Into Account in Estimating the Costs and Benefits of Worker Health and Safety Regulation," *Proceedings of the European Conference on Costs and Benefits of Occupational Health and Safety 1997*, The Hague, Holland, 28–30 May 1997, J. Mossink and F. Licher (eds.), 1998, pp. 69–78.

Ashford, Nicholas A., and Miller, Claudia S., 1998, *Chemical Exposures: Low Levels and High Stakes*, Second Edition, John Wiley. New York.

Josephson, John R., and Josephson, Susan G., 1996, *Abductive Inference*, Cambridge University Press, Cambridge, England.

Chapter 12

ALTERNATIVES ASSESSMENT: PART OF OPERATIONALIZING AND INSTITUTIONALIZING THE PRECAUTIONARY PRINCIPLE

Mary O'Brien

Imagine a woman standing by an icy mountain river. A team of four risk assessors stands behind her, reviewing her situation. The toxicologist says that she ought to wade across the river, because it is not toxic, only cold. The cardiologist says she ought to cross the river, because she looks to be young and not already chilled. Her risks of cardiac arrest, therefore, are low. The hydrologist says she ought to cross the river, because he has seen other rivers like this, and probably this one is not more than four feet deep and it probably has no whirlpools at this location. Finally, the EPA policy specialist says that the woman ought to cross the river, because compared to global warming, ozone depletion, and loss of species diversity, the risks of her crossing are trivial.

The woman refuses to wade across. "Why?" the risk assessors ask. They show her their calculations, condescendingly explaining to her that her risk of dying while wading across the river is one in forty million.

The woman refuses to cross. "Why?" the risk assessors ask again, frustrated by this woman who clearly doesn't understand the nature of risks.

The woman points upstream and says, "Because there is a bridge."

In the above story, the risk assessors are doing risk assessment of only one option: Wading across an icy river. The woman is doing alternatives assessment, and one of the alternatives involves crossing the river, dry and warm, on a bridge.

The last sentence of the Precautionary Principle as developed at Wingspread on January 23–25, 1998 states, "The process of applying the Precautionary Principle must also involve an examination of the full range of alternatives, including no action." This chapter addresses that crucial process: examination of the full range of alternatives.

ALTERNATIVES ASSESSMENT AND THE PRECAUTIONARY PRINCIPLE

Alternatives assessment, the systematic analysis of a reasonable range of options, is a positive and necessary element of long-term use of the Precautionary Principle. Without the presentation of alternatives to hazardous activities, repeated invocation of the Precautionary Principle can be attacked as limitless nay-saying, that is, leading only to bans, phaseouts, and denials of permits.

Alternatives assessment consists of publicly examining a full range of alternatives to a potentially damaging human activity or social arrangement. Since essentially every human activity carries with it the potential to cause some harm (and therefore to trigger consideration of the Precautionary Principle), alternatives assessment allows those who invoke the Precautionary Principle to present attractive, feasible alternatives that show a track record or promise of avoiding, minimizing, or redressing harm.

Absent the articulation of alternatives to a potentially hazardous activity, the implementation of the Precautionary Principle will devolve into a risk assessment exercise, either explicitly or implicitly. Risk assessment will ask, "Is this activity of sufficient potential hazard to invoke precaution?"

The integration of a decision on precaution with alternatives assessment, by contrast, allows more fundamental questions to be asked: "Is this potentially hazardous activity necessary?" and "What less hazardous options are available?"

Alternatives assessment is, therefore, a simple mechanism for providing a set of concrete alternatives on which the Precautionary Principle can be exercised. These concrete alternatives can help dispel theoretical arguments about whether the Precautionary Principle relies on science, because scientific evidence is often available to demonstrate that some of the alternatives are more protective of the environment than the activity being challenged and are feasible. Indeed, when vastly different options are being compared, qualitative comparative analysis, rather than convoluted risk assessment, may suffice to reveal which alternatives are markedly precautionary or hazardous.

Participants at this conference are all familiar with the use of the Precautionary Principle within different contexts. Bo Wahlström (1998), for example, describes the Swedish Act on Chemical Products, which includes "avoiding chemical products for which less hazardous substitutes are available." Andrew Jordan and Timothy O'Riordan (1997) noted the West German concept of *Vorsorge*, which grew out of the belief that "the state should seek to avoid environmental damage by careful forward planning."

My personal acquaintance with a precautionary framework originated with use of the U.S. National Environmental Policy Act (NEPA) regulations (CEQ, 1992), which require, for all federally proposed, funded, and permitted actions, consideration of the potential adverse impacts of that action. If the impacts could be significant, an environmental impact statement is prepared, which analyzes impacts (both beneficial and adverse) of a full range of options for that action (including no action). The NEPA regulations constitute a "look before you leap" approach to decision making, to use the words of Nick Yost, one of the regulations' authors. The approach is applicable in all decision-making processes regarding potentially hazardous processes that could impact the environment or public health.

Through NEPA and courts, I and numerous other citizens, forest workers, and scientists were able to force the U.S. Forest Service to consider, in the early 1980s, a fundamentally less herbicide-dependent approach to vegetation management in the Pacific Northwest (O'Brien, 1990). Once having considered this approach, the Forest Service decided that it was a sound one and selected it as their preferred alternative.

The NEPA regulations do not require the selection of the most precautionary alternative. However, commitment to the Precautionary Principle would result in the selection of alternatives that avoid at least some of the hazards posed by others. For instance, rough rankings of alternatives according to short- and long-term environmental criteria (including uncertainty of benign outcomes) can be made, and then alternatives restricted to those indicating more, rather than less, overall precaution are selected.

That multiple-year process of developing an alternative to herbicide dependency with the Forest Service revealed to me another reality about alternatives assessment, one that may help explain why it is sometimes left behind in the rush by environmental advocates to invoke "precaution": Development and implementation of less hazardous alternatives are often slow, requiring much collaborative work. Often those who are knowledgeable about the hazards posed by some technology or process are much less knowledgeable about the technologies, economics, or social processes that may be needed to replace the hazardous activity.

Moreover, when alternatives are proposed, they are generally resisted, along with acknowledgment of hazard. Business-as-usual may be hazardous, but it brings power, money, and control to certain people, corporations, and networks, and change is resisted.

In the end, however, precautionary decisions are more likely to be made when attractive, least-harm alternatives are represented than when a potentially harmful activity is challenged in the absence of reasonable alternatives. This is illustrated by the story at the beginning of this chapter: Although the risk of mortality from wading across an icy river might be extremely low, even minor discomfort (e.g., becoming cold and soaked) seems less acceptable in the face of the reasonable alternative of crossing the river on a bridge.

A key element of alternatives assessment is its public nature, whether at a local or international level, because the environmental consequences of the decision making are public. The alternatives assessment process is necessarily enriched by broadly based public participation, because a full range of alternatives is more likely to be considered when diverse publics determine the range of alternatives examined and suggest specific reasonable alternatives, as well as their short- and long-term benefits and drawbacks.

Much of the text that follows is based on the NEPA process but is not in any way limited to this procedure. Alternatives assessment, in many forms, can facilitate implementation of the Precautionary Principle locally, nationally, or globally.

This chapter describes the elements of the alternatives assessment process, so that the reader can consider its potential importance as an explicit part of implementing the Precautionary Principle.

ESSENTIAL ELEMENTS OF AN ALTERNATIVES ASSESSMENT

There are several essential elements to an alternatives assessment that must be considered. A discussion of these elements follows.

Presentation of a Full Range of Options

A range of reasonable choices for behaving must be presented. The options must include those that seem to promise the least adverse impact on the environment and public health and/or those that seem to promise the greatest environmental and public health advantages.

Key to the presentation of a full range of options is a sufficiently large scope for decision making. If the scope of the decision is how to incinerate hazardous wastes, for instance, the range of options will be far narrower than if the scope of the decision is how to address the problems of hazardous waste disposal or the production of hazardous waste.

Presentation of Potential Adverse Effects of Each Option

The different types of adverse environmental consequences considered for the different choices must be as varied as the types of adverse environmental consequences humans are causing on earth. The NEPA regulations clearly define adverse effects. (The regulations sometimes refer to the effects as "impacts," and other times as "effects.")

First, the NEPA regulations correctly note that our environment is more than simply ecological or biological functioning. This means that effects to be considered include more than illness or death. The NEPA regulation 1508.8 ("Effects") notes:

> Effects include ecological (such as the effects on natural resources and on the components, structures, and functioning of affected ecosystems), aesthetic, historic, cultural, economic, social, or cumulative.

Second, the NEPA regulations talk of the varied scales of time and space of different effects, and of their combinations:

1. Direct effects, " . . . are caused by the action and occur at the same time and place" (40 CFR 1508.8).
2. Indirect effects " . . . are caused by the action and are later in time or farther removed in distance, but are still reasonably foreseeable. Indirect effects may include growth inducing effects and other effects related to induced changes in the pattern of land use, population density or growth rate, and related effects on air and water and other natural systems, including ecosystems" (40 CFR 1508.8).
3. Cumulative effects are effects "which [result] from the incremental impact of the action when added to other past, present, and reasonably foreseeable future actions regardless of [who] undertakes such other actions. Cumulative impacts can result from individually minor but collectively significant actions taking place over a period of time" (40 CFR 1508.7).

Example: The cumulative effects of draining one wetland. Using the fictitious example of a proposed action to drain a five-acre wetland in Oregon's Willamette Valley, environmental attorney Terence Thatcher (1990) asked what should be considered when assessing the cumulative impacts of draining that wetland. He wrote eloquently that the cumulative effects should include a discussion of the filling, draining, and polluting of wetlands, acts that are occurring along the Pacific Flyway from the Arctic to Central America by myriad "minor" human activities.

While such an analysis might seem unreasonable to the developer who wants to build one factory, draining a five-acre wetland *does* make a difference. This is because the damages that are occurring in the world *are* cumulative. Pintail ducks, for instance, *are* in trouble because of thousands of small activities that have eaten at their habitat throughout the Pacific Flyway. If the private landowner who wants to drain the five-acre wetland in Oregon is *not* responsible for the Pacific Flyway and pintail ducks, who *is*?

An alternatives assessment, if it examines a broad range of options, presents the public with starkly different effects of different options. An alternative that requires no wetland draining and no pollution, for instance, might avoid all cumulative adverse effects on the Pacific Flyway, requiring mention that the option that *would* drain the wetland *would* negatively affect the Pacific Flyway. The no-drain alternative might thus (correctly) be seen as carrying enormous value.

An alternatives assessment must also consider short-term effects versus long-term effects; less severe effects that affect large numbers of people or wildlife; severe effects that affect small numbers of people or wildlife; and irreversible effects.

When each option is associated with some type of adverse effect, there is no magic method of knowing which of the effects is more "acceptable." One set of people might argue for an alternative that may severely affect only one fern species, for instance, while another set of people, who know the tenuous hold that rare fern still has on Earth, may argue for another alternative that will cost a developer more money or that even might impact several populations of three more common fern species.

One alternative might adversely affect workers, while another alternative reduces nonrenewable resources of future generations.

When all kinds of adverse effects are being acknowledged, it becomes important to see if certain options pose markedly fewer hazards while meeting fundamental human needs. Necessity (as opposed to convenience, consumptive habits, protection of industry, or mere preference) may be clarified through the alternatives assessment process.

Presentation of Potential Beneficial Effects of Each Option

We may almost automatically think of direct, indirect, and cumulative effects as being adverse effects. Logically, however, "effects" includes beneficial effects, as well. Like adverse impacts, beneficial effects may vary greatly among the alternatives being considered in precautionary decision making.

And like adverse impacts, the direct and indirect benefits of an option can be "ecological, aesthetic, historic, cultural, economic, social, or cumulative."

Example: Social benefits of solar or wind power. In his book, *The Whale and the Reactor,* Langdon Winner (1986) wrote of the social and cultural choices we are making when we adopt particular technologies. The choice between nuclear power and solar energy technologies, for instance, includes more than issues of waste disposal, accidents, worker and community exposure to radiation, and economics. Nuclear power involves centralized control of energy, while solar and wind energy allow dispersed, local control of energy and energy costs. Small Danish cooperatives, for example, are currently generating 6 percent of Denmark's electricity through wind. As a recent *New York Times* article noted, "For wind, supply starts at the windmills, which do not require mines, oil fields, tankers, dumps, refineries and combustion" (Simons, 1997).

Nuclear power necessarily involves police state powers and secrecy in order to protect radioactive materials against theft for nuclear weapons making. Is there a larger environmental issue than nuclear war? One of the major environmental benefits of solar energy then is its lack of dependence on complex policing.

If a risk assessment of a nuclear power facility were being developed without considering solar power as an alternative, the consideration that the nuclear facility would require police state powers and centralized control of energy could theoretically arise. However, this drawback of nuclear power is more likely to be *noticed* when listing the *benefits* of solar power. Decentralization of energy is often stated as a benefit of solar technologies. Benefits, then, can be direct, indirect, cumulative, short term, and long term. Unfortunately, all our efforts to allow ecosystems and organisms to retain or regain integrity are reversible. Only *adverse* consequences of our activities seem to have the potential for being essentially irreversible, at least short of geologic time. The willful extinction of species, for instance, is irreversible; the erosion of soil to bedrock is irreversible within the timelines we generally consider. Precaution and alternatives are needed.

These, then, are the three elements in an alternatives assessment: a wide range of options, consideration of the benefits of each option, and consideration of the disadvantages of each option.

WHEN THE PUBLIC MIGHT BE AFFECTED: THE ASSESSMENT HAS TO BE A PUBLIC PROCESS

Because environment-impacting activities inevitably impact public resources, the public has a right to be involved in the assessment of alternatives to those activities. How is meaningful involvement ensured? We can turn once again to NEPA regulations for a model of public participation.

The Public Must Be Able to Affect the Breadth of the Alternatives Assessment

Under NEPA, when a federal agency is going to prepare an environmental impact statement (EIS), the agency must notify the public and solicit ideas for the necessary scope of the environmental analysis. At this point, the public can indicate that particular alternatives to the proposal are reasonable and must therefore be described and that particular direct, indirect, and cumulative impacts must be considered. This process, in which the public indicates the alternatives and impacts that need to be included in the EIS, is called "scoping."

A crucial step in the NEPA process is that the federal agency must respond to public scoping comments either by addressing each issue raised or by explicitly stating why the issue is not significant (40 CFR 1501.7(a)(3)). In other words, the agency must do more than simply "hear" the public; it must consider the proposed alternative or impact in the EIS or publicly explain why it is not significant enough to be included.

It is at this point that the public must try to widen the decision-making context if the agency has inappropriately narrowed the scope of the proposed decision, thereby limiting the range of alternatives that can feasibly be considered.

In the case of national or international initiatives, for example, a notice could be offered nationally or worldwide, requesting brief, clear proposals for how a particular problem should be stated, and thereby approached. This is not ordinarily done; however, it would help shed light on the choices that are being made when decision makers eventually decide to limit the scope in particular ways.

Decision Makers Must Respond to Public Comments on Drafts of a Decision-Making Document on the Basis of Alternatives Assessment

Under NEPA, after a federal agency has received comments from the public regarding the breadth and depth of an EIS, it then prepares a draft EIS. This draft EIS presents the pros and cons of the different options and usually suggests a preferred alternative.

The agency must make this draft EIS publicly available and give the public time to comment on its sufficiency and accuracy. The public can also comment on the proposed decision. Once again, the agency must publicly acknowledge each comment (40 CFR 1503.4(a)):

> An agency preparing a final environmental impact statement shall assess and consider comments both individually and collectively, and shall respond by one or more of the means listed

below, stating its response in the final statement. Possible responses are to:

1. Modify alternatives including the proposed action.
2. Develop and evaluate alternatives not previously given serious consideration by the agency.
3. Supplement, improve, or modify its analyses.
4. Make factual corrections.
5. Explain why the comments do not warrant further agency response, citing the sources, authorities, or reasons which support the agency's position and, if appropriate, indicate those circumstances which would trigger agency reappraisal or further response.

Scoping and a process requiring responsiveness to public comments on a draft decision document allow any person to suggest alternative behaviors that are reasonable and to note adverse and beneficial effects of each alternative considered. These processes yield three major social benefits:

• Benefit #1: Those who are or may be adversely affected by environment-impacting activities are allowed to bring up potentially better alternatives and options. This means that those who are managing the assessment process cannot limit the process to a few unsatisfactory options. Those who are managing the process may not be the ones who would suffer the adverse consequences, and so they might not be thinking or caring about such consequences. They might lack crucial information about the adverse or beneficial impacts of certain options. They might want to stack the deck in favor of an option that protects their power or profits, even though that option disadvantages people with less power or access to the managers of the process.

These scoping and comments processes do not depend on generating large numbers of commenters or having personal access to the managers of the assessment process. Rather, these processes give greater power to ideas and information according to their validity and feasibility.

• Benefit #2: The scoping and comments processes draw on numerous sources of experience and alternative ways of doing things. Those who are managing the assessment process might be unaware of a number of good alternatives. Alternatively, they might be afraid to suggest options that challenge the status quo or might be resistant to options that challenge their own power. When assessment processes are public, everyone learns of options and information that had not been considered before.

- Benefit #3: The public becomes more conscious of the fact that impacts on the public environment are *choices* about behavior, not inevitabilities. They become aware that caring for the environment in a precautionary manner is technically and economically feasible.

The term *public* cannot mean only those humans who live within a few social steps of the decision makers. What, then, is the limit of public involvement in an alternatives assessment? Practically, not all affected publics will have the time, interest, or ability to participate in all alternatives assessments. Rather, the best alternatives assessment processes are those that actively reach out to members of diverse publics, with different perspectives, so that a full range of alternatives and a broad range of impacts and benefits of those alternatives is considered.

It is a matter of how broadly the scope of inquiry is drawn—An entire community? A surrounding state? Humans *and* wildlife? A nation? The globe? Ideally, alternatives assessment embodies the saying, "Think globally, act locally." But global thinking will not take place if the globally knowledgeable are not encouraged to participate.

Decision Makers Must Be Accountable to a Public Alternatives Assessment Process

A key aspect of the U.S. federal NEPA process is that the public has the ability to ensure that the assessment process is done with integrity: Citizens can sue if it is not. They can sue a federal agency if the agency has not discussed reasonable alternatives, adverse impacts, or benefits that were brought to their attention during the scoping and commenting phases.[1] Without this recourse, many federal agencies would merely go through the motions of the public commenting processes and would discuss only minor adverse impacts and a narrow scope of alternatives.

Entrance into court is not necessarily the only option for accountability. If a group were entering into an alternatives assessment process with a federal government, for instance, up-front provisions could be established for independent arbitration in the event some public members have reason to believe that the managers of the assessment process might evade accountability to the agreed-on public process. Provisions for binding arbitration might be possible. Under binding arbitration, the parties to the assessment process would be required to submit to the decision of the arbitrator regarding public accountability rather than go to court.

Alternatives assessment processes need to provide for some form of meaningful public recourse if an unreasonably narrow range of alternatives is considered or significant information is ignored. Otherwise, the public's involvement can be time consuming, resource intensive, and utterly ignored.

But how can arbitrators or courts avoid having to consider an infinity of alternatives claimed by participants to be reasonable? A bottom line for arbitrators or courts to find a legally required alternatives assessment inadequate might be failure to analyze an alternative that is already utilized somewhere in the world for reasonably similar programs and that had been brought to the attention of the business, industry, agency, or decision-making body during the assessment process, or would clearly have been known to those entities.

Arbitration or litigation of precautionary-based decisions could be based on criteria of feasibility or overriding public health, transgenerational, or environmental value. Economic considerations, generally the most socially powerful argument offered in favor of environmentally destructive behaviors, could at times be challenged in a court that is interpreting Precautionary Principle legislation. As Jordan and O'Riordan (1997) noted, "If a possible outcome is potentially destabilizing to the natural order or to social equity, can it truly be regarded as a realistic option to the point where lost 'benefits' ought to constitute a 'sacrifice'?"

CONCLUSION

When joined to alternatives assessment, Precautionary Principle commitments are transformed from mere nay-saying and reaction to positive action on behalf of hope, vision, and feasible ways to behave as humans within a long-term, interdependent, multispecies world.

Alternatives assessment is made up of only a few working parts, which are sturdy:

- A set of options for ways to behave in relation to the environment (including most-precautionary options);
- Consideration of the environmental and social advantages of each option; and
- Consideration of the adverse environmental and social consequences of each option.

When an alternatives assessment is public, the public will bring good ideas and information. When a public alternatives assessment process is enforceable, the likelihood of responsiveness to public input is fundamentally increased.

Finally, a public alternatives assessment process places before everyone the reality that each business, industry, public agency, society, and international community is making *choices* to behave in a hazardous or precautionary manner toward the environment. A public alternatives assessment process helps the public become aware that current rules regarding the rights of private corporations to burden and damage communities and the environment are choices some sectors of our trade communities or societies may have made in the past, but the rules can be altered, if we decide to alter them.

The Precautionary Principle helps us acknowledge how hazardous we humans are to each other and to all of life on Earth; alternatives assessment helps us acknowledge the opportunities we have to behave much more decently.

NOTE

1. The ability to sue, of course, depends on having access to an attorney who cares about the case and who has the skills and ability to carry the case. In some areas of the country, public interest environmental attorneys are scarce. Sometimes a group does not have money to hire an attorney, and if no attorney will carry the case pro bono, then the group may in reality not have access to courts. The Equal Access to Justice Act provides for attorneys to recover their fees from the federal government if they win a public interest suit against the government (Axline, 1991). This allows some attorneys to take public-interest environmental cases on a pro bono basis. The importance to our democracy of "equal access" provisions cannot be overemphasized.

REFERENCES

Axline, Michael D. 1991. *Environmental Citizen Suits*. Salem, NH: Butterworth Legal Publishers.

Blake, Tupper, and Peter Steinhart. 1987. *Tracks in the Sky*. San Francisco: Chronicle Books.

CEQ (Council on Environmental Quality). 1992. *40 Code of Federal Regulations*, Parts 1500–1508.

Jordan, Andrew, and Timothy O'Riordan. 1997. "The Precautionary Principle in Contemporary Environmental Policy and Politics." Draft paper for presentation to the Wingspread International Conference on Implementing the Precautionary Principle, Racine, Wisconsin, 23–25 January 1998.

O'Brien, Mary. 1990. NEPA as It Was Meant to Be: NCAP v. Block, Herbicides, and Region 6 Forest Service. *Environmental Law*, 20:735–745.

Ozone Secretariat, United Nations Environment Programme. 1993. *Handbook for the Montreal Protocol on Substances That Deplete the Ozone Layer*. [No city given.]

Simons, Marlise. 1997. "In the New Europe, a Tilt to Using Wind's Power." *New York Times*, December 7, 1997, Section I, p. 20.

Thatcher, Terence. 1990. "Understanding Interdependence in the Natural Environment; Some Thoughts on Cumulative Impact Assessment under the National Environmental Policy Act." *Environmental Law,* 20:611–647.

Wahlström, Bo. "The Precautionary Approach to Chemicals Management— Swedish Views and Experiences." Draft paper for presentation to the Wingspread International Conference on Implementing the Precautionary Principle, Racine, Wisconsin, 23–25 January 1998.

Winner, Langdon. 1986. *The Whale and the Reactor: A Search for Limits in an Age of High Technology.* Chicago: University of Chicago Press.

Chapter 13

ENVIRONMENTAL BONDS:
IMPLEMENTING THE PRECAUTIONARY
PRINCIPLE IN ENVIRONMENTALLY POLICY

Laura Cornwell and Robert Costanza

In the broadest sense, the Precautionary Principle provides guidance on how scientific information should be utilized in formulating policy under uncertainty. It is often wrongly assumed that science can provide certainty if only enough research is undertaken. The truth is that while science can continue to improve our knowledge base, it can rarely, if ever, provide certainty. Environmental policy must be formulated on the basis of the best, *currently available*, scientific information, while erring on the side of caution. As Hey (1993) eloquently summarized, "The Precautionary Principle assumes that science does not always provide the insights needed to effectively protect the environment and that undesirable effects may be caused if measures are taken only when science does provide such insights."

It is precisely because of this scientific uncertainty that the Precautionary Principle has emerged in environmental policy and has been incorporated into numerous international agreements and treaties on environmental matters. Indeed, the Precautionary Principle was the main recommendation of the conference on Sustainable Development, Science and Policy in Bergen in 1990 (Perrings, 1991) and has been advocated by several other recent, international conferences including the Second International Conference on the North Sea (Lauck et al., 1998), the Second World Climate Conference in Geneva in November 1990, The International Conference on an Agenda of Science for Environmental and Development into the Twenty-first Century in Vienna in November 1991, and the United Nations Conference on Environment and Development—The Earth Summit—in Rio de Janeiro in June 1992 (Barbier et al., 1994). The principle is so frequently incorporated into international environmental resolutions that it is seen as a basic normative principle of international environmental law (Cameron and Abouchar, 1991).

The Federal Republic of Germany's *Vorsorgeprinzip*, which included several principles related to environmental stewardship, is often cited as the conceptual origin of the Precautionary Principle (cf. Gray and Brewers, 1996). The principle recommends formulating policies that have provisions for preventing possible environmental damage (particularly when threats of irreversibility exist), even though it is not known that damages will occur or to what extent (Glasser et al., 1994). Though the principle advocates the upfront commitment of resources to safeguard against potentially damaging, future outcomes, it offers no guidance as to what precautionary measures should be taken, the amount of resources to be committed, or which future outcomes should be avoided. Bodansky (1992) noted that "though the Precautionary Principle provides a useful overall orientation, it is an insufficient basis for policy and largely lacks legal content."

ENVIRONMENTAL BONDS

In order to operationalize the Precautionary Principle in environmental policy, we advocate environmental bonding. An environmental bonding system requires those seeking to use society's resources to post a bond in advance of any potentially, environmentally damaging activity. Bonding instruments also have the many added advantages that economic incentives can offer environmental policy (we have discussed these advantages at length elsewhere in this volume, see Cornwell, 1997; Cornwell and Costanza, 1992, 1994). The most important aspect of the bond amount is that it provides the incentive to behave in an environmentally responsible manner. If resource users could demonstrate that damages to the environment were less than the amount of the bond (over a predetermined length of time, specified in the bond), this difference and a portion of earned interest would be refunded. Thus, the environmental bonding system ensures that the funds available for protecting the environment are roughly equal to the potential harm facing its resources.

In addition to direct charges for known environmental damages, a bond equal to the current best estimate of the largest potential future environmental damages would be levied and kept in an interest-bearing escrow account for a predetermined length of time. The bond plus a portion of the interest would be returned at a specified time if and only if the firm or other agent could demonstrate that the suspected damages had not occurred or would not occur. If damages did occur, the bond would be used to rehabilitate or repair the environment and to compensate injured parties. By requiring the users of environmental resources to post a bond adequate to cover potential future environmental damages, the burden of proof and the costs associated with the burden are shifted from the public to the resource user. Additionally, a strong economic incentive is provided to research the true costs of environmentally damaging activities and to develop innovative, cost-effective pollution control technologies.

Neither the principle nor the bonding instrument is new. The environmental bond has its roots in the "material use fees" first proposed by Mill (1972) and Solow (1971), the simplest working example of which is the refundable deposit on glass bottles. The aim of the deposit is to encourage the users to dispose of the commodity in the most desirable way (by recycling) and to avoid its disposal in the least desirable way (as litter). The deposit may not be sufficient to cover the cost of the worst possible method of disposal, but it is generally set at a level high enough to make returning the bottle privately profitable. The important feature of the fee is that by insisting that consumers pay in advance for the costs they might inflict on society if they adopted the most harmful method of disposal, it reverses the usual presumption of innocence over guilt as applied to environmental damages. The innocent-until-proven-guilty argument is not applicable in the case of firms using societal resources as receivers of privately generated waste since there is no question that the act is being committed. It is the amount of damage that is uncertain, and society should not bear this risk.

Another precedent for environmental bonds are the producer-paid performance bonds often required for federal, state, or local government construction work. For example, the Miller Act (40 U.S.C. 270), a 1935 federal statute, requires contractors performing construction contracts for the federal government to secure performance bonds. Performance bonds provide a contractual guarantee that the principal (the entity which is doing the work or providing the service) will perform in a designated way. Bonds are frequently required for construction work done in the private sector as well.

Performance bonds are frequently posted in the form of corporate surety bonds that are licensed under various insurance laws and, under their charter, have legal authority to act as financial guarantee for others. The unrecoverable cost of this service is usually 1 to 5 percent of the bond amount. However, under the Miller Act (FAR 28.203-1 and 28.203-2), any contract above a designated amount ($25,000 in the case of construction) can be backed by other types of securities, such as U.S. bonds or notes, in lieu of a bond guaranteed by a surety company. In this case, the contractor provides a duly executed power of attorney and an agreement authorizing collection on the bond or notes if he or she defaults on the contract (Johnson, 1986). If the contractor performs all the obligations specified in the contract, the securities are returned to the contractor and the usual cost of the surety is avoided.

Environmental bonds would work in a similar manner (by providing a contractual guarantee that the principal would perform in an environmentally benign manner) but would be levied for the current best estimate of the largest potential future environmental damages. A quasi-judicial body would be necessary to resolve disputes about when and how much refund on the bonds should be awarded. This body would utilize the latest independent scientific information on the worst-case ecological damages that could result from a firm's activities, but with the burden of proof falling on the economic agent that stands to gain from the activity. Protocol for worst-case analysis already exists within the U.S. Environmental Protection Agency. In 1977, the U.S. Council on Environmental Quality required worst-case analysis for implementing the National Environmental Policy Act of 1969. This required the regulatory agency to consider the worst environmental consequences of an action when scientific uncertainty was involved (Fogleman, 1986). A specific example includes owners and operators of deepwater ports who are required to "establish and maintain evidence of financial responsibility sufficient to meet the maximum liability to which the responsible party could be subject to" (33 U.S.C. § 2716(d)).

CHARACTERISTICS OF BONDS

On the following pages we discuss various aspects of bonds and how these characteristics influence their use in precautionary policy making.

Three-Party Guarantee

Bonds are used extensively in government and business contracting, both nationally and internationally, when a guarantee of indemnity is required if one party of the contract fails to complete their obligations (Johnson, 1986). In legal terms, such bonds are considered accessories to the contract and are often required at several steps along the way toward project completion (Rowe, 1987). Bonds fall into a category of legal guarantees that involve three parties. The party required to post a bond, or regulated entity, is termed the *principal* (or sometimes the *obligor*) in such legal agreements. The entity that is protected by the bond (in the case of application to the environmental field, the regulatory authority on behalf of the public) is termed the *obligee*, "beneficiary" or sometimes *employer*. The third party in such agreements is the bonding company that posts bonds on behalf of the principal. In the United States, the main issuers of bond guarantees are specialized firms known as *surety companies*.[1] Third-party bonds are therefore referred to as *surety bonds*.

Two-party legal guarantees are sometimes called *personal bonds* or *self assurance mechanisms* but are most often referred to by the specific assurance mechanism used, such as a trust fund (Rowe, 1987). Two-party guarantees are not true bonds but could be utilized in some cases where bonding is an appropriate regulatory instrument and is considered briefly, later in this chapter. It is important to note that self-assurance mechanisms do not carry all of the advantages that suretyship provides. These advantages are discussed next.

Guarantee Types

Bonds are utilized as a legal guarantee of contractual obligation. For example, environmental bonds would guarantee fulfillment of an environmental regulation, which the principal is subject to under law. Bonded guarantees can generally be placed into two categories: financial and performance. A financial bond guarantees that a monetary payment will be made in the amount specified by the bond, if the contractual obligation in the bond is not met. A performance bond, on the other hand, guarantees completion of the obligation (Alber, 1992). For example, in the case of a waste water treatment plant, the contractual obligation would require the plant to meet certain effluent standards. If those standards were not met, a third-party financial bond would require the plant to pay a fine. If the plant was unable to pay the fine (termed *default*), the surety company that guarantees their bond would pay the fine on behalf of the treatment plant. A third-party performance bond would require the plant to come into compliance with the effluent standards designated in their contract (or permit, in the case of waste water treatment plants). In the case of default, the responsibility of installing the necessary abatement equipment would rest with the surety company.

The Underwriting Process

Companies seeking third-party bonds are required to undergo an extensive underwriting process. This process is similar to the credit evaluation given by commercial lending institutions to potential borrowers, but this process is more comprehensive. Underwriting ensures that the principal has the financial and managerial strength and expertise to fulfill the obligation the surety is guaranteeing (Holton, 1987). Underwriting is an important characteristic of the surety bonding process to understand because of the benefits it could provide the regulatory authority and, eventually, the public. These benefits would not be realized in two-party agreements. In general, surety underwriters evaluate the principal's capacity, capital, and character (Gaunt et al., 1982; Johnson, 1986).

Capacity in the surety business refers to the principal's technical and managerial ability to complete the terms set forth in the contract. The underwriter may examine several factors related to capacity, including: previous experience, current work load, organization, general business practices (e.g., personnel policies, distribution of employee salary, working conditions, short- and long-range plans, purchasing policies), academic background and adequacy of staff, quality of outside services, type of equipment owned versus that required, technical expertise, awareness of risk, and knowledge of risk reduction (Gaunt et al., 1982; Holton, 1987; Johnson, 1986; Planning Research Corporation, 1992).

As the name implies, capital evaluations focus on the principal's financial qualifications. Financial qualifications evaluated by sureties are similar to those evaluated by lending institutions. These include factors related to capital resources, financial strength, and credit standing (Johnson, 1986). Capital resource evaluations are often based on three-year certified public accountant audits. The level of financial strength needed to underwrite a bond is dependent on the value of the bond being requested and the type of industry in which the principal is operating (Severson et al., 1994). For example, some industries traditionally operate at a much higher debt-to-equity ratio than others. In the agricultural business, ratios exceeding 40 percent are not unusual (Johnson, 1986). This ratio would be considered severe financial stress in most other industries (Shogren et al., 1993). Barometers of financial strength may include quality of assets, profit and loss reports, and dollars spent on company maintenance and improvements. Surety companies make use of Dun and Bradstreet reports for credit checks (Johnson, 1986). These reports contain a history of the principal and its organization and owners, as well as supplier payment records. The surety also requests information from the principal's lending institution, such as account balances, loan amounts and payment history, as well as any issued letters of credit (Holton, 1987; Johnson, 1986; Planning Research Corporation, 1992).

Character evaluation is a subjective measure. Assessed attributes are a function of the particular industry the contractor is working in and the experience of the underwriter. In general, the underwriter is interested in measures of the principal's overall integrity and reliability and may include factors such as continuity of ownership and management and the financial condition of stockholders (Johnson, 1986). Often, previous obligees are contacted to attest to the character of a bond applicant (Planning Research Corporation, 1992).

Perhaps the most important underwriting factors for environmental bonds are technical expertise, awareness of risk, and knowledge of risk reduction. Technical expertise determines if the principal's management team has the knowledge required to complete the proposed project, if they have complied fully with the terms of contracts in the past, if they are complying with the terms of other contracts they may hold, and if the company has been cited for any violations in the past (Planning Research Corporation, 1992). By verifying compliance with past permits or leases and evaluating a principal's risk aversion, the surety would serve as an extension of the regulatory authority. Under an environmental bonding program, the surety would ensure that only those firms that were qualified and were taking the appropriate measures to avoid putting the public at risk would operate in the environmental arena. The regulated firm, the entity receiving rents from such operation, would bear the costs of such assessments through the payment of a bond processing fee. Assessments conducted by the regulatory authority, on the other hand, are financed by the public.

Another aspect of suretyship underwriting that would be particularly advantageous in the environmental field is the periodic review process. After issuing a bond, the surety assumes responsibility for the principal's actions. It is in the interest of the surety to ensure that the principal does not default on a contract. The surety, therefore, performs periodic assessments of work progress and financial condition during the course of contract completion. If the surety has any indication that the principal is at risk of defaulting on a project, the surety may provide financial or other assistance to prevent or reduce their losses (Johnson, 1986). Therefore, sureties would not only screen principals wanting to operate in the environmental arena, but they would also monitor their compliance.

Surety Versus Insurance

Surety bond companies are licensed and regulated by state and federal government insurance regulators and are often subsidiaries of insurance companies (Rowe, 1987). Though they share some of the same properties, there are important distinctions between third-party bond agreements and insurance agreements. Table 13.1 outlines some of the differences between surety bond contracts and insurance policies. In short, insurance is used to protect against property or casualty loss or damage. Insurance policies are two-party agreements where insured parties pay a premium to transfer the liability from potential loss to the insurer, while the insurer expects that a certain percentage of premiums collected will be paid out for losses. Bonds, on the other hand, are used to guarantee the fulfillment of an obligation. Surety bond contracts are three-party agreements in which liability does not transfer to

TABLE 13.1. Properties of Surety Bonds Versus Insurance Policies

Properties	Insurance	Surety
Contractual Specifications	Protects against property or casualty loss or damage.	Guarantees the fulfillment of an obligation.
Number of Parties	Two: the insurance company and the insured. A third party only enters into the contractual agreement if it is part of the property or casualty loss.	Three: the surety company, the principal, and the obligee.
Premium	Used by the insurer to assume risk for any one party by spreading the known probability of risk across all policy holders. The premium for an individual policy holder is assessed on the basis of the cost of the loss, weighted by the probability that the loss will occur.	Used by the surety company as a service fee to process the bond application.
Liability	Full liability for loss or damage is assumed by the insurance company.	Responsibility for honoring the conditions of the contract are held jointly by the surety company and the principal. If costs are incurred by the surety company they may subrogate, that is, they may pursue full cost recovery from the principal. Indeed, the principal's responsibility to indemnify the surety is granted in common law. Many bond contracts also contain specific clauses of indemnity (PRC, 1992).
Expectation of Loss or Default	The insurance company expects that a certain percentage of premiums collected will be paid out for losses. The probability of losses occurring is generally known with some certainty.	Default on the obligation is not expected. Before being bonded, the principal undergoes a rigorous underwriting process by the surety company. Bonding agreements compare closely with financial loan agreements in this sense. Surety companies do not expect defaults on bonds in much the same way as banks do not expect defaults on home mortgages.
Duration of Contract	Duration of the policy is based on premium periods. The policy can be canceled by either party.	Bond contracts cannot be canceled. The contract is based on an obligation and only expires when the obligation is complete.

Sources: Dawson, 1993; Planning Research Corporation, 1992; Rowe, 1987.

the surety company and default on the obligation is not expected (Planning Research Corporation, 1992). The distinctions are noteworthy for the different incentives they provide and will be addressed later in the chapter.

Subsidiary Contract

Bond guarantees are subsidiary contracts to a primary obligation. The bond is applicable only to the obligations specified in the contract to which it is attached and does not provide for compensation for other debts or obligation the principal may owe the obligee. If, for some reason, the primary contract is void, the bond also becomes unenforceable (Rowe, 1987).

Penal Sum

All bonds carry an overall monetary limit, termed *penal sum* in the United States, regardless if they are performance or payment bonds. In the case of performance bonds, the penal sum is an estimate of the cost to complete the entire contractual obligation, in the case of default. The surety guaranteeing performance or payment is never liable beyond the penal sum of the bond. U.S. courts, however, have not recognized this limit of liability and have decided in favor of completion of the performance even in cases where costs exceed the maximum monetary value set in the bond (Rowe, 1987; Estes and Connolly, 1990; Johnson, 1988; cf. also Ryan and Wright, 1990).

Cost

Principals are charged a premium by the surety company to issue a bond. As outlined in Table 13.1, premiums are utilized by the surety company to finance the underwriting process and cover other fees associated with issuing a bond. This fee may be charged on an annual basis or as a one-time payment. In the United States, premiums are generally single payments between 0.4 and 0.7 percent of the total contract value (Rowe, 1987). In addition to premium fees, principals may be required to post collateral to ensure the surety will be reimbursed in the case of default (Planning Research Corporation, 1992). Collateral requirements are imposed at the discretion of the surety and are based on several factors, including complexity of the project, both technical and legal; the ability of the principal; the nature of the guarantee; the maximum amount of the guarantee; the duration of the guarantee; as well as several other factors specific to a particular industry (Rowe, 1987). Collateral is returned upon completion of the primary contract.

TRADITIONAL USES OF BONDS

Bonds are widely utilized as legal contracts for several types of national and international business and finance guarantees. Traditional reliance on bonding mechanisms in the United States, Canada, and Europe dates back to at least the early 1700s (Planning Research Corporation, 1992). In the international arena, bond guarantees are utilized for everything from aircraft leasing and machinery purchase to sugar delivery and electronic communication (Rowe, 1987). The most common use within the United States is in the construction business.

Bond types are often defined by the type of guarantee. Guarantee types can be broadly categorized as fidelity, payment, and performance. Fidelity bonds guarantee the honesty or integrity of the person or subject of the agreement. Payment bonds guarantee monetary compensation in the case of default. Performance bonds guarantee completion of work or other actions specified in the primary contract, in case of default.

Use of Bonds for Environmental Regulation

In addition to their traditional uses, existing bond types have potential for application in environmental regulation. Surety bonds could supplement, or in some cases, replace, the current command and control regulation in the United States, while providing a needed margin of caution for the environment. Bonds have several advantages for the regulatory authority and public, the regulated community, and the surety industry.

Potential Benefits to the Regulatory Authority and the Public

Third-party bond guarantees could be advantageous to the regulatory authority, and ultimately the public, for several reasons. In general, bond guarantees could be utilized to ensure that a regulated entity would comply with existing environmental regulation. A bond would expand the regulatory authority's monitoring and enforcement capabilities by including the resources of the surety. Requiring a bond would increase the likelihood that the regulated entity would be qualified to complete the bonded activity in accordance with existing regulation. The surety would evaluate the regulated entity's financial status, workload, experience, and ability to fulfill its obligations. Requiring bonds would provide the public with a secure source of private-sector funding to restore environmental damage caused by noncompliance, or a guarantee of performance for bonded activities, even in the event of bankruptcy of the principal. The underwriting process would help to ensure that the regulated entity has the experience and resources to complete the terms of the bonded activity. After issuing the bond, the surety would evaluate the regulated entity during the life of the bond to ensure proper performance. Thus, the regulatory authority could reduce resources devoted to monitoring. Because sureties are regulated at the state and federal levels, regulatory authorities would have minimal need to oversee the surety's involvement in the environmental field. This may help to lessen the burden on agency legal and cost recovery personnel (Planning Research Corporation, 1992).

Finally, in the case of performance bonds, the regulatory authority would not be required to be expert in every industry under its jurisdiction. Sureties that enter a particular bonding market become expert in that market in order to complete the underwriting process. Sureties would ensure that companies for which they were holding guarantees were not only environmentally benign, but were also using all the means available to them to ensure they stayed that way. As a result, regulatory authorities could focus on mandating levels or conditions of environmental acceptability rather than the processes used to achieve those levels or conditions.

Potential Benefits to the Regulated Community

Besides the advantages to the firm of economic incentive instruments in general and bonds in particular, surety bonds provide the regulated community with some additional benefits. Because of the underwriting process, requiring bonds in environmental regulation could exclude principals that were not qualified to complete the bonded activity. In the private sector, this could free up work for firms that have a history of compliance and could provide incentives for companies to maintain high standards for achieving environmental acceptability (Mariotti, 1996).

The underwriting process also prevents principals from over committing their resources and accepting work in areas in which they do not have the experience, resources, or financial capability to perform properly (Planning Research Corporation, 1992). While a benefit to the regulatory authority and the public, this also serves the function of strengthening industry sectors and provides incentives to follow good business practices, while avoiding bankruptcy. The underwriting process helps economic agents subject to environmental regulation to maintain an awareness of the environmental risks associated with their business and provides additional incentives for the principal to take actions to reduce risks in order to increase its ability to be bonded. Because it is in the interest of the surety for the principal not to default on a bond, sureties can provide expertise and funding to the principal in the event of performance or financial problems (Planning Research Corporation, 1992).

Finally, for activities requiring financial guarantees, surety bonds could provide firms with less expensive options than they might otherwise have. In the case of trust funds, for example, the firm would be required to initiate an account for the full amount of the financial guarantee required. This could tie up a considerable portion of a firm's assets and would limit the types of investments that could be made (Mussatti, 1991; Shogren et al., 1993). Sureties, on the other hand, promise the full amount of the financial guarantee required, while only costing the principal a fraction of a percentage of its total guarantee (cf. Rowe, 1987).

Potential Benefits to the Surety Industry

The potential benefits of environmental bonds to surety companies are fairly straightforward. Environmental bonds would provide a new and expanding market for surety companies. Many traditional surety bond clients are expanding into environmental projects (Planning Research Corporation, 1992). Surety companies could offer traditional clients an additional service. In general, the underwriting required for bonds would also provide surety companies the opportunity to diversify their underwriting capabilities for hybrid projects as well as those particular to the environmental field.

Potential Limits of Bonds

Potential limits to utilization of bonds in the environmental field would have to be minimized before a successful program could be implemented. Bonding requirements could create cost and underwriting barriers to entry for some market participants, particularly small businesses (Planning Research Corporation, 1992). There are, however, programs that could accommodate some of these concerns. The Small Business Administration (SBA) does have a Surety Bond Guarantee Program for "small and emerging contractors who cannot get surety bonds through regular commercial channels" (SBA, 1991). Participants need to meet SBA's size eligibility standards, which require that average annual receipts from the previous three fiscal years not exceed $3.5 million. Performance and payment bonds up to $1.25 million are available through SBA. Similar organizations could potentially be formed to accommodate companies that did not meet SBA's size eligibility requirements but were new entrants into a market requiring environmental bonds.

Imposing a bonding requirement on the regulated community could increase the cost of doing business over traditional command and control mechanisms. Reserved assets for collateral reduces investment opportunities for the regulated firm. It is important to note, however, that this reduction is from an initial level of investment that may have been suboptimal. If a regulated industry is not internalizing the social costs of its potentially environmentally harmful activity, its level of investment is artificially high (Mussatti, 1991). The imposition of a bonding scheme could actually bring investment closer to a level that is optimal.

Bonds could create barriers to innovative technology use, depending on the standards in the industry and the requirements of the surety. The surety industry may be unwilling to issue bonds involving untested technologies since it would not be able to assess the risk of default. Depending on the cost savings to the firm from utilization of more efficient technology, one option is to require higher collateral until the effectiveness of the technology can be assessed. This collateral could be reduced as the technology became more proven. The regulated entity would have incentive to implement the innovation so long as the cost savings over time from innovation were greater than the collateral requirements. Other possibilities include grace periods granted by the regulatory authority for innovative instruments that have passed some type of initial screening process.

MATCHING APPROPRIATE INCENTIVE MECHANISMS WITH ENVIRONMENTAL DAMAGE TYPES

For the purposes of environmental regulation, the type of regulatory tool chosen for a particular application would depend on the goal of the regulation. Careful consideration would need to be given to the type of incentive the mechanism would provide the principal, as well as the level of precautionary control the principal may or may not have. Though bonding is certainly not a panacea, the list of potential applications to environmental regulation is practically limitless. Design considerations are as numerous as the application potential and are not the topic of this chapter. We do, however, consider some broad categories of environmental damages and propose which guarantee mechanism or mechanisms might be most appropriate by utilizing some examples. For illustrating bonding mechanisms, the broad environmental damage categories we use are expected, accidental, and continuous.

Expected Damages

We are using the term *expected* to define those damages resulting from an environmentally altering or damaging activity (i.e., damages are expected) that has a limited time horizon. Provisions for remediation or mitigation of expected damages are arranged before permits or allowances for the damaging activity are granted. For example, firms conducting mining activities on public lands are required to mitigate the site after project completion. Expected damages are perhaps the category of environmental damages best suited for regulation through bonds since firms have precautionary control.

When remediation activities are expected (with a limited time horizon) rather than as an ongoing production process of the firm, the incentive for the firm is to avoid the cost of repairing damages. During permit negotiations while securing rights to exploit a resource, the incentive for the economic agent is to make promises about environmental restoration. Once the profits have been exhausted on such a project, however, fulfilling the restoration promise is strictly a cost to the firm. Profit maximization points the firm toward seeking a less costly alternative to land restoration, despite earlier restoration promises (Mussatti, 1991). A guarantee that such activities will take place is therefore warranted. In order to maintain its ability to be bonded and regain any collateral that may have been required, an economic agent would have the incentive to repair damages.

Expected damages are generally associated with a project that has a defined beginning and end over a relatively short time horizon (say, 10 years or less) and requires securing a permit or lease before its undertaking. The subsidiary nature of surety bonds lends itself well to such a legal arrangement. Limited time horizons are also important for assessing damage cost and extent. The upper boundaries of damages are fairly well understood over short time horizons. The longer the time between establishment of bond amounts and completion of the project, the more difficult it becomes to establish potential damage cost and extent (Shogren et al., 1993).

Damages are inevitable in the case of economic activities that create expected damages and their extent can be estimated. It is therefore known, prior to commencing such activities, that mitigation measures will be necessary. One of the arguments against utilizing bonding in environmental programs is the conviction in the system of jurisprudence in the United States that innocence is the assumption and guilt requires proof. In the case of economic activities that create damages ex-post, the probability of damage is one. Prior to this type of activity being permitted, assurances of the probability of remediation to societies' resources should also be one.

Finally, in the case of ex-post damages, damages are observable and attributable to a particular economic agent and their activities. A particular action is required of that agent. The emphasis of the action is generally on the outcome rather than on its process. That is, the desired outcome is mitigation of the damaging activity and the tools utilized to mitigate are (generally) less important. Damages that can be observed and attributed to a particular economic agent are good candidates for performance bonds. The interests of the regulatory authority and the public are in damage remediation not monetary compensation. Payment bonds, while preferred to no financial guarantee, would only represent an intermediary step. The regulatory authority would then need to enter the business of remediation or hire others to complete the task.

Logging companies that are interested in cutting a road through public lands for short-term use to access timber on a leasehold provides an example. The regulatory authority that has jurisdiction over the land may determine that, from an ex-post social welfare standpoint, benefits gained outweigh the costs of the road being built. However, the authority also determines that long-term use and access into the area by the general public is not desirable. As a condition of the permit to build the road and gain right-of-way access, the timber company must remediate the road site after project completion. After completing a mitigation plan and design (contract), a performance bond would provide a precautionary measure to guarantee completion of the plan.

Accidental Damages

We are using the term *accidental damages* to define unintended, event-oriented damages caused by an economic agent during its daily operations. Accidental damages may have a degree of probability associated with their occurrence or may be truly uncertain events. Risk factors can be those that the firm has control over and can reduce or may not be controllable.

If the probability of a damaging event is known for a regulated industry and this risk can be controlled by individual firms (i.e., the firm has precautionary control), bonds are an appropriate regulatory tool. Bond underwriting evaluates each principal individually. Firms would have the incentive to employ risk-averse strategies to ensure they could maintain their ability to be bonded and reduce their collateral requirements to the lowest possible cost. Firms that maintained a history of risk aversion and compliance would have minimal or nonexistent collateral requirements. The transport of crude oil provides an example. Based on past history and factors such as weather, port characteristics of entry and exit, size and draft of the ship, experience of crew, and so forth, the probability of an oil tanker grounding can be calculated. If the tanker involuntarily releases its cargo into the environment, the damages are catastrophic and subject to long time horizons. However, the oil transport company has control over the environmental risk (of crude oil release, at least—physical damage to benthic communities is another matter). Tankers that are double-hulled may ground but will not release their contents. If companies in the oil transport business were required to be bonded to guarantee against such environmental disasters, all tankers would (most likely and for the sake of this example) be double-hulled because surety companies would make this a requirement of their guarantee. Rather than the onus being on the Coast Guard (the regulatory authority that oversees oil tankers) to mandate that certain technologies be put in place, surety companies that entered the oil tanker bonding business would (theoretically) become expert in the oil transport business and would ensure that transport companies were taking all precautionary measures available to prevent default on the bonded activity, in this case, a guarantee that no crude would be released into the environment.

If the probability of an accidental damaging event is known for a regulated industry but cannot be controlled by individual regulated entities, third-party bonds are probably not the best instrument to utilize. Since any given firm would not have control over noncompliant events, and could do nothing to prevent events from occurring, spreading the risk across the regulated community would reduce the burden for any given firm. Insurance instruments that evaluate the industry as a whole rather than at the firm level and that spread risk would be more cost efficient. An example would be environmental damages caused by aperiodic, natural events. Waste water treatment plants that exceed their capacity during extreme flooding events would have no control over the flood event, though its risk may be known. Plants in a given region (EPA's regional designations, for example) could be insured against such damages.

For truly uncertain accidental damaging events that have the potential to cause catastrophic or irreversible environmental damages, third-party bonds are not appropriate tools under the current system of bond markets. The surety industry would not enter into such an agreement (Johnson, 1988; Ryan and Wright, 1990). Self-assurance mechanisms, however, such as trust funds would be appropriate. Trust funds (also termed *self bonds*) for environmental regulation would name the regulatory authority as beneficiary and would be established for an amount equal to the maximum potential damage. If it was in the regulatory authority's interest to have the regulated entity mitigate or remediate damages, the financial assurance could be subsidiary to a performance contract. In the case of noncompliance or bankruptcy, the regulatory authority would have the trust fund to hire mitigation measures. Self bonds for highly uncertain, catastrophic damages have received much criticism, particularly for their liquidity constraints. The responses, however, have maintained that economic activities that place societies resources at such risk should not be permitted if financial guarantees cannot be levied. In the case of damages that have the potential to be mitigated or at least off set, this argument is potentially justified. In the often cited example of species extinction, however, it would not be possible to establish a bond amount or remediate damages. Ultimately these are value-based questions and need to be addressed by society—not by private industry or bureaucracies. The point here is that one type of environmental bond could be utilized and would provide the degree of precaution that should be afforded such environmental activities.

Continuous Damages

The term *continuous damages* is being used here to define damages that are the result of ongoing production processes of an economic agent. These damages are expected and require a permit that limits their quantity over temporal and spatial scales or their concentration relative to ambient conditions. Examples of these include the many "end-of-the-pipe" type pollutants in the air and water media that are regulated by EPA.

Continuous damages could have many of the characteristics conducive to an environmental bonding program. The incentives to avoid detection rather than implement mitigation equipment in line with permitted levels are well documented for economic agents emitting continuous waste streams, both theoretically and empirically (cf. Ackerman and Stewart, 1985; Hartford, 1987; Vig and Kraft, 1990). Requiring a guarantee is therefore justified. Damages are expected and often require some degree of precautionary measure and also a permit. Bonds could be subsidiary guarantees to the conditions of such permits. The extent of maximal potential damages is generally known. Because damages are being emitted via a "pipe," damages are most often observable (or detectable) within the limits of interest and are attributable to a particular economic agent. Though continuous damages are, by definition, not subject to a limited time period, bonds could be issued over a given time horizon. At the end of the bond period, the economic agent could be reevaluated and another bond could be issued.

The type of bond most appropriate for economic agents in continuous damage activities would depend on the nature of the damage. Performance bonds could be utilized to guarantee that permitted levels were not exceeded at the point of emission. In the case of solid waste, performance bonds could also be utilized to remediate past offenses. However, for the air and water media, a type of hybrid bond would be needed. A performance bond could be utilized for guaranteeing compliance at the end of the pipe. Sometimes firms are required to off-set damages that have already occurred and are not attributable to any one agent. However, it may be in the interest of the regulatory authority to require a payment bond for damages already released into the air or watershed to initiate a type of mitigation trust fund. This fund would pool resources from all offenders and be earmarked for a larger mitigation effort than any one agent could perform independently.

Nonpoint sources are a type of continuous damage that probably does not lend itself well to bonding because most often damages are not attributable to any one source. Siltation in watersheds located in agricultural areas is an example. There are, however, certain actions that farmers can take to reduce nonpoint source runoff. In the case where these actions are deemed to be necessary, performance bonds could be utilized to guarantee certain tillage practices, the planting of buffer vegetation, or the decommissioning of certain acreage.

BONDS IN ENVIRONMENTAL POLICY

Examination of U.S. environmental legislation and regulation reveals that bonds are currently used as regulatory instruments under current policy. At least five governmental agencies or commissions currently make use of bonds in their regulatory programs, including Department of the Interior, Department of Transportation, Environmental Protection Agency, Nuclear Regulatory Commission, and Department of Commerce. In a couple of cases, bonds are required by either statute or regulation. In other programs, they are named or implied as an option among several regulatory tools. Programs that utilize bonds or have the potential for utilization are contained in the Appendix at the back of this volume. The list includes the statute that applies to the program, the specification of the statute with respect to environmental protection and protection mechanisms, the regulatory authority that administers the program, and the regulatory authority's interpretation of the program.

Bonds in Statutes

At least 10 statutes pertaining to environmental protection allow for the utilization of bonding instruments. Indeed, the Surface Mining Control and Reclamation Act (SMCRA) (30 U.S.C. §§ 1201–1211, 1231–1328) mandates the use of performance bonds to guarantee reclamation of mining sites for coal and other minerals. The stated goal of the reclamation is "to prevent or mitigate adverse environmental effects of present and future surface coal mining operations" (30 U.S.C. § 1201(d & k)). The penal sum of the bond is not specifically stated but is required to "assure the completion of the reclamation" (30 U.S.C. § 1259 (a)).

Bonding is specifically named in statutory language as one of several options for establishing financial responsibility in at least six other environmental programs (33 U.S.C. § 2716 (e) with 43 U.S.C. § 1331 and 33 U.S.C. § 1503 (c), 33 U.S.C. § 1321 (p) with 43 U.S.C. § 1331 and 33 U.S.C. § 1503 (c), 30 U.S.C. § 226 (g), 42 U.S.C. § 9608 (b), 42 U.S.C. § 6991c (c)).

In at least three other programs, statutory languages require an assurance or establishment of financial responsibility and also require the regulatory authority to determine the appropriate tools (42 U.S.C. § 2210 (a) with (42 U.S.C. § 10141 (b)(1)(A) (i & ii, iii) and 42 U.S.C. § 10131(a)). Bonds are therefore allowable under these programs, and establishing proof of responsibility is required.

Bonds in Regulation

In the case where bonds are named as an option under the statute, the regulatory authority's interpretation requires bonds in at least one regulatory program. The Bureau of Land Management specifically requires the posting of a bond under their administration of oil and gas leases on public lands (43 C.F.R. Part 3104). The bond guarantees the complete and timely plugging of the well(s), reclamation of the lease area(s), and restoration of any lands or surface waters adversely affected by the lease operation. The penal sum of the bond is not specified but is required to be the minimum amount necessary to ensure compliance.

Where bonds are not specifically named as an option under statute but establishment of financial responsibility is, the interpretation of regulatory authorities names bonds in all three programs. These programs are all administered under the Nuclear Regulatory Commission and include regulations pertaining to decommissioning of nuclear facilities, storage of nuclear material in repositories, and land disposal of radioactive waste (10 C.F.R. Part 61, 10 C.F.R. Part 70, 10 C.F.R. Part 72).

In some cases, the regulatory authority requires guarantees in their interpretation when they are not called for in statutory language. Bonds are named as options in at least three programs. These programs are all administered by the EPA and include regulations pertaining to disposal of PCBs; treatment, storage, and disposal facilities for hazardous waste; and underground injection wells (40 C.F.R. Part 761, 40 C.F.R. Part 264, 40 C.F.R. Part 144).

Finally, some regulatory programs contain vague language that could allow for the utilization of bonds. Two examples of these include deep seabed exploration and recovery (15 C.F.R. 970, 15 C.F.R. 971). Both programs are administered by the Department of Commerce under the National Oceanic and Atmospheric Administration. Regulatory language indicates that applicants for deep seabed mining exploration "must show to the Administrator's satisfaction that he is reasonably capable of committing or raising sufficient funds to carry out [the entire program outlined in his plan, including protection of the environment and monitoring the effectiveness of that protection]" (15 C.F.R. Part 970). The regulation does not specify how applicants obtain these funds.

Specific details of an environmental bonding system would vary, depending on its application. These details are critical to the success of the proposal and need to be given careful consideration. There are several examples of well-intentioned performance bonding systems that have failed because of simple oversights in the implementation of the system. For example, mine reclamation bonds have, in some cases, been set so low that it was cheaper for the mining companies to default than to reclaim the site. It is of prime importance that the bond be large enough to cover the worst-case damages so that malfunctions like this do not occur.

CONCLUSION

Because of their logic, fairness, efficiency, and ability to implement the Precautionary Principle in a practical way, environmental bonds hold significant promise. They use legal and financial mechanisms with long and successful precedents and therefore might be both practical and politically feasible. They can potentially solve the problem of "how precautionary should we be?" and allow the Precautionary Principle to play an increasingly central role in environmental policy. They are certainly not panaceas, but they can be an important component in improved environmental policy.

NOTE

1. Internationally, the bonding business is not exclusive to a particular type of organization but varies from one country to another. Most banks in the U.S. are not authorized to issues bonds, though in Europe this is common practice (Rowe, 1987).

REFERENCES

Ackerman, B., and R. Stewart. 1985. "Reforming Environmental Law: The Democratic Case for Market Incentives." *Columbia Journal of Environmental Law* 13:178–199.

Alber, P. 1992. "Making Sense Out of Performance and Payment Bonds." *Michigan Bar Journal* 71:1–8.

Barbier, E., J. Burgess, and C. Folke. 1994. *Paradise Lost?* Earthscan Publications, London.

Bodansky, D. 1992. "The Precautionary Principle." *Environment* 34:4–5.

Cameron, J., and J. Abouchar. 1991. "The Precautionary Principle: A Fundamental Principle of Law and Policy for the Protection of the Global Environment." *Boston College International and Comparative Law Review* 14:1–27.

Cornwell, L. 1997. "Policy Tools for Environmentally Sustainable Development." Ph.D. thesis. University of Maryland, College Park.

Cornwell, L., and R. Costanza. 1994. "An Experimental Analysis of the Effectiveness of an Environmental Assurance Bonding System on Player Behavior in a Simulated Firm." *Ecological Economics* 11:213–226.

———. 1992. "The 4P Approach to Dealing with Scientific Uncertainty." *Environment* 34:12–20.

Dawson, A. 1993. "Understanding Commercial Suretyship: Bids, Performance, and Payment Bond." *National Bar Association* 7(3):10.

Estes, J., and M. Connolly. 1990. "Avoiding Exposure to Environmental Liabilities Concerns for Sureties." American Bar Association, Forum Committee on the Construction Industry.

Fogleman, V. 1986. "Worst Case Analysis: A Continued Requirement Under the National Environmental Policy Act?" *Columbia Journal of Environmental Law* 13:53.

Gaunt, L., N. Williams, and E. Randall. 1982. *Commercial Liability Underwriting, Second Edition.* Insurance Institute of America.

Glasser, H., P. Craig, and W. Kempton. 1994. "Ethics and Values in Environmental Policy: The Said and the UNCED." In: J. van den Bergh and J. van der Straaten (eds.) *Toward Sustainable Development: Concepts, Methods and Policy.* Island Press for the International Society for Ecological Economics, Washington, D.C.

Gray, J., and J. Brewers. 1996. "Towards a Scientific Definition of the Precautionary Principle." *Marine Pollution Bulletin* 32 (11):768–771.

Harford, J. 1987. "Self-Reporting of Pollution and the Firm's Behavior Under Imperfectly Enforceable Regulation." *Journal of Environmental Economics and Management* 14:293–303.

Hey, E. 1993. "The Precautionary Principle." *Marine Pollution Bulletin* 26 (1):53–54.

Holton, R. 1987. *Underwriting Principles & Practices,* Third Edition, The National Underwriter Company, Cincinnati, Ohio.

Johnson, M. 1986. "Performance Bonds." A final report prepared for the U.S. Environmental Protection Agency, Office of Waste Programs Enforcement, contract number 15-4040-00. Washington, D.C.

———. 1988. "An Analysis of Alternative Cleanup Financing Mechanisms for Their Potential Application to CERCLA Settlements." Study prepared for the United States Environmental Protection Agency, Office of Waste Programs Enforcement.

Lauck, T., C. Clark, M. Mangel, and G. Munro. 1998. "Implementing the Precautionary Principle in Fisheries Management Through Marine Reserves." *Ecological Applications* 8(1) Supplement: S72–S78.

Mariotti, J. 1996. *The Power of Partnership.* Blackwell Business Press, Boston.

Mill, E. 1972. *Urban Economics.* Scott Forseman, Glenville, IL.

Mussatti, D. 1991. "Springing the Social Trap: Bonding Schemes as a Regulatory Alternative." Unpublished manuscript.

Perrings, C. 1991. "Reserved Rationality and the Precautionary Principle: Technological Change, Time and Uncertainty in Environmental Decision Making. In R. Costanza (ed.) *Ecological Economics: The Science and Management of Sustainability.* Columbia University Press, New York.

Planning Research Corporation. 1992. "Guide to the Use and Implementation of Surety Bonds in Environmental Programs." Study prepared for the United States Environmental Protection Agency, Office of Policy Analysis, Regulatory Innovations Branch.

Rowe, M. 1987. *Guarantees, Standby Letters of Credit and Other Securities.* Euromoney Publications, London.

Ryan, W., Jr., and R. Wright. 1990. *Hazardous Waste Liability and the Surety.* American Bar Association, Tort and Insurance Section, Washington, D.C.

SBA. 1991. "The Facts About Surety Bond Guarantee Program." Circular.

Severson, G., J. Russell, and E. Jaselskis. 1994. "Predicting Contract Surety Bond Claims Using Contractor Financial Data." *Journal of Construction Engineering and Management* 120(2):405–420.

Shogren, J., J. Herriges, and R. Govindasamy. 1993. "Limits to Environmental Bonds." *Ecological Economics* 8:109–133.

Solow, R. 1971. "The Economist's Solution to Pollution Control." *Science* 173:498–503.

Vig, N., and M. Kraft. 1990. *Environmental Policy in the 1990s.* Congressional Quarterly Press, Washington, D.C.

Chapter 14

THE PRECAUTIONARY PRINCIPLE AND CORPORATE DISCLOSURE

Sanford Lewis

The growing recognition of the need for precautionary action on environmental issues presents new challenges. Issues such as global warming and exposure to endocrine-disrupting compounds from production systems and products demand dramatic readjustment of industrial and other activities, even in the absence of conclusive proof about how likely the most probable or even worst impacts could be.

The Precautionary Principle states that "when an activity raises threats of harm to human health or the environment, precautionary measures should be taken even if some cause and effect relationships are not fully established scientifically." Various existing and proposed mechanisms attempt to advance this precautionary approach, ranging from promoting additional study, to shifting burdens of proof in litigation or administrative proceedings, to providing incentives for preventive behavior, to action forcing measures such as bans and phaseouts of substances suspected of causing the harms. In this chapter, we examine the potential for strategies to promote precaution on the basis of disclosure of corporate information—including both local, community-based action and public policy approaches.

THE ROLE OF CORPORATE DISCLOSURE IN PROMOTING THE PRECAUTIONARY PRINCIPLE

Economic theory holds that marketplaces and economic activities are organized around information. Decisions based on a balance of costs and benefits require perfect information both about the impacts of an activity as well as the availability of alternatives to prevent harm (the latter easier to provide than the former). The existence of "imperfect" information is cited by economists as one pivotal reason why markets work imperfectly to meet consumers' and society's needs, as well as why environmental and public health degradation continues to occur. This lack of perfect access to information clearly can work to undercut society's responsiveness in taking precautions on issues of social concern.

For instance, experience with societal responses to lead, asbestos, and tobacco, as well as a host of other hazardous technologies, demonstrates a pattern of corporate denials, or inflated corporate assertions of scientific uncertainty, when the short-term (or long-term) profits of industries could be affected negatively by decisive societal responses. Corporate-sponsored scientific denials are well highlighted in recent tobacco litigation. The absence of disclosure of "what they knew when they knew it" and the fabrication of quasi-scientific "expert" views on safety, led to years of delay in corporate accountability and societal responsiveness to the hazards posed by cigarette smoking. The same seems equally true in other long-standing hazards, such as the dangers posed by asbestos and lead.

This denial and obscuring of dangers is not a victimless offense—it has led to countless unnecessary deaths and illnesses. In each instance, if the government and public had adequate, untainted information, it might have acted many years earlier and avoided decades of damage to public health.

The experience of society in these matters points to the need to clarify what we mean when we refer to the "uncertainty" surrounding a particular issue. There is surely "real uncertainty," both reducible and irreducible, in many technical arenas, but just as often there may be what we can refer to as "smokescreen uncertainty"—uncertainty created by corporate (and sometimes government) science for the purposes of forestalling decisive action. Experience shows that when affected by demands for precaution, many industries (and sometimes the government) engage in an array of tactics to create a false sense of uncertainty (or certainty that there is no problem):

• Concealment of what the management and corporate personnel already know;

• Hiring of scientists inside and outside the corporation both to create "contrary" opinions and to promote the notion of uncertainty;

- Development of third-party "independent" research organizations that represent an industry's interests but are far enough removed that they appear to provide an unbiased, expert opinion of the matter; and
- Manipulation of public relations and political "spins" to create an atmosphere of uncertainty and a "consensus" need for more study before action is taken.[1]

While there is bona fide uncertainty in some scientific matters that should be dealt with in public policy decision making, specific attention is needed in precautionary strategies to minimize and control "smokescreen uncertainty."

A New Initiative Against Disclosure: Building New Corporate Smokescreens

Public policy can either foster or reduce the potential for smokescreen uncertainty to undercut needed public responses on issues requiring precautionary action. An example of a public policy approach that targets and prevents corporate smokescreen efforts is contained in recent court decisions that denied industries the right to use attorney–client privilege to conceal scientific documentation of health impacts of tobacco and, in environmental cases, of other scientific studies on remediation.

Abuses of Attorney–Client Privilege in Tobacco and Environmental Litigation

In the tobacco litigation in Minnesota, Ramsey County District Judge Kenneth Fitzpatrick wrote that the tobacco industry falsely claimed attorney–client privilege to keep documents private. "Upon review of randomly selected documents, it has been determined that defendants have in numerous instances claimed privilege where none is due and blatantly abused the categorization process," Fitzpatrick wrote.[2] The attorney–client privilege is intended to protect communications between lawyers and their clients, but the tobacco companies had attempted to process all sorts of scientific and marketing studies through their lawyers in order to provide a veil of secrecy. The tobacco companies routinely marked scientific research papers "attorney work product" or attorney–client privilege, even when they had not been created for use in litigation.

The same attorney–client rationale has been applied by industry in environmental cases in an attempt to conceal needed environmental data. For instance, the Phelps Dodge Corporation attempted to use attorney–client privilege to conceal documentation regarding remediation of a waste site. Phelps Dodge, Inc., sold property to the U.S. Postal Service (USPS) in

Maspeth (Queens), New York, in the mid-1980s for construction of a postal distribution building. Phelps Dodge agreed to clean up the former copper refining site, but as the cleanup process continued it became apparent to Phelps Dodge and its contractors that heavy metal contaminants onsite (e.g., arsenic, cadmium, and lead) were more widespread, resulting in a much more expensive cleanup than anticipated. In response, Phelps Dodge officials—under the leadership of the company president—apparently employed a strategy of concealment in an attempt to strap the USPS and the U.S. taxpayers with the costs of the cleanup. One major strategy was to claim attorney–client privilege for studies conducted by consultants that revealed information relevant to the extent of contamination and costs of the cleanup. In 1994, the court reviewed these attorney–client privilege claims, document by document, and found about 80 percent of the documents ineligible for such treatment. The court issued an explicit ruling, with a six-page list showing the numerous studies, letters, and evaluations that the company inappropriately attempted to keep out of government hands.[3] The lengthy list of documents that the company had attempted to cover as "privileged" included many documents that had merely been copied to attorneys and others in which attorneys had no real role. The court noted that the data were "generated through studies and collected through observation of the physical condition of the Property. . . . Such underlying factual data can never be protected by attorney–client privilege and neither can the resulting opinions and recommendations."[4]

Had these documents remained out of public view in the Phelps Dodge matter, the government may have been incapacitated from winning its later court victory, in which the company was required to resume ownership of the site. The 1997 court decision in favor of the government found that Phelps Dodge had breached its contract with the U.S. Postal Service by delaying and declining to fulfill its contractual responsibility to excavate all of its contamination.[5] Without disclosure of documents showing the extent of contamination and remedial needs, the court might never have reached this result, and either the environment might have remained contaminated or the taxpayers would have been forced to pick up the tab.

Environmental Audit Laws: A New Corporate Concealment Approach

These recent court rulings against corporate secrecy through attorney–client privilege allow society to scrutinize corporate activities to support precaution and accountability. However, in the legislative arena the law has been shifting elsewhere "in favor" of allowing more "smokescreen" uncertainty to be applied by corporations. In the past few years, corporations have won new rights of environmental concealment in twenty states, on the basis of the principle that "honest" self-appraisals—termed "self-critical analysis" or "environmental audits"—merit a privilege against forced disclosure in private or public litigation. Such laws encompass vast new rights of concealment of information relevant to the issue of precaution, as the "audits" are broadly defined to include an array of corporate environmental studies.

For instance, the Michigan environmental audit law enacted in March 1996 is broad in the documentation that is rendered privileged against forced disclosure. An "audit" can include studies designed to identify "historical or current noncompliance . . . or to identify an environmental hazard, contamination, or other adverse environmental condition. . . ." The Michigan law also provides, as many other such laws do, that concealed data may include "field notes, records of observations, findings . . . photographs, computer generated and electronically recorded information, maps, charts, graphs, and surveys. . . ."[6] The Michigan law and many of the others may block government and citizen access to studies of environmental hazards, such as company studies of actual and historical contamination, dioxin and other toxic chemical emissions, and evaluations of pollution prevention opportunities.

As an example of how the laws may be abused, consider the attempt of an Ohio landfill operator to use that state's new audit law to conceal environmental data:

Waste Management, Inc., has operated a landfill in Cincinnati's lower-income Winton Hills neighborhood for 25 years. Activists have long suspected that gas leakage and other environmental hazards contributed to residents' health problems, like persistent headaches and respiratory difficulties.

In 1994, when the company wanted to build a 106-foot vertical expansion to the landfill, Communities United For Action (CUFA)—like any self-respecting community group—launched a campaign to deny the expansion and close the landfill. CUFA participated in an adjudication hearing before the Ohio EPA to urge the agency to deny a permit needed for the expansion.

The Ohio EPA in 1995 denied the permit, and the expansion was never built. Waste Management spokesperson Kathy Trent says the proposal "didn't meet the siting location restrictions," which include limits on how close a landfill can be built to a residence. Waste Management appealed the decision, and again CUFA intervened with others to close the facility. To support its case, CUFA accessed Waste Management documents, including groundwater and gas contamination information, some from corporate consultants' never-implemented environmental recommendations. "This community has had a health epidemic for the last 15 years," says Linda Briscoe, president of Winton Hills Citizen Action Association and CUFA member. "Everything we had been saying, now we have the documents to prove it."

However, Waste Management, one of the country's largest waste haulers, argued that a pending Ohio "audit privilege" law—passed in 1996—made some papers privileged information, not available to the government or citizens. Waste Management insisted the "audited" documents, including air emissions data and compliance reviews, were private, even though some dated back to 1988, well before the law was passed. In reaffirming the permit denial, still based on the siting restrictions, the Ohio EPA ruled the controversial Waste Management documents were not private and admitted the data into the record.[7]

If the same Waste Management documents had been generated *after* the law was enacted, they may well have been treated as concealable. It is most likely that many future cases will be affected by these laws and that the more protective environmental path may never be taken because of a lack of publicly available information.

A PROACTIVE RESPONSE TO THE SMOKESCREEN PROBLEM: NEW PRINCIPLES OF DISCLOSURE FOR RESPONDING TO SECRECY ON PRECAUTIONARY MATTERS

Fighting over access to information in the courts and fighting the trends such as audit privilege that can undercut the courtroom rights to such information are important parts of precautionary advocacy. If corporations become less able to conceal hazardous activities, they face stronger economic and public accountability incentives because of realistic threats of potential liability and adverse publicity. This can help to discourage activities that jeopardize health and the environment.

The Need for Early, Far-Reaching Rules on Disclosure

However, because disclosure in the courts typically comes after the damage has been done and someone has mounted the effort and resources to sue over it, judicially focused disclosure strategies must be accompanied by other, earlier disclosure requirements that allow more preventive interventions. Earlier anti-smokescreen strategies can head off the avoidance of precautionary action "at the pass."

The following are some starting principles for a public policy program using corporate disclosure to advance the Precautionary Principle. These rules should be lodged in the usual places, such as state and federal environmental rules, but are also suited to adoption in other innovative areas—corporation laws and corporate charters, corporate codes of conduct, public scorecards and evaluation tools, shareholder reporting requirements, local ordinances, and contracts (such as lawsuit settlements, good neighbor agreements, and vendor contracts).

Designating Precautionary Matters

A listing of "precautionary matters" should be developed. This can identify the types of activities and risks that trigger the Precautionary Principle. (An example of this is the International Joint Commission's targeting of persistent toxic compounds.) These will be referred to here as "precautionary matters." This targeting may be very specific—for example, a list of potentially endocrine-disrupting compounds or more general, such as a focus on all activities in the vicinity of the habitat of a specific endangered species, or all polluting activities in an area that is already overburdened with pollution. Society can also establish principles that guide public and private institutions, acting in the absence of specific listing of their activities, to know when facts before them, or activities they are engaged in, constitute a "precautionary" matter. These can include both activities that *could* pose environmental harm as well as matters for which we are sure of the detrimental impacts.

Disclosure Rules

If an activity is a precautionary matter, certain baseline disclosure rules can be applied:

- *An affirmative duty to study.* Public and private institutions can be given an affirmative duty to study the risks (uncertainties, complexities, and magnitude of harm) and to adopt and advance state of the art preventive alternatives. For instance, the book *Our Stolen Future* suggests that we "require companies selling products, especially food but also consumer goods and other potential sources of exposure, to monitor their products for contamination." The authors of that book note that the existing system run by the Food and Drug Administration (FDA) is misdirected, placing the responsibility of testing on an underfunded and understaffed body. The financial burden for this testing should be shifted to the manufacturer and distributor to fund truly independent testing and analysis, with the FDA instead responsible for oversight to ensure compliance.

- *An obligation of rapid disclosure and transparency.* The institutions that conduct such studies can be obliged to promptly disclose all data generated regarding precautionary matters. Categories of disclosure can be clarified and include as a minimum the availability and feasibility of safer alternatives, environmental and health impacts, sustainable jobs, and job-transition issues.
- *Data pooling.* Data relative to precautionary matters can be pooled via the internet for real-time access of the public and government.
- *An override of inconsistent privileges.* Certain corporate privileges of concealment can be treated as subordinate to precautionary disclosure needs. For example, attorney–client privilege, environmental audit privilege, freedom of information act exemptions, and trade secrecy would be preempted where there is an overriding need for disclosure on precautionary matters.[8]
- *Triggering action and socially accountable third-party review.* Citizen oversight boards should be empaneled to ascertain when available information triggers action requirements (e.g., restrictions, bans, or phaseouts of substances or activities) under new or existing statutory mechanisms or third-party review mechanisms, such as independent technical review processes that can distinguish genuine uncertainty from artificial and other smokescreen elements.
- *Broadening the reach of the Toxics Release Inventory.* The Toxics Release Inventory is an established model, requiring manufacturing corporations to disclose a wide array of emissions of certain listed substances to air, water, and land. This right-to-know law can be expanded to include known endocrine-disrupting and persistent organic compounds—both those emitted to air and water, and those incorporated into products sold or transferred from each facility.

Acting Locally

Although policy reforms on a state and national level are necessary, grassroots advocates can also be effective in promoting the Precautionary Principle through efforts focused on local industries. For instance, they can encourage local industries to enter good neighbor agreements—legally binding commitments made by companies to local community, labor, or environmental organizations to establish sustainable practices and launch processes for community engagement and oversight.[9] These contracts typically stem from resolutions of local citizens' challenges to companies' environmental or zoning permits, as well as from other well-organized efforts focused directly on the local industries. These agreements often reflect a "win-win" approach to problem solving on thorny environmental and economic development issues. These agreements generally have two elements:

- Concrete commitments, such as ending the use of a specific substance, or providing funding for local health institutions; and
- Process requirements, such as the establishment of citizen oversight processes and disclosure requirements.

With these arrangements, the community benefits from a safer, healthier environment, while corporations may develop positive relationships with the community, identify safer and cleaner process and product alternatives, avert lawsuits, and safeguard against potential disasters. The disclosure rules described in the public policy section, earlier in this chapter, can easily be folded into the framework of such a local agreement. For instance, an agreement could establish a framework for oversight, study, and disclosure on the use and emission of endocrine-disrupting compounds.

CONCLUSION

Disclosure is an important part of an overall strategy for promoting precaution. In our age in which decision making and policy making are driven by information technologies, including the internet, a combination of public policy changes and local advocacy can go a long way toward advancing the kinds of corporate disclosures that will promote the Precautionary Principle. The underlying theme of disclosure is the redistribution of power, the elimination of false or smokescreen uncertainties in critically important issues, and the correction of imperfect information for political and economic decision making. Strong advocacy will be needed to protect public health and the environment through a broadened concept of right to know as delineated in this chapter.

NOTES

1. Boston University Professor David Ozonoff wrote about this phenomenon in "Science for Sale," *Toxic Deception* (Center for Public Integrity, 1997). He describes how the asbestos industry has argued its case in the courts:

    Assertion: Asbestos doesn't hurt your health. Ok, it does hurt your health but it doesn't cause cancer. Ok, asbestos can cause cancer, but not our kind of asbestos. Ok, our kind of asbestos can cause cancer, but not the kind this person got. Ok, our kind of asbestos can cause cancer, but not at the doses to which this person was exposed. Ok, asbestos does cause cancer, and at this dosage, but this person got his disease from something else like smoking. Ok, he was exposed to our asbestos, and it did cause his cancer, but we did not know about the danger when we exposed him. Ok, we knew about the danger when we exposed him, but the statute of limitations has run out. Ok, the statute of limitations hasn't run out, but if we're guilty we'll go out of business and everybody will be worse off. Ok, we'll agree to go out of business, but only if you let us keep part of our company intact and only if you limit our liability for the harms we have caused.

Cited by Ralph Nader, "REAL JUNK SCIENCE," *New Solutions*, Vol. 8, No. 1, 1998, p. 33 at 38.

2. "Judge Wants Tobacco Papers Released," Associated Press, March 8, 1998.
3. *U.S. Postal Service v. Phelps Dodge Refining Corporation*, 852 F Supp. 156 (EDNY 1994). (Phelps Dodge I).
4. Phelps Dodge I at 162.
5. *U.S. Postal Service v. Phelps Dodge Refining Corporations*, 950 F Supp. 504 at (E.D. NY, 1997). (Phelps Dodge II).
6. Section 148 of the Michigan Natural Resources and Environment Protection Act, 14081(B).
7. From Margaret Littman, "Privileged and Immune," *The Neighborhood Works*, March–April 1998.
8. The book *Our Stolen Future* notes include the following:

Trade secret laws have been enacted to prevent business competitors from gaining an unfair economic advantage by adopting a company's methods without having borne the cost of product research and development. In practice, these laws are routinely used by manufacturers to deny the public access to information about the composition of their products. Since a skilled chemist can discover what a product contains, we are skeptical that trade secrets are keeping such information from business competitors determined to find out. One has to ask who is being kept in the dark by trade secret provisions, save consumers, who do not have the money to do the chemical analysis. Until manufacturers provide honest and complete labels for their products, consumers will not have the information they need to protect themselves and their families from hormonally active compounds.

9. Examples of good neighbor type agreements include the following:

• Union Oil Company (Unocal) agreed with community and environmental organizations in Rodeo, California, and provided for a safety audit supervised by local residents, new environmental emissions monitoring, and new economic commitments by Unocal such as local hiring, $4.5 million for transportation infrastructure funding, and the establishment of a $300,000 annual community benefits fund.

- Amalgamated Clothing and Textile Workers Union persuaded the management of the Sheldahl in Northfield, Minnesota, to commit (in collective bargaining) to study and implement alternatives to the use of cancer-causing methylene chloride, a pollutant of concern to the community.
- In Seguin, Texas, a steel mini-mill operated by Structural Metals, Inc. underwent public and government review to identify the best available strategies for preventing pollution.

Although good neighbor agreements (GNAs) are formed at the local level, states can also promote these agreements. For instance, a new law enacted in Iowa in 1996 provides extra consideration in subsidy programs for companies that sign GNAs with community or labor organizations. For other examples of GNAs, see information posted on the web at http://www.envirolink.org/orgs/gnp/gnas.htm or contact the Good Neighbor Project at P.O. Box 79225, Waverley, MA 02479; (617) 489-3686.

Chapter 15

~

PRACTICING THE PRINCIPLE

Richard B. Sclove
and Madeleine L. Scammell

My mother, my grandmothers, and six aunts have all had mastectomies. Seven are dead. The two who survive have just completed rounds of chemotherapy and radiation. I've had my own problems: two biopsies for breast cancer and a small tumor between my ribs diagnosed as a 'borderline malignancy' . . . I cannot prove that [my family] developed cancer from nuclear fallout in Utah. But I can't prove they didn't.

—*Terry Tempest Williams*[1]

People who inhabit and intimately know a single place day-after-day, understand things about the environmental risks they face that no outside or objective perspective can provide. Ordinary middle- and lower-class citizens (e.g., mothers, fishermen, secretaries, and industrial workers) are frequently the first to discern health problems and environmental ills. In the face of a death or loss of livelihood, practicing the Precautionary Principle—taking prudent action steps in the face of incomplete knowledge—is a survival strategy.

In a democracy, it normally goes without saying that decisions affecting all citizens should be made democratically. Decisions that involve scientific and technical complexity stand as grand exceptions to this rule. They certainly affect all citizens profoundly: The world is continuously shaped by advances in telecommunications, computers, materials science, weaponry, biotechnology, home appliances, energy production, air and ground transportation, and environmental and medical understanding. Yet policies incorporating technical complexity are customarily framed by representatives of just three groups: business, the military, and universities. These are the groups invited to testify at congressional hearings, serve on government advisory panels, and prepare influential policy studies.[2]

Institutionalizing the Precautionary Principle requires open and informed processes that involve all potentially affected parties in making and implementing responsible decisions in the face of uncertainty. In this chapter we describe two models of how the Precautionary Principle is being exercised in practice by people participating in deciding their own fate. The first model, community-based research, provides a direct, bottom-up avenue for popular engagement in setting research agendas and conducting community-driven research projects. Community-based research empowers workers, grassroots, and other community groups to take action in the face of uncertainty and institutional resistance. The second model, the consensus conference, is an innovative method for introducing informed citizen concerns into high-level and complex policy-making processes—processes in which a high degree of uncertainty about future impacts is endemic.

COMMUNITY-BASED RESEARCH

> I'd rake a bath and break out, like chicken pox. Take another and there's the pox again. I took a water sample to the health department; they said nothing's wrong with it. I thought they was good people, smarter than I was. But they wasn't.
>
> —*Victim of toxic waste poisoning in Woburn, Massachusetts*[3]

The recently popularized story of toxic waste in the town of Woburn, Massachusetts, in the United States includes a dramatic example of community-based research: Two decades ago children in Woburn were contracting leukemia at alarming rates. Other childhood disorders such as urinary tract and respiratory disease were also unusually common, as were mothers' miscarriages. The families of the leukemia victims were the first to discern a geographical pattern in the proliferation of disease.

Anne Anderson, a Woburn housewife whose son, Jimmy, had leukemia, began gathering information about other sick children on the basis of chance meetings with other victims' families and word of mouth. She theorized that the proliferation had something to do with the town water supply and asked state officials to test the water. She was rebuffed.

The affected families responded by initiating their own epidemiological research. Eventually they were able to establish the existence of a cluster of leukemia cases and then relate it to industrial carcinogens leaked into the water supply. Their civil suit against the corporations responsible for the contamination resulted in an $8 million out-of-court settlement and provided the major impetus for federal Superfund legislation that provides resources to clean up the country's worst toxic waste sites.[4]

Two key factors led to this outcome: victims and their families organized and worked together; and victims and their families were able to enlist the help of several scientists at the Harvard School of Public Health, and at other research institutions in the area, who conducted crucial research both with and on behalf of the affected families. The Woburn case is an example of what community-based research can accomplish.

In contrast with the prevailing undemocratic model of research, where expert concerns or market incentives drive research agendas, community-based research is rooted in the community serves a community and encourages participation of community members at all levels. For instance, the Woburn case involved citizens collaborating with university experts committed to helping citizens conduct research. It was an interdisciplinary effort with researchers and experts collaborating across several disciplines and with community members.

Scientists working with Woburn also took into consideration the observations and knowledge of the Woburn residents. Community-based research recognizes the indigenous expertise of communities and aims not merely to advance understanding, but also to ensure that their knowledge contributes to making a concrete and constructive difference in the world. Community-based research seeks solutions—even before scientific certainty proves a causal relationship.

In the Woburn case, it may not have been necessary for residents of Woburn to initiate their own study had the state officials in Massachusetts practiced the Precautionary Principle. By practicing the Precautionary Principle, the burden of proof would not have been on the families of leukemia victims. Were it not for the assistance of sympathetic researchers, Woburn might still be in the company of communities across the country on the front lines of toxic waste litigation where the burden of proof continues to make victims of innocent residents.

Examples of Community-Based Research and Its Practical Results

Following are three mini-case studies of community-based research and how results are used widely in the United States to empower communities, prevent harm, and effect social change.[5]

Jacksonville Community Council, Inc., Jacksonville, Florida: Assessing the Fairness of Public Service Distribution

The Jacksonville Community Council, Inc. (JCCI), is a broad-based civic organization that performs research intended to improve the quality of life in northeast Florida. Study topics are selected by a committee of JCCI members—which includes all citizens who express an interest—after soliciting input from public officials, nonprofit organizations, human service agencies, labor leaders, minority community leaders, and the public at large. Members of the JCCI and the community study the selected topics, reach consensus on key findings, compile a list of recommended solutions, and then establish a task force to promote implementation.

In the early 1990s, some residents suspected that government services were not being provided equally across Jacksonville's several distinct neighborhoods. A 1994 JCCI study examined public services in Jacksonville—including streets and drainage, parks and recreation, and police and fire services—to determine their geographic distribution and to evaluate whether needs were being met throughout the city. On the basis of its research, the JCCI recommended better communication between city functionaries and the public, more citizen involvement in decisions about the distribution of public services, improved monitoring of public service distribution, and adherence to standards of distributional fairness. These recommendations resulted in developing an annually updated "Equity Index" that measures the fairness of public service distribution in the Jacksonville area. One early result is that the Sheriff's Office implemented a new system for more equitable police patrol services.

Neighborhood Planning for Community Revitalization, Minneapolis, Minnesota: Planning to Revitalize the South East Industrial Area

Neighborhood Planning for Community Revitalization (NPCR) is a consortium of colleges and universities in Minneapolis and St. Paul, Minnesota, that works with neighborhood organizations on pressing urban issues. Community organizations apply for assistance in conducting neighborhood research projects by working closely with NPCR's project director to develop a project proposal. After NPCR approves a proposal, the community organization uses the money awarded to hire a student researcher. NPCR works closely with the organization and the student from the initiation of a research project to completion.

For example, residents and business owners in the South East Industrial Area (SEIA), just outside Minneapolis, were concerned that their area's viability was threatened by increasing pollution, overstrict zoning laws, crime, and the lack of sidewalks, bike paths, and park space. In addition, various groups affected by the SEIA had a contentious history and had not worked together for years. The SEIA community appealed to NPCR for assistance. Researchers, working jointly through NPCR and the SEIA community members, conducted a research project that established that an urban area can compete with the suburbs in quality of life and still retain industrial and heavy commercial business. As a result the city, county, and state agencies formed a steering committee to prepare a master development plan for the area. The original research project was funded by NPCR and involved 960 hours of time committed by graduate student researchers.

Childhood Cancer Research Institute, Worcester, Massachusetts: Preventing Radiation Contamination Risks in Native-American Communities

The Childhood Cancer Research Institute (CCRI) is a small nonprofit organization located at Clark University in Worcester, Massachusetts. The CCRI's mission is to prevent childhood cancer by investigating the causes of disease and educating the public on its findings. The CCRI specializes in epidemiological studies of radiation and other related causes of cancer and in promoting public participation in radiation and public health risk assessment. The CCRI staff have made particular efforts to develop long-term collaborative relationships with Native-American communities.

In 1993, the CCRI responded to the concerns of the Native-American Western Shoshone and Southern Paiute communities. In collaboration with several tribal groups and Clark University, the CCRI developed a model for working in partnership with communities to improve public health protection from environmental contaminants. The partnership prepared community exposure profiles, trained community members on matters of environmental health, strategized on nuclear hazards' management, and provided outreach to other Native-American communities. Funding for these projects came from private foundations and small programs in two federal government agencies. As a result, the Native-American communities established a sustainable infrastructure for community planning and group decision making. This provides the participating communities with a sense of ownership in the risk management process and in epidemiological and radioactive-dose-reconstruction studies conducted by government and academic researchers. This infrastructure also requires scientists to work through Native-American communities when they do research, first by securing community permission and then, often, a community's knowledgeable input.

The preceding examples attest to citizens' ability to act constructively before it is too late. Community-based research offers a tested and relatively economical means for addressing a wide variety of social, economic, and environmental problems. Insofar as communities identify potential problems, initiate research to reduce uncertainty, or formulate action plans in the face of uncertainty and take remedial or preventive action before problems develop into full-blown catastrophes, community-based research exemplifies one type of precautionary approach. It can also begin to counterbalance the undemocratic processes that currently determine most national research agendas.

CONSENSUS CONFERENCES: CITIZEN-BASED TECHNOLOGY ASSESSMENT

According to conventional wisdom, ordinary citizens are excluded from most decisions involving scientific knowledge and technology because non-experts are ill-equipped to comment on complex technical matters, and they probably wouldn't want to anyway. The earlier examples of community-based research contradict these suppositions. And the success of an innovative European process, dubbed the "consensus conference," has also begun to shed new light on the subject. Pioneered during the late 1980s by the Danish Board of Technology, a parliament agency charged with organizing technology assessment activities, the process is intended to stimulate broad and intelligent social debate on policy issues involving technical complexity and scientific uncertainty.

Not only are laypeople elevated to positions of preeminence in a consensus conference, but a carefully planned program of reading and discussion—culminating in a forum open to the public—ensures that they become well-informed prior to rendering judgment as well. Both the forum and the subsequent judgment, written up in a formal report, become a focus of intense national attention—usually at a time when the issue at hand is due to come before Parliament. Though consensus conferences are hardly meant to dictate public policy, they do give legislators some sense of where the people who elected them might stand on important questions. Involving laypeople in policy deliberations empowers them to act on the elements of the Precautionary Principle. Consensus conferences can furthermore help industry steer clear of new products or processes that are likely to cause public harm or spark public opposition.

Since 1987 the Danish Board of Technology has organized about 20 consensus conferences on topics ranging from genetic engineering to educational technology, food irradiation, air pollution, human infertility, sustainable agriculture, telecommuting, and the future of private automobiles. Ironically, the process began gaining popularity and diffusing to other nations (including to date The Netherlands, the United Kingdom, Norway, Switzerland, France, and Japan) just as the U.S. Congress abolished its Office of Technology Assessment (OTA), whose establishment in 1972 helped motivate Europeans to develop their own technology assessment agencies. But the truth is that when the OTA faced the chopping block in 1995, those rallying to its defense were primarily a small cadre of professional policy analysts or other experts who had themselves participated in OTA studies—hardly a sizable cross-section of the American public. By contrast, a consensus conference format, which engages a much wider range of people, holds the potential to build a broader constituency familiar with and supportive of technology assessment.

The Danish consensus conference is an institutionalized manifestation of the Precautionary Principle. It is a democratic process that helps ensure informed consent and that publicly examines research agendas and new technologies during their early stages, when there is normally both great uncertainty about outcomes but also relatively great latitude to steer resource and development agendas down alternative, *socially preferred* paths.

Framing the Discussion

To organize a consensus conference, the Danish Board of Technology first selects a salient topic—one that is of social concern, pertinent to upcoming parliamentary deliberations, and complex, requiring judgment on such diverse matters as ethics, disputed scientific claims, and government policy. The board has also found that topics suited to the consensus conference format should be intermediate in scope—broader than assessing the toxicity of a single chemical, for instance, but more narrow than trying to formulate a comprehensive national environmental strategy. The board then chooses a well-balanced steering committee of knowledgeable stakeholders to oversee the organization of the conference; a typical committee might include an academic scientist, an industry researcher, a trade unionist, a public-interest group representative, and a project manager from the board's own professional staff.

With the topic in hand and the steering committee on deck, the board advertises in local newspapers throughout Denmark for volunteer, layparticipants. Candidates must send in a one-page letter describing their background and their reasons for wanting to participate. From the 100 to 200 replies that it receives, the board chooses a panel of about 15 people who roughly represent the demographic breadth of the Danish population and who lack significant prior knowledge of, or a specific material interest in, the topic. Groups include homemakers, office and factory workers, and garbage collectors as well as university-educated professionals. They are not, however, intended to comprise a random scientific sample of the Danish population. After all, each panelist is literate and motivated enough to have responded in writing to a newspaper advertisement.

At the outset of a first preparatory weekend meeting, the laygroup, with the help of a skilled facilitator, discusses an expert background paper, commissioned by the board and screened by the steering committee, that maps the political terrain surrounding the chosen topic. The laygroup next begins formulating questions to be addressed during the public forum. On the basis of the laypanel's questions, the board goes on to assemble an expert panel that includes not only credentialed scientific and technical experts but also experts in ethics or social science and knowledgeable representatives of stakeholder groups such as trade unions, industry, and environmental organizations.

The laygroup then meets for a second preparatory weekend, during which members, again with the facilitator's help, discuss more background readings provided by the steering committee, refine their questions, and, if they want, suggest additions to or deletions from the expert panel. Afterward, the board finalizes selection of the expert panel and asks its members to prepare succinct oral and written responses to the laygroup's questions, expressing themselves in language that laypeople will understand.

The concluding public forum, normally a four-day event chaired by the facilitator who presided over the preparatory weekends, brings the lay and expert panels together and draws the media, members of Parliament, and interested Danish citizens. On the first day, each expert speaks for 20 to 30 minutes and then addresses follow-up questions from the laypanel and, if time allows, the audience. Afterward, the laygroup retires to discuss what they have heard. On the second day, the laygroup members publicly cross-examines the expert panel in order to fill in gaps and probe further into areas of disagreement.

Laypanel Reports

Once cross-examination has been completed, the experts and stakeholders are politely dismissed. The remainder of that day and on through the third day, the laygroup prepares its report, summarizing the issues on which it could reach consensus and identifying any remaining points of disagreement. The board provides secretarial and editing assistance, but the laypanel retains full control over the report's content. On the fourth and final day, the expert group has a brief opportunity to correct any outright misrepresentations of its testimony, but not otherwise to comment on the documents substance. Directly afterward, the laygroup presents its report at a national press conference.

Laypanel reports are typically 15 to 30 pages long, clearly reasoned, and nuanced in judgment. The report from the 1992 Danish conference on genetically engineered animals is a case in point, showing a perspective that is neither pro- nor anti-technology in any general sense. The panel expressed concern that patenting animals could deepen the risk of their being treated purely as objects. Members also feared that objectification of animals could be a step down a slippery slope toward objectification of people. Regarding the possible ecological consequences of releasing genetically altered animals into the wild, they noted that such animals could dominate or out-compete wild species or transfer unwanted characteristics to them. "Also," they wrote, "there could be risks which one is unable to foresee and therefore cannot assess." On the other hand, the group saw no appreciable ecological hazard in releasing genetically engineered cows or other large domestic animals into fenced fields and endorsed deep-freezing animal sperm cells and eggs to help preserve biodiversity.[6]

Portions of laypanel reports can be incisive and impassioned as well, especially in comparison with the circumspection and dry language that is conventional in expert policy analyzes. Having noted that the "idea of genetic normalcy, once far-fetched, is drawing close with the development of a full genetic map," a 1988 OTA study of human genome research prepared by experts concluded blandly that "concepts of what is normal will always be influenced by cultural variations."[7] In contrast, a 1989 Danish consensus panel on the same subject recalled the "frightening" eugenic programs of the 1930s and worried that "the possibility of diagnosing fetuses earlier and earlier in pregnancy in order to find genetic defects creates the risk of an unacceptable perception of man—a perception according to which we aspire to be perfect." The laygroup went on to appeal for further popular debate on the concept of normalcy. Fearing that parents might one day seek abortions upon learning that a fetus was, say, color blind or left-handed, 14 of the panel's 15 members also requested legislation that would make fetal screening for such conditions illegal under most circumstances.[8]

A Precautionary Approach to Social Issues

This central concern with social issues becomes much more likely when expert testimony is integrated with everyday citizen perspectives. For instance, while the executive summary of the OTA study on human genome research states that "the core issue" is how to divide up resources so that genome research is balanced against other kinds of biomedical and biological research,[9] the Danish consensus conference report, prepared by people whose lives are not intimately bound up in the funding dramas of university and national laboratories, opens with a succinct statement of social concerns, ethical judgments, and political recommendations. And these perspectives are integrated into virtually every succeeding page, whereas the OTA study discusses ethics in a single discrete chapter on the subject. The Danish consensus conference report concludes with a call for more school instruction in "subjects such as biology, religion, philosophy, and social science"; better popular dissemination of "immediately understandable" information about genetics; and vigorous government efforts to promote the broadest possible, popular discussion of "technological and ethical issues."[10] The corresponding OTA study does not even consider such ideas.

When the Danish laygroup did address the matter of how to divide up resources, the members differed significantly from the OTA investigators. Rather than focusing solely on balancing different kinds of biomedical and biological research against one another, they supported basic research in genetics but also called for more research on the interplay between environmental factors and generic inheritance and for more research on the social consequences of science. They challenged the quest for exotic technical fixes for disease and social problems, pointing out that many proven measures for protecting health and bettering social conditions and work environments are not being applied. Finally, the group members recommended a more "humanistic and interdisciplinary" national research portfolio that would stimulate a constructive exchange of ideas about research repercussions and permit "the soul to come along."[11] The consensus conference format enabled the Danish laypanelists, when pondering policy decisions entailing far-reaching but uncertain impacts, to express the characteristic predisposition of citizens who are treated with dignity and respect to act on principles of ethical sensitivity, caution, and prudential action.

Not that consensus conferences are better than the expert-driven OTA approach in every possible way. While less accessibly written, arid, and less attentive to social considerations, a traditional OTA report did provide more technical detail and analytic depth. But OTA-style analysis can, in principle, contribute to the consensus conference process. For example, a 1993 Dutch consensus conference on animal biotechnology used a prior OTA study as a starting point for its own more participatory inquiry.

Timeliness and Responsiveness

Once the panelists have announced their conclusions, the Danish Board of Technology exemplifies its commitment to encouraging informed discussion by publicizing them through local debates, leaflets, and videos. In the case of biotechnology, the board has subsidized more than 600 local debate meetings. The board also works to ensure that people are primed for this whirlwind of postconference activity. For example, the final four-day public forums are held in the Parliament building, where they are easily accessible to members of Parliament and the press.

Nor is it any accident that the topics addressed in consensus conferences are so often of parliamentary concern when the panelists issue their findings. The board has developed the ability to organize a conference on six months notice or less, largely for the purpose of attaining that goal. This timeliness represents yet another advantage over the way technology assessment has been handled in the United States: Relying mostly on lengthy analysis and reviews by experts and interest groups, the OTA required, on average, two years to produce a published report on a topic assigned by Congress.

In fact, one complaint leveled by the congressional Republicans who argued for eliminating the OTA was that the process it employed was mismatched to legislative timetables. Upon learning about consensus conferences and their relatively swift pace, Congressman Robert S. Walker—at the time, Republican chairman of the House Science Committee—told a March 1995 public forum that if such a process can "cut down the timeframe and give us useful information, that would be something we would be very interested in."[12] This suggests that at least some of the political resistance to practicing the Precautionary Principle might be reduced to the extent that it can be institutionalized in effective, timely, and economical ways.

The Danish Board of Technology's efforts do seem to be enhancing public understanding of technically complex issues. A 1991 study by the European Commission discovered that Danish citizens were better informed about biotechnology, a subject that several consensus conferences had addressed, than were the citizens of other European countries, and that Danes were relatively accepting of their nation's biotechnology policies as well.[13] Significantly, too, Dr. Simon Joss of the University of Westminster, who has conducted interviews on consensus conferences with Danish members of Parliament, has found the legislators to be generally appreciative of the process—indeed, to the point where several eagerly pulled down conference reports kept at hand on their office shelves.[14]

And although consensus conferences are not intended to have a direct impact on public policy, they do in some cases. For instance, conferences that were held in the late 1980s influenced the Danish Parliament to pass

legislation limiting the use of genetic screening in hiring and insurance decisions, to exclude genetically modified animals from the government's initial
biotechnology research and development program, and to prohibit food
irradiation for everything except dry spices.[15] Manufacturers are taking heed
of the reports that emerge from consensus conferences as well. According to
Professor Tarja Cronberg in reports issued by the Technical University of
Denmark, Danish industry originally revisited even the idea of establishing
the Board of Technology but has since had a change of heart. The reasons
are illuminating.

In conventional politics of technology, the public's first opportunity to
react to an innovation can occur years or even decades after crucial decisions
about the form that innovation will take have already been made. In such
a situation, the only feasible choice is between pushing the technology forward or bringing everything to a halt. And no one really wins: Pushing the
technology forward risks leaving opponents bitterly disillusioned, whereas
bringing everything to a halt can jeopardize jobs and enormous investments
of developmental money, time, and talent. The mass movements of the
1970s and 1980s that more or less derailed nuclear power are a clear example of the phenomenon.

By contrast, implementing the Precautionary Principle with early public
involvement and publicity—as a consensus conference permits—can facilitate more flexible, socially responsive research and design modifications all
along the way. This holds the potential for a fair, less adversarial, and more
economical path of technological evolution.[16] A representative of the Danish Council of Industry relates that corporations have benefited from their
nation's participatory approach to technology assessment because "product
developers have worked in a more critical environment, thus being able to
forecast some of the negative reactions and improve their products in the
early phase."[17]

For example, Novo Nordisk, a large Danish biotechnology company,
reevaluated its research and development strategies after a 1992 panel
deplored the design of animals suited to the rigors of existing agricultural systems but endorsed the use of genetic engineering to help treat incurable diseases. The firm now wants to concentrate on work more likely to win popular approval, such as animal-based production of drugs for severe human
illnesses.

CONCLUSION

At least in the abstract, we Americans are fiercely proud of our democratic
heritage and our technical prowess. But it is striking how little we do to
ensure that these twin sources of national pride are in harmony with one

another. Community-based research and consensus conferences are not
magic cures for all that ails democracy or for ensuring that science and technology become fully responsive to social concerns. But they do reawaken
hope that, even in a complex technological age, democratic principles and
procedures, incorporating a precautionary approach to decision making
under uncertainty, can prevail.

NOTES

1. Terry Tempest Williams, *Refuge: An Unnatural History of Family and Place* (New York: Pantheon Books, 1991).
2. Richard E. Sclove, "Better Approaches to Science Policy," *Science*, Vol. 279 (27 Feb. 1998), p. 1283; also available under the "Publications" section of the World Wide Web pages of the Loka Institute at <www.loka.org>.
3. Quoted in Phil Brown and Edwin J. Mikkelson, *No Safe Place: Toxic Waste, Leukemia, and Community Action* (Berkeley: University of California Press, 1990), p. 145.
4. For more information on the Woburn case, see Jonathan Harr, *A Civil Action* (Vintage Books: New York, 1996); and Brown and Mikkelson, op. cit.
5. These and additional case studies may be found in Richard S. Sclove, Madeleine L. Scammell, and Breena Holland, *Community-Based Research in the United States: An Introductory Reconnaissance, Including Twelve Organizational Case Studies and Comparison with the Dutch Science Shops and the Mainstream American Research System* (Amherst, MA: The Loka Institute, July 1998); this report is available as a free download via the World Wide Web pages of the Loka Institute at <www.loka.org>.
6. "Consensus Conference on Technological Animals: Final Document (prelimi-nary issue)" (Copenhagen: Danish Board of Technology, 1992); also available on the World Wide Web at <http://www.tekno.dk/eng/publicat/92teaneo.htm>.
7. U.S. Congress, Office of Technology Assessment, Mapping Our Genes—Genome Projects: How Big, How Fast?, OTA-BA-373 (Washington, DC: U.S. Government Printing Office, April 1988), p. 85.
8. "Consensus Conference on the Application of Knowledge Gained From Map-ping the Human Genome: Final Document" (Copenhagen: Danish Board of Technology, 1989).
9. *Mapping Our Genes*, p. 10.
10. "Consensus Conference on the Application of Knowledge Gained From Map-ping the Human Genome," pp. 28–29.
11. Ibid., pp. 7, 17–25.
12. Robert S. Walker, "Democratizing R&D Policymaking," lecture and discussion during the 10th Annual Meeting of the National Association for Science, Technology & Society, Arlington, Virginia, March 2, 1995.
13. INRA (Europe) and European Coordination Office SA/NV, Eurobarometer 35.1: Biotechnology (Brussels: European Commission; Directorate-General; Science, Research, Development; "CUBE"—Biotechnology Unit, June 1991).

14. Simon Joss, telephone interview with Richard E. Sclove, July 14, 1995.
15. Lars Kluver, "Consensus Conferences at the Danish Board of Technology," In: *Public Participation in Science: The Role of Consensus Conferences in Europe*, eds. Simon Joss and John Durant (London: Science Museum, 1995), p. 44.
16. Richard E. Sclove, *Democracy and Technology* (New York and London: Guilford Press, 1995), esp. pp. 183–184.
17. Quoted in Tarja Cronberg, "Technology Assessment in the Danish Socio-Political Context," *Technology Assessment Texts* No. 9 (Lyngby, Denmark: Unit of Technology Assessment, Technical University of Denmark, no date), p. 11.

Chapter 16

HOW MUCH INFORMATION DO WE NEED BEFORE EXERCISING PRECAUTION?

Gordon K. Durnil

How much information does a policy maker need before he or she can make a decision? The question is not intended as an academic exercise, but as a pragmatic reality check. The threshold of evidence needed to trigger a decision depends on the nature, background, and training of each policy maker. Those of us who would attempt to convince policy makers to make decisions in our favor must know and understand the ethos of each individual charged with such decisions. We need research to determine what kind of information, and how much, is needed to tip the decision-making scales.

I am neither scientist nor academic. I am a politician, a lawyer, an environmentalist, and a decision maker. The ability to make decisions, usually correct decisions, has been a talent by which I have been favorably judged over the past forty years. I gather the evidence, weigh it, and make a decision. Since I have never had the opportunity to be a dictator, my decisions require me to motivate others, usually a majority, to accept my way of thinking.

How much information do I need before making a decision? As a lawyer I sometimes need a preponderance of the evidence, but at other times I need proof beyond a reasonable doubt. As a political leader I need to hear many views on a subject before I make a decision. Sometimes the decision is that no decision is necessary, but I make decisions based on a set of facts presented to me or discovered by me, and I can usually determine when I have enough information to decide. Often my political decisions are based on what the people want or believe to be true, as exemplified by the science of randomly testing their attitudes and beliefs, because, in a government of the people, it is the view of the majority that most often moves elected leaders to action. Policy makers outside of government make decisions in similar fashion.

The typical decision-making chief executive of a large corporation calls in a trusted staff of experts and listens to what they recommend regarding the use of a particular substance or process. He likely does not invite the view of his competitor, nor does he solicit the opinions of the environmentalists who are picketing at his gate because he thinks they may not be interested in the success of his business. He studies what his trusted staff or vendors present to him. He weighs it and makes his decision. The CEO and his advisors don't want to cause harm to anyone or anything, and they sleep well at night because they think they are being cautious, but they do have a preeminent purpose. They want to make a profit (a motivation that encourages people to exercise and preserve individual freedom—a good thing). The CEO's advice probably included antagonistic information that was laced with uncertainty, while the protagonists were much more certain in their presentations, saying "the environmentalists can't prove what they say." The CEO then presents his recommendation to a board of directors who ratify or overrule his decision.

The decision-making process is more likely to be laced with tedium than it is with some heavenly revelation of "the right thing to do." We can agree or disagree with why decisions are made, but, if we are to encourage the implementation of a Precautionary Principle, we must not lose sight of how decisions are made. To affect those decisions, we need to influence the public and we need a clear message to carry to the board rooms.

In 1989, President George Bush appointed, and the U.S. Senate confirmed, me as the U.S. Chairman of the International Joint Commission (IJC), a binational organization, unique in this world.[1] The commission has three U.S. commissioners and three Canadian commissioners acting as a unitary body, not bound by the policies of their two governments. Decisions are made by consensus, with a common goal of finding the best answer to the problem studied, irrespective of the politics of the two nations, but other nations resist the idea of a binational watchdog assessing their actions. In my speech to the plenary session of the United Nations Conference on Environment and Development in Rio de Janeiro a few years ago, I pointed out a simple fact—no other nations work as collegially in transboundary matters as do the United States and Canada. I noted that our two governments, especially around the Great Lakes, had done a pretty good job of dealing with basic environmental concerns. But in listening to world leaders in Rio, I found that the rest of the world, the so-called developing nations, are still concerned about the basics—such as population, poverty, and sewage—while in the industrialized world, our concerns are now more subtle and quite likely more serious over the long term.

Beginning in 1989, I found myself enveloped in international scientific discussions about the dangers of onerous substances being discharged into our air, land, and water. I had heard about cancer as an end-point effect of pollution, but I didn't know if I believed it. I hadn't heard about persistent toxic substances, nor had I heard about subtle human health effects such as learning disabilities, reproductive problems, and immune suppression in new born babies. The IJC gathered volunteer scientists from both the United States and Canada—from government, industry, environmental groups, and academia—to serve in their personal capacities, not bound by the views of their employers. Those scientists from diverse backgrounds and disciplines, from the Gulf of Maine to the Juan de Fuca Straits, convinced me that we now know enough about the subtle adverse human health effects of such onerous discharges to begin virtually eliminating their use.

The United States and Canada came to that same conclusion a decade earlier in their Great Lakes Water Quality Agreement.[2] My initial reaction was to examine how governments and industry were dealing with the possibility that some environmental practices might be putting our children at risk. What I found was a system so disingenuous that it spurred me on to further interest in the realm of environmental protection. I found a system where government, industry, academia, and even hard-core environmentalists had come to agreement on the thought that we humans can manage the unmanageable through permits and regulation. I found a system where governments stocked fish into the Great Lakes and then issued warnings to humans not to eat those very same fish for fear of adverse health effects. I found a system that only rarely practiced prevention as a tool for environmental protection, a system that called waste reduction "pollution prevention" and a system that promoted recycling as the epitome of environmental stewardship. I found a system that was adverse to priority setting and a system that used scientific uncertainty as proof that no harm was possible. Ironically, one of the critical elements in establishing the need for a Precautionary Principle—scientific uncertainty—is also a prime tool for those who believe the principle is unnecessary. The antagonists say, and many in the public believe, if you can't prove it is true, then it is not true.

In its *Fifth Biennial Report on Great Lakes Water Quality*,[3] the IJC recognized that solutions to many environmental problems may not always be found in government and it encouraged a stronger participation by individuals, businesses, and nongovernmental organizations. We once again recommended a coordinated strategy for the virtual elimination of persistent toxic substances, and we said that an essential part of that strategy was preventive measures to keep new, harmful chemicals from entering the marketplace through the principle of reverse onus, "That is, when approval is sought for the manufacture, use or discharge of any substance which will or may enter the environment, the applicant must prove, as a general rule, that the substance is not harmful to the environment or human health."

In that same 1990 report, the IJC recommended to the governments of the United States and Canada a concept maybe even more important than reverse onus: "To raise the level of knowledge among the general public about the importance of a clean environment and what individuals can do to prevent, avoid and remediate degradation of the ecosystem, the Commission again recommends the Parties prepare and urge the use of a comprehensive public information and education program." My thought then, as it is now, was that the small percentage of our population who indulge themselves in environmental discussions tend to presume that the entire population is as well informed as they are. That is a false presumption. We make little progress when we tell people to change a process, an industrial feedstock, or a lifestyle, without clearly informing them why they should do so.

If we are to convince humans that they need to exercise precaution, we must give them reasons. As hard as it is for many of us to accept, we have not yet crossed the awareness threshold and we have credibility problems. When we present evidence that a discharge might be causing an adverse health effect, many think we have nonenvironmental motives. When a child comes home from school fully cognizant about the intricacies of recycling, but deficient in the 3 Rs, the parents suspect an environmentalist conspiracy. There is a cultural gulf between people on opposites sides of the environmental debate that gets in the way of success. In the privacy of their minds they think, people like me have difficulty dealing with people like you, and vice versa.

My audiences are quite diverse, from the very conservative to the very liberal, and one thing has become quite clear—neither side of the environmental debate believes that the other side is sincere. It is a problem that needs solving. It's a threshold we must cross, if precaution is our goal. Too often Republicans believe that all environmental claims are false, and too often Democrats believe all such claims are true. Partisanship gets in the way and one thing is certain—as long as the environmental issue is the province of but one political party, neither political party will give it the attention it deserves. So we must broaden our discussion if we are to implement the Precautionary Principle. We cannot restrict our debate to the technical arena. We must enter the political arena where policy decisions are made, and we need to improve our public relations.

In August of 1997, I attended my first Republican meeting, ever, where the environment was formally on the agenda. I moderated a panel at the Midwestern Republican Leadership Conference on how Republicans handle or mishandle the environmental issue. On the panel were leaders of two Republican environmental organizations—REP America and the Coalition of Republican Environmental Activists—neither of which existed two years ago when I published my book *The Making of a Conservative Environmentalist*.[4] Hundreds of Republican leaders from the thirteen midwestern states attended and participated in that session. I talked about the normal political rhetoric heard at most Republican meetings, such as the breakup of the family—but then I talked about adult wildlife abandoning their young. We talked about our educational deficits and learning problems, and then I mentioned the dulling syndrome in wildlife around the Great Lakes. We talked about anxiety and behavior problems among young humans and other youthful mammals, and we talked about female birds trying to nest with other female birds. We also clarified that human health is an environmental issue of prime importance. That might sound a little silly in this forum, but few people think of health when the word environment is mentioned. We talked about "bad science" as point of view or as a dilatory tactic for lobbyists. We broadened the discussion, so we can now suggest that we have enough evidence to exercise precaution, among a group of Republican leaders who would have previously simply believed those antagonists who charge that environmentalists promote bad science and are "a bunch of anti-free market liberals." And we discussed public opinion surveys that indicate that Republicans lose female votes over environmental issues.

It is important to bring policy makers into our discussions, but we must do more. To get the attention of policy makers we must prioritize the where, what, why, and when of how we would apply the Precautionary Principle. If we tell a decision maker that we want the principle to apply everywhere as soon as possible, we will lose the attention of the policy maker. We can bemoan the state of our society, but we also need to recognize that a good message is not good enough in today's society. We need good messengers.

To get from science to policy, the science must be communicated, and scientists tend not to be good messengers. Scientists need to come forward, not only with what they know as technicians, but with what they have concluded based on a great deal of experience.

We need to think about how scientists are trained and how they practice their profession. The average person believes that scientific truth exists for most questions. They believe science can be as absolute as math. They believe that scientists are broad thinkers, unfettered in their search for the truth. But we all know that with most research grants comes a framework that often limits thinking. The Toronto City Council was considering a restriction on the purchase of PVC pipe, and after I made my presentation, a company scientist leaned over to tell me that he agreed with what I said, but that he must promote the opposing view. Recently, a new study indicated that PCBs and DDT were not factors in breast cancers. It contradicted another recent study that reported just the opposite, and the very next day another study said that the breast cancer culprit was weight. Obviously, when scientists come to opposite conclusions about the same subject, the public sees no need for applying precaution. So we must discuss the contemporary practice of science as a barrier to implementing the Precautionary Principle. We need to decide if our goal is defending the way we do things or preventing adverse health effects.

As a decision maker I want to hear all sides of an issue, weigh the evidence, and then make the decision. Those who would have me decide in their favor should be focused on an identifiable goal. They should be able to clearly enunciate what it is they want to accomplish. Environmentalists tend to be of a different mind. Environmentalists are a diverse group, so diverse it often appears we don't have a clear goal. Too often we give the perception that we believe all human actions are bad, which is a losing proposition for our concerns, and the "Chicken Little" tactic is an unqualified barrier to environmental success. Some of us can get as excited about healthy trees as we do about healthy humans, we don't see a conflict, but that can be a confusing conundrum for decision makers. For a decision maker to conclude that the Precautionary Principle is needed, the science should be clear, the communication must be clear, and the priorities should be obvious.

The agreement by the governments of the United States and Canada in the Great Lakes Water Quality Agreement mandated that: The discharge of toxic substances in dangerous amounts be prohibited and the discharge of any or all persistent toxic substances be virtually eliminated.[5] It was the formal proclamation of a Precautionary Principle in 1978, but years have passed and not one such substance has been virtually eliminated. So, in 1992, in it's

Sixth Biennial Report on Great Lakes Water Quality,[6] the IJC tried to refine the message and make it more specific. We advocated a weight-of-evidence approach to the identification of substances that may cause harm. We found disingenuous the prevailing thought that science must prove with 100 percent certainty that an exposure would result in an adverse health effect before pulling the trigger on precaution, and we expressed our belief that informed decisions could be made after weighing the available evidence. We said, "The focus must be on preventing the generation of persistent toxic substances in the first place, rather than trying to control their use, release and disposal after they are produced."

To put the principle of prevention into play, we advocated a system of sunseting a prioritized list of toxic substances. We defined sunseting as "a comprehensive process to restrict, phase out and eventually ban the manufacture, generation, use, transport, storage, discharge and disposal of a persistent toxic substance." We knew that some substances would require an orderly sunseting process over a long period of time to protect workers and avoid chaos, while others could be sunset rather quickly. On our list for sunseting, we included PCBs, DDT, dieldrin, toxaphene, mirex, and hexachlorobenzene. We also suggested, in consultation with industry and others, that production process and feedstock chemicals be altered so that dioxins, furans, and hexachlorobenzene would no longer result as by-products. We also recommended that the governments "review the use and disposal practices for lead and mercury, and sunset their use wherever possible." Then, because many of the worrisome substances were synthetic chlorinated organics we decided that we could not seriously come to grips with toxics unless we took a good hard look at how society uses chlorine. We recommended that governments and industry work together to develop an orderly and reasonable timetable to sunset the use of chlorine as an industrial feedstock.

The aftereffects of the chlorine recommendation were interesting. For more information, see *The Making of a Conservative Environmentalist.*[7] Lobbyists came after us politically conservative commissioners with all guns blazing, calling us liberals, saying our science was bad, as they emulated tobacco lobbyists in their exercise of the three Ds—deny, divert, and delay. They denied what we said was true. They said their science didn't show what ours did, even though we often discovered that they had no health effects science. They diverted attention from what we said by calling us names, and they delayed for as long as possible any interference with their normal course of business. And some of my friends believed what the lobbyists said, thinking I had become a liberal. But most often my friends would simply lose interest when the word "environment" was mentioned. Many good friends were reluctant to read my book because the word "environmentalist" was in the title, thinking that the subject matter might be too difficult, too arcane, or too inane. The stimulating interest these friends have in almost every sub-ject matter is somehow automatically lost when the word "environment" is interjected into the discussion, and I fear they represent the many, not the few.

When I travel my state in political campaigns and ask people about major problems they might have with state government, the Indiana Department of Environmental Management always tops the list of horrendous stories. The unreasonableness of bureaucrats obsessed with finding fault, instead of protecting the environment, creates a lot of anger. But such anger at envi-ronmental harassment shouldn't revolve around the existence of real prob-lems, because recognizing the existence of a problem, environmental or not, is not the focus of a political philosophy. If you and I see a broken chair, should I sit down anyway and say that my political philosophy won't allow me to accept the fact that the chair is broken? Sounds silly, I know, but when it comes to matters environmental, broad-based problem recognition is still a major hurdle. How problems are handled, however, is a matter for philo-sophical debate (e.g., governmental cleanups, the polluter pays). Too many of us want to believe an environmental problem is real or false on the basis of what we're told by the definers of our political philosophy. It seems that much of the businessperson's displeasure with environmental dictates has to do with the failure of environmental agencies to set priorities or to provide timely factual information on which to act. Citizen displeasure over envi-ronmental strictures often has to do with governments that declare some-thing too evil for usage, but that sells permits to authorize the use of the evil substance. And Republican anger often has to do with the nonenvironmen-tal partisan political philosophies espoused by those who have traditionally led the environmental debate.

Albert Einstein once said, "Everything has changed, soured our way of thinking." That fitting quote found its way to the cover of the *Seventh Bien-nial Report on Great Lakes Water Quality*,[8] wherein the IJC said, "Precaution in the introduction and continued use of chemical substances in commerce is a basic underpinning of the proposed virtual elimination strategy. It is gen-erally agreed, in principle, that the burden of proof concerning the 'safety' of chemicals should lie with the proponent for the manufacture, import or use of at least substances new to commerce in Canada and the United States, rather than with society as a whole to provide absolute proof of adverse impacts." We also said, "This principle should in the Commission's view, be adopted for all human-made chemicals shown or reasonably suspected to be persistent and toxic, including those already manufactured or otherwise in commerce. The onus should be on the producers and users of any suspected persistent toxic substance to prove that it is, in fact, both 'safe' and necessary, even if it is already in commerce."

If we are to accomplish our goal of implementing the Precautionary Principle, however, we must be prepared to recognize that society is not yet convinced of the need to alter life styles, change processes, or alter policy. Society has not yet come to the majority view that some personal sacrifice is needed, primarily because the environmental message is delivered from such confusing and diverse perspectives. Environmental claims still seem far fetched to many. Reporters give us options in their environmental stories— believe or don't believe, either is okay. Scientists love to debunk the findings of their colleagues.[9] The medical profession is geared toward curing and not toward prevention, and environmentalists in support of each other's pet project present confusing and diverse messages. Somehow, we must consider how we get to a point of some agreement on where precaution is needed, where priorities can be set, and where rationality can prevail over the claims of the lobbyists who say bad science and a desire to destroy capitalism are at the root of environmentalist claims. If we advocate the exercise of a principle that requires sacrifice without making the case for its need, we are wasting time and effort.

I have my best luck in attracting support for precaution when I talk about breast cancer. I tell my audiences that when my wife and I were married, one woman in twenty contracted breast cancer. When my wife had her double mastectomy, it was one woman in twelve. Now it's one woman in eight, but we still formally declare that we don't know what causes breast cancer nor do we know how to cure it. We brag that we have figured out how to keep more survivors alive for a longer time, but we dedicate most of our efforts toward looking for a cure after the woman has suffered through the mental and physical pain. After that simple example, and after telling them that only one penny out of each cancer research dollar is spent on prevention, the audience is usually ready for a discussion about how to prevent some breast cancers in the next generation of girls.

The public will support preventive measures to environmental problems, but only if we explain the problem and convince them of its dangers to them. My experience with public opinion surveys since 1967—when the sample is truly random, the questions properly framed, and the interviewers professionally trained—is that the results can be relied on as of the day they are taken. Surveys give us some good news relative to public attitudes about the environment, but only if we offer potential answers with a menu style of questioning instead of open-ended questions. If we ask if we can have a clean environment and economic growth at the same time, 67 percent say yes (as opposed to 50 percent in 1978).[10] On the other hand, nearly 90 percent say they have never considered contacting an elected official about an environmental matter.[11] The survey questions that move policy makers, however, are more often those that reflect the off-the-cuff thinking of the public, such as, "What is the most important problem facing your community?" On that question, the environment has traditionally been in the 1 percent range. In a recent Indiana survey, it was 0.7 percent. In those off-the-cuff citizen responses, the environment shrinks to the level of unconcern when compared to issues such as education, crime, taxes, the decline of morality, health care costs, drug abuse, government corruption, economic matters, the way young people think and act, government debt/budget deficit, welfare, immigration, medicare, and other senior issues. All those issues, and more, reveal greater public concern and interest than do environmental issues.

I have watched those interested in such ideas as the Precautionary Principle give the matter great thought and then relate that thought to mid-level persons in government, hoping that government will change. I have observed the environmental community debate among themselves and come to full agreement on various proposals, but somehow fail to get their message to policy makers in a fashion that causes policy makers to sit up and take notice. On the other hand, I notice that lobbyists for business and labor spend very little time with mid-level governmental workers or with nongovernmental organizations interested in change. Instead they employ two primary tactics—they communicate with policy makers and they prod the public to demand action from their elected officials. If we are to convince policy makers that the Precautionary Principle should be implemented, then we need to adopt proven tactics that can generate positive results.

Do we have sufficient evidence to warrant the implementation of the Precautionary Principle? What is that evidence? Is scientific uncertainty a sufficient reason to implement a Precautionary Principle, or is it proof that no such principle is needed? What will happen if we don't implement a Precautionary Principle? As we enter into the deeper intellectual aspects of this volume, let's not presume that the American people think these questions have been answered satisfactorily.

NOTES

1. International Joint Commission—United States and Canada, created by The Boundary Waters Treaty of 1909 between the United States and Canada.

2. Great Lakes Water Quality Agreement with Annexes and Terms of Reference between the United States and Canada signed at Ottawa, November 22, 1978 as amended by Protocol, 1987.
3. International Joint Commission, *Fifth Biennial Report Under the Great Lakes Water Quality Agreement of 1978 to the Governments of the United States and Canada*, 1990.
4. Gordon K. Durnil, *The Making of a Conservative Environmentalist* (Indianapolis: Indiana University Press, 1995).
5. Great Lakes Water Quality Agreement.
6. International Joint Commission, *Sixth Biennial Report Under the Great Lakes Water Quality Agreement of 1978 to the Governments of the United States and Canada*, 1992.
7. Durnil, *The Making of a Conservative Environmentalist*.
8. International Joint Commission, *Seventh Biennial Report Under the Great Lakes Water Quality Agreement of 1978 to the Governments of the United States and Canada*, 1994.
9. Gordon K. Durnil, *Is America Beyond Reform?* (Bend, OR: Sligo Press, 1997).
10. Cambridge Reports/Research International, September 1994, for *Times Mirror Magazine*.
11. Roper Organization and Cantril Research, Inc., for the President's Council on Environmental Quality, December 1990.

Part IV

The IV

THE PRECAUTIONARY PRINCIPLE IN ACTION

The chapters in this part all apply the Precautionary Principle to a specific problem. The authors vividly describe what precautionary action would be when applied to agriculture, manufacturing, gasoline additives, and endocrine disruption. This section includes discussions on a broad array of issues in order to think through problems beyond toxic chemicals.

As far as we know, Fred Kirschenmann's chapter is the first to apply the Precautionary Principle to agriculture, one of the most destructive of all human undertakings, particularly when it is globalized and industrialized. Kirschenmann asks a question that is key to the Precautionary Principle: Can we say "yes" to human industry, including producing food? It will seem to naysayers that the Precautionary Principle stops all activity. Kirschenmann takes one of the absolute essentials of human life and thinks through how we might eat without destroying the natural world.

Chapters 18 and 19 should be read together because they illuminate the issues around gasoline additives and opportunities (won and lost) to apply the Precautionary Principle. Peter Montague tells the story of tetraethyl lead and how close the United States came to actually applying the Precautionary Principle in the 1920s. He explains the consequences of that decision through to the late 1980s. Ted Schettler reviews the status of oil companies' efforts to add manganese to gasoline. Manganese is a necessary trace element for humans but too much can be toxic. The question is, Have we learned anything from the experience of tetraethyl lead and the human health consequences?

In chapter 20, Ken Geiser wrestles with clean production techniques in manufacturing and whether they are truly precautionary and how to make them part of precautionary action. He urges us not to make the glib arguments that something that is good for the environment is really good for the financial bottom line. Perhaps most interesting is that consumers, not just producers, are challenged to adopt the Precautionary Principle. What is precautionary clean production and what is precautionary consumption?

The final chapter in this part ties many themes together. Peter deFur's examination of endocrine disruption raises the disturbing idea that humans have so destroyed the environment that the only possibility is to apply the Precautionary Principle: There is no turning back. DeFur explains the problem of endocrine disruption—how we know and what we know. He goes on to apply the Precautionary Principle and a handful of subsidiary principles on the basis that the scientific uncertainty around endocrine disruption is vast, but the threat looms even larger.

Chapter 17

CAN WE SAY "YES" TO AGRICULTURE USING THE PRECAUTIONARY PRINCIPLE: A FARMER'S PERSPECTIVE

Frederick Kirschenmann

This chapter asks the question: Can human industry and ecological health co-exist? Indeed it asks whether humans can live in their ecological neighborhoods[1] in ways that can *improve* the health of those neighborhoods. It asks that question primarily with respect to agriculture.

Agriculture has been chosen as the focus of this inquiry for four reasons:

- It may have been agriculture that shaped our industrial ethic that largely determines our present attitude toward nature.
- Agriculture clearly has been a major cause of environmental degradation, almost from its inception.
- The production of food is different from the production of any other goods. Humans need food to survive, and agriculture is essential for food production.
- The author is a farmer and has been struggling with this question on his own farm for over twenty years.

Niles Eldredge, evolutionary biologist with the Museum of Natural History, argues that it may have been agriculture, invented some 10,000 years ago, that started us down the path of believing that we could all "step out of" our local ecosystems. He believes that the practice of domesticating certain

279

species of plants and animals for food led us to conclude that we could sur-vive and solve all of our problems through human cleverness without the aid of nature. "We told Mother Nature that we didn't need her anymore; that we could take care of ourselves" (Eldredge, 1995, 93).

This sense of having been emancipated from our bondedness with our local ecological neighborhoods led us to believe that we could bend nature to our will and design systems of human industry that could transcend the diversity, complexity, and native nuances of local ecologies.

Eventually that led us to adopt an agriculture that mirrors the industrial paradigm. The principle features of that paradigm are specialization, rou-tinization, and control (Boehlje, 1995). This system of agriculture runs farms like factories, relying heavily on fossil fuel energy and other exogenous inputs to achieve its goals. It tends to ignore both the resources and the lim-itations of local ecological neighborhoods.

In a brief booklet published by the U.S. Department of Agriculture's Soil Conservation Service (SCS) over forty years ago, W.C. Lowdermilk (for-merly assistant chief of the SCS) asserted that destructive agricultural prac-tices helped to topple empires and wipe out entire civilizations, for over 7,000 years. Soil erosion, deforestation, and overgrazing all contributed to the demise of civilizations (Lowdermilk, 1953). Dr. Lowdermilk came to these conclusions based on a personal study (conducted in 1938 and 1939) of the regions in question. Ecological destruction, caused by agriculture, is not a new phenomenon.

At the same time Lowdermilk's study revealed that some farming prac-tices enabled societies to flourish. He discovered fields that had been farmed for thousands of years without soil or environmental deterioration.

AGRICULTURE AND THE PRECAUTIONARY PRINCIPLE

Since agricultural practices have caused both degradation and stability, and since agriculture is an activity that is essential to human survival, agriculture is a good place to test the application of the Precautionary Principle.

At its heart, the Precautionary Principle proposes that we should act to protect the health of our ecological neighborhoods *in advance* of scientific certainty. In other words, we should act to avoid risk, rather than external-izing risk or weighing risk against the benefits.

The Precautionary Principle has been described in two different formats. The Rio Declaration uses the principle in a reactive manner. Scientific uncertainty should not *prevent* action to protect the environment. More recently the principle has been stated proactively. We *should* act to protect the environment before scientific proof of harm can be established. Using this principle, what are the benchmarks that we can use to determine when we can say "yes" to human industry? At least four benchmarks come to mind.

- We could say "yes" when we can demonstrate that the magnitude of potential harm is limited. If the effect of the introduction of a material or practice could last for generations—if it doesn't disappear in one generation—we should say "no." If the potential harm is limited to one generation, we may say "yes."
- If the geography of potential harm is limited, we could say "yes." The larger the area affected by the introduction of a material or practice, the more precaution we must use. For example, if toxins to be introduced are air borne, water borne, bioaccumulative, or in other ways ubiquitous to the environment, we should say "no."
- If the biology of potential harm is limited, we could say "yes." Since all species in any ecological neighborhood have coevolved, the Precautionary Principle must take the welfare of all species, not just the human species, into consideration. If the introduction of a material or practice potentially threatens harm to the stability and integrity of the biotic community, we should say "no."
- If the social cost is limited, we could say "yes." Often we say "yes" to the introduction of materials or practices because the short-term economic gain is attractive, or because of the potential for economic gain for one sector of society. But sometimes these gains are not weighed against the long-term economic costs. If the introduction of a material or practice compromises future economic well-being or is achieved in one sector of society at the expense of another sector, we should say "no."

These benchmarks are useful for determining when to say "yes" to agriculture. Indeed, they may be essential to the survival of industrial agriculture. In fact, failure to use such benchmarks invites the collapse of industrial agriculture.

Derrick Jensen gives us a graphic example of such impending collapse in a brief article in the October 13, 1996, issue of the Sunday *New York Times Magazine*. The article, entitled "Hush of the Hives," provides a poignant picture of what happens when agriculture ignores precautionary benchmarks.

Jensen called attention to the well-known fact that European honey bees throughout this country are dying—80 percent loss in Maine, 55–75 percent loss in Massachusetts, 60 percent loss in Michigan. But Jensen tells us it is pointless to blame the Varroa mite or the beekeeper who inadvertently brought the mite into the United States when he smuggled honeybee queens in from South America or Europe hoping that they would increase his honey production. Because, says Jensen, "The collapse was inevitable anyway."

In February the hills surrounding Modesto, Calif., roll with white-blossomed almond trees. Although mono-cropped miles of almond flowers may be beautiful they're as unnatural as Franken-stien's monster; the staggering number of blooms to be pollinated grossly overmatches the capacity of wild pollinators like bumble-bees, moths, wasps and beetles to set fruit, causing almond ranch-ers to pay distant beekeepers up to $35 per hive to bring in bees for the four-week bloom.

Almonds aren't the only crop needing pollination. Apples, cherries, pears, raspberries, cranberries, blueberries, cucumbers, watermelons—each of these densely packed crops requires simi-larly densely packed beehives to set fruit (Jensen, 1996).

When we impose a specialized, monocrop agriculture onto diverse, local ecosystems without attending to precautionary benchmarks, we create brit-tle ecologies that are vulnerable to something as tiny as a mite. Whether it's a fruit-setting monoculture on the West coast, a cereal grain monoculture in the Northern Plains, the genetic uniformity of a single variety of potatoes in Ireland, or a single strain of hi-bred seed corn in the corn belt—we know that specialization, ignores interdependent biotic communities of local ecosystems, invites collapse.

Because farmers have ignored precautionary benchmarks, they have rou-tinely destroyed valuable production resources before they recognized the value of those resources for their own farming operations.

Steve Buchmann and Gary Nabhan give us some poignant examples of such destruction in *The Forgotten Pollinators*. They point out that pollinators that are native to local watersheds are often more efficient than imported European honeybees.

While honeybees can pollinate alfalfa flowers, they have no predilection for "tripping" these complex legume blossoms—that is they seldom "unhinge" the keel and wing petals held under tension to release the stamens and stigma protected within. The alkali bee, though, is a master at this maneuver, busily tripping more flowers per legume than an individual plant can mature as ripe fruits (Buchmann and Nabhan, 1996, 186).

Nabhan and Buchmann go on to point out that in the Great Basin older Idaho farmers discovered this ecosystem-service of the alkali bee and actively cultivated their nesting habitat. As a result, their alfalfa seed yields jumped from 300–600 pounds per acre to well over 1,000 pounds per acre.

But later, the increasing demand for seed and the evolution of a genera-tion of farmers who did not understand the connection between the alkali bees and high alfalfa yields led farmers to plow out the highly alkaline soils (to plant more alfalfa) that had provided the habitat for the alkali bees. Sub-sequently dieldrin and parathion, which are very toxic to alkali bees, were also applied to alfalfa crops. These practices resulted in the disappearance of the alkali bees and a dramatic loss of income from alfalfa seed production.

SUSTAINABILITY VERSUS REGENERATION

The precautionary benchmarks of not doing too much harm are still re-active in the sense that *they attempt to determine what we can get away with, rather than asking how we might use human industry to restore the health of our ecological neighborhoods.* This point is critical and perhaps suggests the crucial focus where we need to make a fundamental shift in our thinking. Usually when we think about making the environment safer, we are thinking of ways to "green up" our technologies and practices without reconsidering our place in the biotic communities in which we evolved.

A more positive and effective way of implementing the Precautionary Principle with respect to agriculture would be to hook agriculture into the evolutionary process of nature and make better use of the ecosystem services that nature provides. In other words, to practice agroecology[2] (Altieri, 1987).

Agriculture provides us with some interesting examples of this way of thinking about the Precautionary Principle. When green manure, legumi-nous cover crops, are used as a source of nitrogen, rather than purchasing industrial nitrogen, it not only reduces soil erosion and enables the farmer to have better control of her costs, it is also less likely to overload the environ-ment with nitrogen and more likely to prevent nitrate leeching, and the decline of species richness, as well as improve soil quality. When a diversity of crops is grown in a diverse crop rotation that is compatible with the eco-logical neighborhood in which the farm exists, rather than a single crop monoculture, similar dual benefits become evident. Not only does such diversification reduce the farmer's risks, it also provides diverse habitat for wildlife, creates opportunities for nutrient cycling, and dramatically reduces crop loss due to insects, disease, and weeds. Consequently, it also reduces the need for excessive amounts of pesticides.

Biointensive Integrated Pest Management practices have demonstrated that such a shift in thinking can also be beneficial (both environmentally and economically) in industrial agriculture systems. And farmers who have taken this initial step into the world of ecological farming sometimes find that the leap into a full-blown agroecology paradigm seems less foreboding.

When we extend this model of ecological farming to the watershed level and begin redesigning watershed landscapes to restore native plants in hedgerows or other field boundaries as well as uncultivated areas, and fold agricultural fields into that landscape, the resultant habitat can invite the return of many native species that could simultaneously restore the ecological health of the watershed and enhance agricultural production.[3]

NEW DIRECTIONS

Shifting agriculture from an industrial to an ecological paradigm, and adopting farming practices that make use of the ecosystem services of local watersheds, will require some fundamental changes in the way we do agricultural research and in the way we approach farming practices.

The research model suggested by Raoul Robinson may be better suited to these farming methods than the research designs that have become common in the industrial paradigm. Robinson proposes the development of "farmer clubs," consisting of farmers and researchers, working together to breed plants with diverse genetic mosaics to fit into diverse cropping systems that are suited to specific ecological neighborhoods (Robinson, 1996). One can easily imagine farmers and researchers working together in such clubs also learning how to adapt ecosystem services in local watersheds to farming goals.

Farming in concert with the ecological processes of local watersheds will also mean that we will need to shift from exclusively reductionist research (designed to produce single, linear effects) to contextual research that attempts to understand the complex interactions of organisms in their watershed environments. This will mean that we need to view both farm plants and animals *and* research designs quite differently. Craig Holdrege implies, for example, that we will no longer be able to think of a cow as a bioreactor that we can induce to produce more milk by tweaking some single part of the cow's system, assuming it has no impact on the rest of the cow, the farm, or the watershed (Holdrege, 1996). The fact that we can still assume that increasing a cow's milk production by one quart per day without affecting the rest of the organism, when an additional 300 to 500 quarts of blood must flow through the udder to produce that extra quart of milk, suggests just how far we have drifted from any kind of contextual thinking.

Ecological farming will also change the way we view our farms. Pete G. Kevan and his colleagues suggest that there are at least four "ecological processes in agriculture systems" that are important to understanding "agriculture as applied ecology." These ecological processes are no different in agriculture than they are in any other ecosystem, but they are an important part of understanding how we need to change our perceptions about farming practices if we are going to make creative, pro-active use of the Precautionary Principle (Kevan et al., 1997).

The four processes identified by Kevan are nutrient cycling, competition, symbiosis, and succession.

- *Nutrient cycling.* Industrial agriculture is basically an input–output system. Nutrients are brought into the farm to meet the production output goals of the farm. In ecological systems, nutrients are recycled within the system and the goal is to "reduce the deficit between exports and imports by using intrinsic recycling . . ." and reducing dependence on exogenous inputs.
- *Competition.* Ecological agriculture views competition quite differently from industrial agriculture. Industrial agriculture seeks to *eliminate* competition. Ecological agriculture seeks to understand the complex interrelationship of all organisms in the ecosystem to determine how they can be made beneficial and generally only seeks to *limit* (not eliminate) the competition of those organisms that threaten the farming operation.
- *Symbiosis.* Many organisms that co-evolve in a local ecosystem are mutually beneficial since they have adapted to one another over their long evolutionary journey. Recognizing and enhancing mutualisms is one of the most fruitful areas of exploration for ecological agriculture practices at the watershed level. Soil microbes may yield especially rich resources for ecological agriculture in this regard. Since industrial agriculture seeks to impose a control system onto the watershed, it has little inclination to seek out or understand such mutualisms.
- *Succession.* Managing succession appropriately presents ecological agriculture with one of its most interesting challenges. In nature, succession tends to reach a climax in which biotic communities achieve a dynamic stability. Such stability gives local ecological neighborhoods their character as prairie ecologies, woodland ecologies, and alpine tundra ecologies, for example.

All agriculture tends to disrupt succession and hold it at an early stage. This is one of the principle reasons that weeds are a problem for farmers. Industrial farmers try to hold succession at an early stage with exogenous inputs. Ecological farmers use various strategies to *mirror* succession. Most prevalent have been crop rotations, orchard floor management, slash-and-burn agriculture, and perennial cropping systems. Perhaps one of the most elaborate research experiments designed to overcome the problem of succession inter-ruption is the work carried out at the Land Institute in Salina, Kansas, where efforts are underway to replace annual cereal grains with perennial polycul-tures well adapted to prairie ecologies.

In all of these strategies, the goal is to fit agriculture into the watershed, rather than to impose an alien agriculture onto the natural ecology of the watershed. Ecological agriculture recognizes that the biotic community that has evolved in each ecological neighborhood is a rich resource for farming, and that therefore it just makes good farming sense to preserve its stability and integrity.

For ecological farmers, therefore, Aldo Leopold's ecological standard for a land ethic, is much more than an ethical imperative—it is a practical neces-sity. "A thing is right when it tends to preserve the integrity, stability, and beauty of the biotic community. It is wrong when it tends otherwise" (Leopold, 1949). Perhaps that is another way to proactively state the Pre-cautionary Principle.

BEYOND AGRICULTURE

Does the way we might say "yes" to agriculture have implications for other sectors of human industry? Very likely.

William McDonough, world renowned eco-architect, for example, uses principles for the design and construction of buildings for human shelter that are similar to the principles for agriculture suggested earlier. Since shelter, like food, is essential to human survival, this is another important sector of human industry for testing the viability of the Precautionary Principle.

McDonough proposes an overarching ecological standard for the design and construction of shelter.

> . . . the things we make must not only rise from the ground but return to it, soil to soil, water to water, so everything that is received from the earth can be freely given back without causing harm to any living system (McDonough, 1996, 71).

Applying this principle McDonough suggests three "defining characteris-tics" that should guide the design and construction of human shelter: (1) All waste must become food. McDonough suggests that nature has no concept of waste as we understand it. "Everything is cycled constantly with all waste equaling food for other living systems." (2) All energy must be current. The thing that allows " . . . nature to continually cycle itself through life is energy, and this energy comes from outside the system in the form of per-petual solar income." Nature, consequently, operates entirely on current energy. (3) Biodiversity is essential. Biodiversity gives nature's system its resilience. "What prevents living systems from running down and veering into chaos is a miraculously intricate and symbiotic relationship between millions of organisms, no two of which are alike (McDonough, 1996, 72).

On the basis of his many successful years of experience, McDonough asserts that these principles for designing human shelter are not only an ideal to guide our work, but are also eminently practical and economically viable. As a farmer, it strikes me that McDonough's ecological standard, and his defining characteristics, are directly applicable to agriculture. The fact that these approaches are interchangeable among various types of human industry suggests that a common approach exists.

On our own farm, the "waste should become food" principle provides us with economies of scope that are, in many respects, more important to the financial viability of our operation than economies of scale. The waste from the cropping system (crop residues) is fed to livestock, and the waste from the livestock (manure) is recycled back to the soil and becomes food for the crops. This reduces the need for both purchased livestock feed *and* purchased crop inputs. Also, on our farm, eight diverse crops are produced in a complex crop rotation that, together with the livestock and field boundary trees and hedgerows, insures at least a modicum of diversity. This diversity not only reduces economic risks but also provides habitat for wildlife, much of which in turn helps to control crop pests (e.g., ladybugs to control aphids).

Our energy sources are far from current, which is the "characteristic" that most needs development and where the sustainability of our farm is most vulnerable. While our farming methods have reduced our energy consumption by about 20 percent over the past twenty years, we still need alternative technologies to reduce our reliance on "borrowed" fossil fuel energy. But McDonough's principles are sound and may be universal. It would be interesting to see how they might be applied to all human industry.

TECHNOLOGY AND ETHICS

Perhaps the greatest danger in the application of the Precautionary Principle is that it could lend itself to the notion that all we need to do is assess the potential for damage of specific technologies and practices and then get on with business as usual. The prevailing assumption seems to be that if we switch from polluting technologies (e.g., endocrine-disrupting chemicals) to nonpolluting technologies (e.g., genetic engineering), then we will have solved the problem of ecosystems health. Nothing could be further from the truth. We can only come to such conclusions from the perspective of our having "stepped out of" our local ecosystems.

The technological switch approach ignores fundamental lessons of biology and ecology. What makes the ecological standards proposed by Leopold and McDonough attractive as guiding principles for human industry is that they are grounded in a sophisticated understanding of the way natural systems work. And they use that understanding to determine how human industry can be folded into nature's systems, rather than how human industry can be imposed on those systems without doing too much harm. In other words, they operate out of a different ethical posture.

This ethical shift is perhaps the most important challenge confronting the human species. Having "stepped out of" our local ecosystems, we have lost both the vision and the language for thinking appropriately about the Earth and our place in it. We no longer have any sense of the Earth as "comprehensive community," as Larry Rasmussen puts it. "We do not . . . have a common language, even for something as basic as earth community. We are temporarily stuck in the awkward space between worlds we trusted and ones strangely new to us" (Rasmussen, 1996, 9).

Accordingly our predominant approach to solving the problems confronting us is to develop new and more clever ways to redesign our complex technological systems, rather than rethinking our place in the intricate, interconnected biological world of living organisms—from fungi and viruses to insects and vertebrates.

Increasingly thoughtful observers from many disciplines are concluding that this rethinking is essential. Even food safety analysts are now concluding that unless we begin thinking about our food system from an ecological perspective, all our technical food safety strategies will simply put us on a treadmill of continually creating the problem we are trying to solve. Indeed, Nichols Fox argued that "the considerations that apply to the ecology of other environments apply equally to food. . . . Whenever there is a lack of diversity, when a standardized food product is mass-produced, disease can enter the picture" (Fox, 1997, 76).

In agriculture, architecture, and food processing and distribution, it all seems to boil down to the same awareness. We are ultimately an integral part of a very complex biotic community that has coevolved over billions of years, and there can be no long-term health or survival of our species without the combined health of all the other species-neighbors with whom we share our ecological neighborhood. Our task is to fit human industry into that neighborhood in ways that enhance the health of the whole system.

(Note: Portions of this chapter were presented at a meeting of the Soil and Water Conservation Society, University of Georgia, April 1997.)

Appendix to Chapter 17

In June of 1998, participants in a meeting convened by the International Forum on Agriculture drafted and signed the following document.

THE VANCOUVER STATEMENT ON THE GLOBALIZATION AND INDUSTRIALIZATION OF AGRICULTURE

We believe that the industrialization and globalization of food and fiber imperils humanity and the natural world. Reducing farming to a monocultural, synthetic, transnational corporate business threatens the health, nourishment, right livelihood, and spirituality of communities and the earth. It is insane to believe that we must poison land and water and waste the soil in order to feed and clothe ourselves. Five decades of the so-called Green Revolution have not only led to the destruction and contamination of water, soil, biodiversity, and human communities, but exacerbated hunger worldwide. One of the most critical impacts of industrial agriculture is climate change, which will destroy the natural basis of agriculture itself. The patenting of life, corporate ownership, and manipulation of our genetic heritage is one of the greatest threats ever imposed by industrial agriculture: the human right to feed, clothe, and shelter ourselves and our families is at stake. Institutions and treaties such as the World Trade Organization, the General Agreement on Tariffs and Trade, Codex Alimentarius, North American Free Trade Agreement, the Food and Agriculture Organization, and the European Union have accelerated the process of agricultural industrialization and globalization while promoting the rights of corporations over those of people.

We know that there are nontoxic and nondestructive alternatives to global industrial agriculture, and we know that these alternatives can provide more food. Farmers around the world are farming in ways that respect their unique ecological and cultural communities. Building on their wisdom, all farms of the twenty-first century can be ecologically regenerative, community sustaining, biologically and culturally diverse, as well as energy conserving. We must not only build on the existing knowledge and vision of farmers, but we must expand partnerships and create coalitions that serve to re-empower them.

In order to rescue our food system, we need more skilled farmers who have access to land, seed, and the knowledge of local biological systems. Also essential to a healthy food system, is clean land, air, water, and soil and the right to save seeds to ensure future harvests.

Scientific organizations and transnational corporations that are experimenting with and releasing poisons, synthetic compounds, and genetically modified organisms into the biosphere should be held fully accountable for the safety of their practices and products. Corporations, scientists, and governments should honor the Precautionary Principle and take preventive action in the face of scientific uncertainty in order to avoid cultural and ecological harm.

We affirm, with the Universal Declaration of Human Rights, that the right to food is sacred. The right to food transcends basic nutrition and hunger and includes the right to produce one's own food. We also affirm that consumers have the right to know where their food comes from, what is in it, and how it was produced.

Furthermore, farmers and consumers have a right to maintain local control over food production, distribution, and consumption.

Our bodies, our plants and animals, our air, water, land, and soil, are not commodities and are not patentable. When a food production system violates the rights of citizens and the natural order of the planet's ecosystems, it is essential that we the people make use of our inalienable freedom to correct those abuses. We stand united on these points.

VANCOUVER STATEMENT SIGNATURES

Hal Hamilton, Center for Sustainable Systems, U.S.

Ronnie Cummins, Pure Food Campaign, U.S.

Tim Lang, Centre for Food Policy, U.K.

Carolyn Raffensperger, Science and Environmental Health Network, U.S.

Candido Gryzbowski, IBASE, Brazil

Mark Ritchie, Institute for Agriculture & Trade Policy, U.S.

Victor Suarez Carrero, ANEC (Asociacion Nacional de Empresas Comercializadoras Campesinas), Mexico

Alejandro Rojas, University of British Columbia, Canada

Steve Shrybman, West Coast Environmental Law Association, Canada

Jose A. Lutzenberger, Fundacao Gaia, Brazil

Miguel Altieri, University of California—Berkeley, U.S.

Jeanot Minla Mfou'ou, Agriculture Peasant & Modernization Network, Cameroon

Herb Barbolet and Kathleen Gibson, Farm Folk/City Folk, Canada

Helena Norberg-Hodge, International Society for Ecology and Culture, U.K./U.S.

Carolyn Mugar, Farm Aid, U.S.

Gregor Robertson, Happy Planet Foods, Canada

Mika Iba, Network for Safe and Secure Food and Environment, Japan

Sigmund Kvaloy, Setreng Insitute for Ecophilosophy, Norway

Will Allen, Sustainable Cotton Project, U.S.

Professor Nanjunda Swamy, Karnataka Farmers' Union, India

Franco Adriano Werlang, Fundacao Gaia, Brazil

Lori Ann Thrupp, World Resources Institute, U.S.

Monica Moore, Pesticide Action Network—North America, U.S.

Nancy Hirshberg, Stonyfield Farm, Inc., U.S.

Wendell Berry, Lanes Landing Farm, U.S.

Moura Quayle, University of British Columbia, Canada

Andrew Kimbrell, International Center for Technology Assessment, U.S.

Jerry Mander, Public Media Center, U.S.

Kate Duesterberg, University of Vermont—Center for Sustainable Agriculture, U.S.

Brewster Kneen, "The Ram's Horn," Canada

Cathleen Kneen, "The Ram's Horn," Canada

Fred Kirschenmann, Kirschenmann Family Farm, U.S.

Flavio Valente, Associacao para Projetos de Combate a Fome, Brazil

Karen Lehman, Institute for Agriculture & Trade Policy, U.S.

Kathy Ozer, National Family Farm Coalition, U.S.

Anuradha Mittal, Food First, U.S.

Peter Rosset, Food First, U.S.

Laurie MacBride, Georgia Strait Alliance, Canada

Shirley Sherrod, Federation of Southern Cooperatives, U.S.

Vandana Shiva, Research Institute for Natural Resource Policy, India

Dan Imhoff, Foundation for Deep Ecology, U.S.

Dena Hoff, National Family Farm Coalition, U.S.

Nettie Wiebe, National Farmers' Union, U.S.

Martin Khor, Third World Network, Malaysia

NOTES

1. The term "ecological neighborhood" is used to denote the sum total of a biotic community that has evolved in a particular local ecosystem—for example, a watershed.
2. The word "agroecology" came into our vocabulary around 1970 and refers to a way of practicing agriculture that attempts to balance all of the environmental and economic risks of farming while maintaining productivity over the long term. Or, put more simply, agroecology is "agriculture as applied ecology."
3. Such an effort is now being considered for the Willamette Valley (Oregon) by researchers at Oregon State University.

REFERENCES

Altieri, M. A. 1987. *Agroecology*. Boston: Cambridge University Press.

Boehljie, Michael. 1995. "Industrialization of Agriculture: What Are the Consequences?" (Unpublished paper given at Industrialization of Agriculture Symposium, Minneapolis, Minn., July, 1995)

Buchmann, Steve, and Gary Nabhan. 1996. *The Forgotten Pollinators*. Washington, D.C.: Island Press.

Eldredge, Niles. 1995. *Dominion: Can Nature and Culture Co-Exist?* New York: Henry Holt & Co.

Fox, Nichols. 1997. *Spoiled*. New York: Basic Books.

Holdrege, Craig. 1996. *Genetics and the Manipulation of Life: The Forgotten Factor of Context*. Hudson, N.Y.: Lindisfarne Press.

Jensen, Derrick. 1996. "Hush of the Hives." *New York Times*, October 13.

Kevan, Peter G., Vern G. Thomas, and Svenja Belaoussoff. 1997. "AgrECOLture Defining the Ecology in Agriculture." *Journal of Sustainable Agriculture*. Vol. 9(2/3).

Leopold, Aldo. 1949. *A Sand County Almanac*. London: Oxford University Press.

Lowdermilk, W.C. 1953. *Conquest of the Land Through Seven Thousand Years*. Washington, D.C.: U.S. Government Printing Office.

McDonough, William. 1996. "Design, Ecology, Ethics, and the Making of Things." *Lapis*. Issue 3. New York: New York Open Center.

Rasmussen, Larry L. 1996. *Earth Community Earth Ethics*. Maryknoll, N.Y.: Orbis Books

Robinson, Raoul. 1996. *Return to Resistance*. Davis, Calif.: AgAcess Press.

Chapter 18

~

PRECAUTIONARY ACTION NOT TAKEN: CORPORATE STRUCTURE AND THE CASE STUDY OF TETRAETHYL LEAD IN THE U.S.A.

Peter Montague

The Principle of Precautionary Action was rejected by U.S. corporations and political authorities seventy years ago, and public health authorities acceded to this decision at the time. The historical record is very clear on this point. The same forces at work then are still at work today.

The history of tetraethyl lead in the United States is the history of a power struggle between a handfull of large, publicly held corporations on the one hand and the federal government's public health apparatus on the other. The large corporations prevailed easily. As we think about the possibility of establishing the Precautionary Principle as standard operating procedure in the United States in the 21st century, we would be remiss if we did not examine the nature of this legal entity, the corporation. As things stand today, the corporation—and not government—will determine whether a precautionary approach is possible.

The transnational corporation is the principle institution of our era, and this has been true for roughly the past century. This institution is as important today as the Christian church and the crown were in Europe during the 15th century, determining and shaping reality for most people.

THE NATURE OF THE CORPORATION

In the United States, corporations were initially created as artificial, subordinate entities, chartered by state legislatures, with no rights of their own. Up until 1886 corporations could only serve the public purposes that they were specifically established to serve (e.g., build a canal, construct and manage a toll road, finance and construct a bridge). Their capitalization was fixed by law; they could not own other corporations; often their board of directors would have to live in the state where they were incorporated. Their lifetime was finite, often twenty years.

After 1886 the situation changed. In a U.S. Supreme Court decision, corporations were given the status of "persons" under the U.S. Constitution. After that, corporations could do anything that any other "person" could do, so long as it was legal. Armed with the Constitutional protections of individuals, but having none of the limitations of individuals, corporations soon ceased to be subordinate entities. Today many corporations are countries. Mitsubishi is larger than Indonesia. General Motors is larger than Denmark. Ford Motor is larger than South Africa and larger than Saudi Arabia. Toyota Motor is larger than Portugal. Wal-Mart stores are larger than Israel, larger than Greece. As a result of unlimited lifetime, in 1995–1996 51 of the world's 100 largest economies were corporations while only 49 were countries. (See the chapter appendix.)

For the most part, corporations are staffed by intelligent, well-meaning people. But the personal motivations of those individuals are not what motivate the corporation. The corporation is driven by its own internal logic.

A corporation has an internal drive that is comparable to a human's "will to live." Once a corporation is publicly traded, it:

- Must return a profit to investors;
- Must grow;
- Must externalize costs to the extent feasible.

These are essential characteristics of the corporate form. If a corporation fails to provide a decent return for investors, those investors can (and do) sue for breach of fiduciary trust. This requirement—to turn a profit—narrowly limits what corporations can do. In general, what is unprofitable cannot be pursued. This means that individuals must put aside their consciences when they make decisions for a corporation. The most well-meaning people in the world are not free to act on their personal philosophies when they are acting on behalf of a publicly held corporation. They must do what is profitable, not necessarily what is right.

Corporations must grow for a variety of reasons. In general, larger size brings stability. It also tends to bring greater market share. It also brings a measure of political power, which allows corporate managers to manipulate the political environment within which the corporation must operate. Size also brings with it the power to create and control the demand for goods, through mass-market advertising. A corporation that is not growing is thought to be in trouble and may therefore lose investors.

After they grow to a certain size, corporations cannot feel any sort of "pain." For example, the Exxon Corporation was fined $5 billion for the Exxon Valdez oil spill. On the day that mammoth fine was announced, Exxon's stock rose because investors realized that Exxon was invincible. No matter how odious its behavior, human institutions have no capacity to curb the excesses of a large transnational. Similarly, the day the government of India imposed an $800 million fine on Union Carbide for its role at Bhopal, Carbide's stock went up.

In the United States, about a dozen of these extraordinary creatures own and operate 90 percent of the mass media—controlling almost all books, magazines, records, videos, TV, radio, newspapers, wire services, photo agencies (see Ben Bagdikian, Media Monopoly, 4th edition, 1992). Thus, the number of people who set the terms of public debate in the United States would easily fit into one small room. To the extent that they are visible at all, corporations use the mass media artfully to give themselves the appearance of benevolence.

This is the creature that we are asking to take precautionary action in the best interests of humanity. Unfortunately, this is not an entity with a conscience (it is, after all, not human) or a sense of social purpose, so it is, in general, incapable of taking precautionary action on behalf of the larger society or the next generation. If society wants these entities to take precautionary action, society will have to build that into the legal requirements of the corporation by modifying the corporate charter—the piece of paper issued by state legislatures giving corporations the privilege of existing. In addition, in the United States, corporations could be denied the privileges of personhood under the Constitution. Our rule of thumb could be: If it doesn't breathe, it isn't a person. Thus, corporations could be brought back to the subordinate status that our grandparents and great-grandparents envisioned for these sociopathic inventions.

TETRAETHYL LEAD IN THE UNITED STATES, 1925–1989

The issue seventy years ago was whether General Motors (GM), Standard Oil of New Jersey, and the DuPont corporation should begin putting tetraethyl lead into gasoline. At that time, the toxicity of lead had been well established for a hundred years, but a new gasoline additive was needed by the automobile and petroleum corporations, and lead suited their purposes.

In 1923, the automobile industry was booming. In 1916, 3.6 million cars were registered; in 1920, the number was 9.2 million; and by 1925, it was 17.5 million. Prior to 1920, Ford had grabbed the lion's share of the market by mass producing the standardized Model T, but GM developed a successful strategy for overtaking Ford. In the words of GM Chairman Alfred Sloan, GM created demand "not for basic transportation but for progress in new cars for comfort, convenience, power and style." In the search for greater horsepower, GM developed higher-compression engines. However, with ordinary gasoline, high-compression engines developed an annoying and damaging "knock" because the gasoline burned explosively. So GM chemists searched systematically for a gasoline additive that would make gasoline burn evenly in high-compression engines, eliminating the "knock." On February 1, 1923, in Dayton, Ohio, leaded gasoline went on sale for the first time.

Leaded gasoline was produced by the Ethyl Corporation—a joint venture of GM, Standard Oil of New Jersey, and DuPont. Tetraethyl lead is at least as toxic as normal metallic lead, but with this difference: Tetraethyl lead is a volatile liquid, readily absorbed through the lungs and skin. Almost immediately, workers began to be poisoned. At Standard Oil's Bayway, New Jersey, refinery, five workers died and thirty-five suffered severe palsy, tremors, hallucinations, and other serious symptoms of nerve damage. Several of these workers spent the rest of their lives confined in insane asylums. One of the supervisors at the Bayway facility told the New York Times that "these men probably went insane because they worked too hard." At DuPont's Deepwater, New Jersey, plant, more than three hundred workers were poisoned by tetraethyl lead. DuPont workers dubbed the plant "The House of Butterflies" because so many workers had hallucinations of insects. The New York Times reported that 80 percent of the workers at DuPont's lead plant were poisoned.

These industrial poisonings created headlines nationwide and public health officials became apprehensive about the prospect of treating billions of gallons of gasoline with tons of tetraethyl lead, which would be released into the air along with the exhaust fumes.

In 1924, GM and DuPont paid the federal Bureau of Mines to investigate the hazards of lead from automobile exhausts. The Bureau of Mines agreed to investigate and accepted a stipulation by Charles Kettering, president of the Ethyl Corporation: " . . . the Bureau [shall] refrain from giving out the usual press and progress reports during the course of the work, as [Ethyl Corporation] feels that the newspapers are apt to give scare headlines and false impressions before we definitely know what the influence of the material will be."

Further, the bureau agreed never to mention the word "lead" in its reports but to use only the trade name "Ethyl." Ethyl Corporation insisted that "all manuscripts, before publication, will be submitted to the Company for comment, criticism, and approval." The Bureau of Mines agreed. During an eight-month period, the bureau exposed monkeys, dogs, rabbits, guinea pigs, and pigeons to automobile exhaust on 188 occasions, half for 3 hours at a time and half for 6.

The bureau reported finding no evidence of lead poisoning and no accumulation of lead in any of the animals. The New York Times reported the bureau's results November 1, 1924, with this headline: "No Peril to Public Seen in Ethyl Gas/Bureau of Mines Reports After Long Experiments with Motor Exhausts/More Deaths Unlikely." The Times also reported that "the investigation carried out indicates the danger of sufficient lead accumulation in the streets through the discharging of scale from automobile motors to be seemingly remote."

Despite this reassuring news, public health authorities remained concerned about the prospect of putting millions of pounds of toxic lead in the form of a fine dust into the streets of every American city and town.

Therefore, the U.S. Public Health Service convened a conference May 20, 1925, to determine whether leaded gasoline could be safely manufactured and whether lead from automobile exhausts would harm the general public. Just before the conference, Standard Oil announced it was temporarily suspending the sale of leaded gasoline.

The morning session on May 20 was devoted to speeches by GM, Standard Oil of New Jersey, DuPont, and their new joint venture, Ethyl Corporation, which they had created to market leaded gasoline. The afternoon was devoted to discussions of health. Here in summary is what the conference revealed:

- Charles F. Kettering, president of the Ethyl Corporation, pointed to the unique properties of tetraethyl lead as an anti-knock additive. Other additives gummed up the engine, but the lead compounds passed out through the exhaust, leaving the engine clean, he said.
- Mr. Kettering said American automobiles would burn 15 billion gallons of gasoline in 1926.

- Lt. Col. E.B. Vedder, chief of the U.S. Chemical War Service, said lead is a cumulative poison.
- Robert Kehoe, a medical consultant to GM and to the Ethyl Corporation, confirmed that "in sublethal dose, lead is cumulative."
- Joseph C. Aub of Harvard University emphasized that "lead is an accumulative poison."
- Robert Kehoe, the industry's consultant, established that lead passed through the placenta of a rabbit and contaminated unborn rabbits with lead if the pregnant mother was exposed. He established that pregnant rabbits exposed to lead had abortions, miscarriages, and premature births. He acknowledged that poisoning by tetraethyl lead is the same as other lead poisoning: "In those cases in which absorption is present over a long period of time the symptoms do not differ strikingly from the symptoms in chronic lead poisoning . . . ," Kehoe said.
- Alice Hamilton of Harvard University—one of the country's acknowledged experts on lead poisoning—said, " . . . lead is a slow and cumulative poison and . . . it does not usually produce striking symptoms that are easily recognized."
- E.R. Hayhurst from Ohio State University made the point that serious lead poisoning "is most apt to occur in cases using lead in the form of a dust."
- R.R. Sayers of the U.S. Bureau of Mines described experiments in which five times the normal amount of tetraethyl lead was added to gasoline and animals were forced to breathe the exhaust fumes. "The dust from the floor of the test chamber contained 10.5 percent of lead within six months without cleaning," Sayers said.
- Professor Joseph C. Aub of Harvard calculated that if all gasoline sold in 1926 were leaded, 50,000 tons of lead would be spewed as a fine dust across America's highways, road, and urban streets.
- David Edsall, dean of the Harvard School of Public Health, summarized as follows:

    "The only conclusion that I can draw from the data presented here today is that in the question of the exhaust . . . I cannot escape feeling that a hazard is perfectly clearly shown thus far by what has been reported here today, that it appears to be a hazard of considerable moment, and that the only way that it could be said that it is a safe thing to continue with that hazard would be after very careful and prolonged and devoted study as to how great the hazard is."

The conference resolved unanimously that the U.S. Surgeon General should appoint a seven-member panel to determine the dangers of leaded gasoline by January 1, 1926, and, until then, the sale of leaded gasoline should remain suspended. At the time, it seemed like a great victory for the Precautionary Principle. But it was not to be.

In summary, various speakers at the conference had established that lead would be emitted from automobile exhausts as a fine dust; lead is a potent brain-damaging poison, and dust is its most dangerous form; when caged laboratory animals were dosed with automobile exhaust, lead dust built up on the bottoms of their cages; lead is a cumulative poison; it passes through the placenta and harms the unborn; it causes low birth weight, spontaneous abortion, and stillbirth. On these points, there was no disagreement.

However, views were split that day in 1925: The corporations wanted to press ahead rapidly, putting about 2 grams (1/14 of an ounce) of lead into every gallon of gasoline. Health officials, on the other hand, urged caution; they wanted to consider the consequences for public health. Without giving it a name, health officials in 1925 were embracing the Precautionary Principle, which says, first, that the burden of proof of safety should be borne by the proponent of a new technology, not by the public; and second, that, where there are threats of serious or irreversible damage, lack of scientific certainty should not be used as an excuse for postponing measures to prevent environmental degradation.

Late in the afternoon, Dr. Yandell Henderson of Yale University summarized what he had heard, as follows: "We have in this room, I find, two diametrically opposed conceptions. The men engaged in industry, chemists, and engineers, take it as a matter of course that a little thing like industrial poisoning should not be allowed to stand in the way of a great industrial advance. On the other hand, the sanitary experts take it as a matter of course that the first consideration is the health of the people."

The American Federation of Labor (AFL) had two representatives at the conference, both of whom embraced the Precautionary Principle:

- Grace M. Burnham, representing the Workers' Health Bureau of the AFL, said, " . . . I think that the United States should be self-respecting enough to realize that, when there is a public health hazard involved which affects the entire population, that hazard ought to be investigated out of public funds and by a responsible public agency. . . . And I believe that until that time, and until the manufacture, distribution, and use of tetraethyl lead has been proved conclusively to be safe, its use should be discontinued."

- Mr. A.L. Berres, representing the Metal Trades Department of the AFL, said, "I feel that, as has been stated here by some of the previous speakers, until such time as it can be definitely determined that there is no hazard in the manufacture and handling of this gas [leaded gasoline], its use ought to be prohibited. . . ."

Dr. Haven Emerson, professor of public health at Columbia University in New York City, summarized, "I presume that it is the inclination of every health officer to urge a continuance of the cessation of the use or sale of the ethyl gasoline that has been voluntarily determined upon by the company."

In sum, in 1925, the public health community, as represented at the May 20th conference, urged the Precautionary Principle: Faced with a known hazard of unknown size, it urged that the hazard be prevented.

The corporations, on the other hand, used arguments that are still common today:

- The dangers have not been proven;
- Animal studies cannot tell us what we need to know about humans;
- Efficiency requires us to adopt new technologies, even though some people may have to be sacrificed; and
- People should act strictly on available facts, not on fears for the future or opinions about what *might* occur.

Sometimes these arguments were combined. For example, Frank A. Howard, representing the Ethyl Corporation, said

> Our continued development of motor fuels is essential in our civilization. . . . Now, as a result of some 10 years' research on the part of the General Motors Corporation and 5 years' research by the Standard Oil Co., or a little bit more, we have this apparent gift of God . . . of tetraethyl lead. . . .
>
> . . . Because some animals die and some do not die in some experiments, shall we give this thing up entirely? . . . I think it would be an unheard-of blunder if we should abandon a thing of this kind merely because of our fears. . . . Possibilities can not be allowed to influence us to such an extent as that in this matter. It must be not fears but facts that we must be guided by. I do not think we are justified in trying to reach a final conclusion in this matter on fears at all; nor are we justified in saying that we will cease this development because of fears we entertain. This development must be stopped, if it is stopped at all, by proofs of the facts.

Dr. Robert Kehoe, a medical consultant to the Ethyl Corporation, gave a similar argument: "I must say, from the standpoint of industry, that when a material is found to be of this importance for the conservation of fuel and for increasing the efficiency of the automobile it is not a thing which may be thrown into the discard on the basis of opinions. It is a thing which should be treated solely on the basis of facts."

Since the "facts" could not include any poisonings until such poisonings had already occurred (until they occurred, they would be nothing more than speculative "fears" or "opinions"), the argument for basing policy strictly on facts produced a policy of experimenting on the public and waiting for the sick and the dead to accumulate. This, then, became the official way of doing business in the United States. Today the language is slightly different; we hear calls for policy based on "sound science" (not on "facts"), but it is the same argument.

Shortly after the May conference, Dr. Emery Hayhurst—a paid consultant to the Ethyl Corporation—wrote an unsigned editorial for the *American Journal of Public Health* titled "Ethyl Gasoline." (He was a member of the journal's editorial board.) In it, he described newspaper advertisements by the Ethyl Corporation that claimed that leaded gasoline was being used around the country with "complete safety and satisfaction." Hayhurst's editorial concluded, "Observational evidence and reports to various health officials over the country, previous to and following the above advertisements have, so far as we have been able to find out, corroborated the statement of 'complete safety' so far as the public health has been concerned."

The May 1925 conference had ended with a unanimous resolution calling on the U.S. Surgeon General to appoint a seven-member, blue-ribbon panel to render an opinion on the dangers of lead by January 1, 1926. For less than six months, the committee studied 252 garage mechanics, filling station attendants, and chauffeurs in Dayton and Cincinnati and concluded, "There are at present no good grounds for prohibiting the use of ethyl gasoline." In sum, the "facts" argument overwhelmed the Precautionary Principle.

In June 1926, GM, DuPont, Standard Oil of New Jersey, and their joint venture, the Ethyl Corporation, started selling leaded gasoline again, and they continued to do so until Congress finally outlawed it completely in 1989. Between 1926 and 1985, 7 million metric tons of toxic lead dust (15.4 billion pounds) were distributed into the environment by the automobile corporations. These corporations still sell their brain-damaging product in third-world nations today.

In 1965, MIT professor Clair C. Patterson examined the situation and concluded that "the average resident of the United States is being subjected to severe chronic lead insult." Patterson went on, "Intellectual irritability and disfunction are associated with classical lead poisoning, and it is possible, and in my opinion probable, that similar impairments on a lesser but still significant scale might occur in persons subjected to severe chronic lead insult." Subsequent studies have confirmed and reconfirmed this view.

The period of greatest lead use was 1945–1971, after which it began to decline. In those years, 165,000 to 275,000 tons of lead dust spewed from the exhaust pipes of American automobiles each year. Americans born during these years have 300 to 1,000 times as much lead in their bodies as pre-Columbian indigenous people had.

The history of tetraethyl lead reveals the outlines of the basic dilemma facing humanity: Can we the people curb the imprudent and reckless actions of corporations in time to avoid public health catastrophe?

Appendix to Chapter 18

GROSS DOMESTIC PRODUCT/REVENUES FOR
NATIONS VERSUS CORPORATIONS, DECEMBER 1996

Country/Corporation	GDP/Revenues for 1996 (in billions of dollars)
1. United States	$6,648,013
2. Japan	4,590,971
3. Germany	2,045,991
4. France	1,330,381
5. Italy	1,024,634
6. United Kingdom	1,017,306
7. Brazil	554,587
8. Canada	542,954
9. China	522,172
10. Spain	482,841
11. Mexico	377,115
12. Russia	376,555
13. South Korea	376,505
14. Australia	331,990
15. Netherlands	329,768
16. India	293,606
17. Argentina	281,922
18. Switzerland	260,352
19. Belgium	227,550
20. Austria	196,546

Country/Corporation	GDP/Revenues for 1996 (in billions of dollars)
21. Sweden	196,441
22. Mitsubishi	175,836
23. Indonesia	174,640
24. Mitsui	171,490
25. Itochu	167,825
26. Sumitomo	162,476
27. General Motors	154,951
28. Marubeni	150,187
29. Denmark	146,076
30. Thailand	143,209
31. Hong Kong	131,881
32. Turkey	131,014
33. Ford Motor	128,439
34. South Africa	121,888
35. Saudi Arabia	117,236
36. Norway	109,568
37. Exxon	101,459
38. Nissho Iwai	100,876
39. Finland	97,961
40. Royal Dutch Shell	94,881
41. Poland	92,580
42. Ukraine	91,307
43. Toyota Motor	88,159
44. Portugal	87,257
45. Wal-Mart Stores	83,412
46. Israel	77,777
47. Greece	77,721
48. Hitachi	76,431
49. Nippon Life Insurance	70,840
50. AT&T	75,094
51. Nippon Telegr./Telepho.	70,844
52. Malaysia	70,626
53. Matsushita Ele. Indust.	69,947
54. Tomen	69,902
55. Singapore	68,949

Country/Corporation	GDP/Revenues for 1996 (in billions of dollars)
56. Colombia	67,266
57. General Electric	64,687
58. Daimler-Benz	64,169
59. Philippines	64,162
60. IBM	64,052
61. Iran	63,716
62. Mobil	59,621
63. Nissan Moto	58,732
64. Venezuela	58,257
65. Nichimen	56,203
66. Kanematsu	55,856
67. Dai-Ichi Mutual Life Ins.	54,900
68. Sears Roebuck	54,825
69. Philip Morris	53,776
70. Chrysler	52,224
71. Ireland	52,060
72. Pakistan	52,011
73. Chile	51,957
74. Siemens	51,055
75. New Zealand	50,777
76. British Petroleum	50,737
77. Tokyo Electric Power	50,359
78. Peru	50,077
79. Volkswagen	49,350
80. Sumitomo Life Insurance	49,063
81. Toshiba	48,228
82. Unilever	45,451
83. Egypt	42,923
84. Algeria	41,941
85. Nestlé	41,626
86. Hungary	41,374
87. Deutsche Telekom	41,071
88. FIAT	40,851
89. Allianz Holding	40,415
90. Sony	40,101

Country/Corporation	GDP/Revenues for 1996 (in billions of dollars)
91. Veba Group	40,072
92. Honda Motor	39,927
93. Elf Aquitaine	39,459
94. State Farm Group	38,850
95. NEC	37,946
96. Prudential Insurance Co.	36,946
97. Oesterreichische	36,766
98. Meiji Mutual Life Ins.	36,344
99. Czech Republic	36,024
100. Daewoo	35,707

From: Ward Morehouse, "Multinational Corporations and Crimes Against Humanity," in Trent Schroyer, editor, A World That Works (New York: The Bootstrap Press, 1997), p. 51. Morehouse attributes the data to these sources: corporation data from "Fortune's Global 500, The World's Largest Corporations," Fortune magazine, August 7, 1995. Country information from The World Development Report (Washington, D.C.: World Bank, 1996).

REFERENCES

"Ethyl Gasoline," American Journal of Public Health, Vol. 15 (1925), pp. 239–240. Rosner and Markowitz, p. 347, identify Hayhurst as the author of the anonymous editorial, based on his correspondence with R.R. Sayers of the U.S. Bureau of Mines.

Bruce A. Fowler et al., Measuring Lead Exposure in Infants, Children, and Other Sensitive Populations (Washington, D.C.: National Academy Press, 1993), pp. 14–15, 107.

David Freestone and Ellen Hey, "Origins and Development of the Precautionary Principle." In: David Freestone and Ellen Hey, editors, The Precautionary Principle and International Law (The Hague, London, and Boston: Kluwer Law International, 1996), pp. 3–15.

David C. Korten, When Corporations Rule the World (San Francisco: Berret-Koehler Publishers, 1995).

Jerome O. Nriagu, "The Rise and Fall of Leaded Gasoline," The Science of the Total Environment, Vol. 92 (1990), pp. 13–28.

Clair C. Patterson, "Contaminated and Natural Lead Environments of Man," Archives of Environmental Health, Vol. 11 (September 1965), pp. 344–360.

David Rosner and Gerald Markowitz, "A 'Gift of God'? The Public Health Controversy over Leaded Gasoline During the 1920s," American Journal of Public Health, Vol. 75, No. 4 (April 1985), pp. 344–352.

Treasury Department, U.S. Public Health Service, Proceedings of a Conference to Determine Whether or Not There Is a Public Health Question in the Manufacture, Dis-

tribution, or Use of Tetraethyl Lead Gasoline (Public Health Bulletin No. 158) (Washington, D.C.: Treasury Department, United States Public Health Service, 1925). Available from William Davis at the National Archives in Washington, D.C.: (202) 501-5350. (National Archives Record Group No. 287; T27.12:158/ 3S1, 24/2316 Box T777. RG 287.)

U.S. Bureau of the Census, *Historical Statistics of the United States, Colonial Times to 1970*, Bicentennial Edition, Part 2 (Washington, D.C.: U.S. Government Printing Office, 1975), Series Q-153, p. 716.

Chapter 19

❧

MANGANESE IN GASOLINE:
A CASE STUDY OF THE NEED FOR
PRECAUTIONARY ACTION

Ted Schettler

A 1995 federal court decision forced the EPA to allow manufacturers to add manganese to gasoline as an octane enhancer, preventing the agency from regulating this additive in a precautionary manner. In a related development, the Ethyl Corporation, the sole producer of the fuel additive, sued the Canadian government for banning manganese in gasoline, citing unfair trade practices under the North American Free Trade Agreement, and arguing that there were no proven adverse equipment or health effects resulting from its use. In July 1998 the Canadian government settled the case, lifting the ban and agreeing to pay Ethyl a reported $13 million for costs and lost profits. The government also agreed to issue a public statement indicating that there are no adverse equipment or health effects resulting from the use of manganese in gasoline. This chapter reviews the current state of understanding the health effects of manganese exposure and asks if laws governing the use of manganese in gasoline are more influenced by concerns for corporate than for public health.

Octane enhancers are added to gasoline to promote the efficient operation of modern, high-compression engines. Organic additives containing oxygen (such as ethanol) or metallic additives, such as tetraethyl-lead or manganese (as methylcyclopentadienyl manganese tricarbonyl or MMT), boost the octane rating to desired levels. MMT is an organic manganese compound produced solely by the Ethyl Corporation at a large plant in South Carolina.[1] It was first added to the commercial gasoline supply as a supplement to tetraethyl lead in 1958.[2] After the phaseout of lead use in gasoline, MMT continued to be used as an octane enhancer. (For additional information about the Ethyl Corporation, or the introduction of tetraethyl lead, see chapter 18.)

However, amendments to the Clean Air Act in 1977 prohibited marketing of fuel additives unless the manufacturer could show that it did not harm emission control systems. That concern prompted the EPA to ban the addition of MMT to unleaded gasoline. Eventually, the Ethyl Corporation submitted evidence showing that equipment was not harmed, but the EPA continued to deny permission to use MMT because of new concerns about the health effects of manganese exposure. In 1995, a federal appeals court found that EPA rules requiring premarket testing of fuel additives, developed in 1994, did not apply since the Ethyl Corporation had applied for permission to reintroduce MMT prior to that date. Moreover, the court found that pre-1994 law permitted the EPA to regulate only currently used additives based on health effects. Since MMT was not in use, the court denied the agency the right to continue the prohibition, despite legitimate reasons for concern. According to the court, the EPA must allow the substance to be manufactured and used—inevitably releasing MMT into the environment.

The task now falls to the EPA to assess the likelihood of harm from MMT use while manufacturers regularly add this compound to an undisclosed portion of the nation's gasoline supply. Placing the burden of proof on the EPA, rather than on the manufacturer, to prove safety denies the agency the right to regulate in a precautionary way and commits public funds to the analysis. Moreover, despite a considerable amount of recent research, there are important gaps in the information necessary to determine whether adding MMT to gasoline poses a significant threat to public health. Filling these gaps will be expensive and time consuming.

Adding MMT to gasoline results in low-level, continuous releases of manganese-containing particles to air, soil, and water. Research shows that excessive manganese exposure has adverse health effects and indicates that children and other members of the general population may be particularly susceptible. But there is considerable uncertainty about the long-term nature of these effects and the level and routes of exposure that may trigger them. MMT use will enable the Ethyl Corporation to recover its investment in the production facility and profit from sales. The more generally realized benefit of MMT use is enhanced engine performance.

Manganese: Essential and Harmful

Manganese is an essential element in trace amounts in plants and animals. It has catalytic properties in several critical enzymatic reactions and is required for photosynthesis. Manganese deficiency may result in inadequate maintenance of connective tissue, cartilage, and bone. In various species, too little dietary manganese causes impaired skeletal development and reproduction, abnormal carbohydrate and lipid metabolism, and movement disorders.

But manganese can also be harmful to the brain, lungs, and reproductive system after excessive exposure. Most manganese comes from food, and regulatory systems in the intestine control the amount of ingested manganese that is actually absorbed. The National Research Council estimates a safe and adequate dietary intake of manganese at 2 to 5 milligrams a day. Although inhaled airborne manganese adds only slightly to total exposure under normal conditions, it bypasses intestinal regulation and is more completely absorbed into the circulation. Consequently, identifying an adequate but nontoxic exposure level requires consideration of multiple sources, various pathways of exposure, and differences in the uptake and distribution of manganese in children and other susceptible members of the population.

Sources of Manganese

The ordinary adult dietary intake of manganese ranges from 0.52 to 5.33 milligrams daily with an average of about 3 milligrams.[3] Infant manganese dietary intake varies considerably depending on the source of food. Human breast milk contains about 6 micrograms of manganese per liter. This is considerably less than infant formula, which contains 77 micrograms per liter if no manganese has been added to it or 99 micrograms per liter if it has been supplemented with manganese.[4] Since soy protein contains relatively high concentrations of manganese, soy-based formula contains 200–300 micrograms of manganese per liter.[5] As a consequence, formula-fed children receive much more manganese than breast-fed children.

Inhaled manganese rarely exceeds 1 percent of the dietary intake, even in polluted areas.[6] However, occupational inhalation exposures may be several orders of magnitude higher than normal in mining and ore-crushing facilities, dry-cell battery production, manufacturing of electrodes, or welding operations. Airborne manganese levels increase when MMT is added to gasoline supplies.

The Fate of Manganese in the Body

In adults, only about 3–5 percent, or approximately 100 micrograms, of ingested manganese is absorbed into the circulation. Much of the absorbed manganese is excreted via bile into the feces so that adults retain only approximately 30 micrograms manganese per day.[7] Animal studies show, however, that young animals absorb much more ingested manganese than adults—about 70 percent for young rats compared to 1 to 2 percent in the adult.[8] Moreover, manganese competes with iron and aluminum for a number of binding sites, and manganese absorption from the intestine is increased by iron deficiency. Manganese balance studies show that infants and young children also absorb more ingested manganese than adults, while they excrete less.[9] And the blood–brain barrier, which keeps many blood-borne chemicals from entering the brain of older children and adults, is immature in infants and allows proportionately more manganese to lodge in the central nervous system.

Airborne manganese is associated with aerosols of varying particle size. Small particles (diameters of less than 5 micrometers) are distributed more deeply into the lungs than larger particles, which tend to be deposited in the upper airways. Manganese particulates resulting from combustion of fuel containing MMT have a mean diameter of 0.2–0.4 micrometers.[10] Animal studies show that these small particles are cleared from the lungs within two weeks and taken into the systemic circulation.[11]

Once in the body, manganese is widely distributed to many tissues. A large portion of body stores (25–40 percent) is found in bone.[12] The liver, kidney, and pancreas also accumulate substantial amounts. In normal individuals, manganese has a total-body half-life of about 37 days.[13]

But the entrance of ingested manganese into the brain is slow, and clearance from the brain is also slow, with a half-life of about 150 days. There is a more direct route to the brain, however, with important implications for manganese toxicity and tissue distribution. Inhaled manganese can be absorbed through nasal tissues and transferred directly to the brain along the olfactory nerve, circumventing the blood–brain barrier. Studies in rats and fish using radiolabeled manganese instilled in the nasal passages indicate that the manganese readily migrates to the olfactory bulbs and then to most parts of the brain and spinal cord.[14] MRI imaging shows that the manganese jet from the olfactory nerve pathway is aimed at the portion of the brain where the most severe pathology is found in autopsy studies of patients with manganese neurotoxicity. Further migration occurs via secondary and tertiary neurons. The relevance of this pathway to humans with long-term, low-level exposures is unknown. In summary:

- Infants and children absorb more and excrete less manganese than adults. Moreover, children less than six months old have an incompletely developed blood–brain barrier, allowing circulating manganese direct access to the brain.

- The absorption and tissue distribution of dietary and airborne manganese differ significantly.
- Manganese inhaled into the nose or lungs bypasses the mechanisms that regulate intestinal absorption and excretion.
- The olfactory nerve provides a direct route for manganese into the brain, bypassing the blood–brain barrier.
- The half-life of manganese in the brain is about four times longer than in other tissues.

THE TOXICITY OF MANGANESE

Although manganese is an essential nutrient in plants and animals, it also has significant toxicity when absorbed and distributed in critical tissues in excessive amounts. Respiratory symptoms, pneumonia, or bronchitis occur in workers with large occupational exposures that are orders of magnitude larger than airborne levels attributable to the use of MMT in gasoline.[15] Adverse reproductive effects include testicular toxicity and reduced testosterone levels in animals exposed to manganese during fetal development at levels that show no other toxic effects but that are considerably higher than normal dietary intake.[16] There is also limited evidence of hormone changes in miners suffering from manganese poisoning.[17]

However, the more critical health effect that may occur at much lower levels of exposure is brain damage. This toxic effect is the most relevant for assessing the safety of adding MMT to gasoline. Obvious neurological effects of manganese were first noted in workers in manganese mines, refineries, and smelters. "Manganism" includes tremor and movement disorders, often preceded by transient "manganese madness," characterized by compulsive behavior that includes running, fighting, and singing.[18] There is considerable overlap between the neurological disorder due to manganese exposure and classical Parkinsonism. This has led to some confusion and disagreement about the relationship between the two conditions and their response to therapy.

The confusion is perhaps best explained by the observation that the areas of the brain affected by Parkinsonism and manganism overlap but are not identical. Moreover, inhaled manganese transported directly along the olfactory nerve may be distributed in the brain somewhat differently than ingested manganese. This implies that the route of exposure to manganese may be an important determinant of the resulting clinical symptoms and response to therapy. Inconsistencies in the response of manganese toxicity to agents useful in treating Parkinsonism may be due to these variables.

ASSESSING SAFETY

Three issues are of primary concern regarding the safety of adding MMT to gasoline. We discuss them in detail below.

Level of Exposure: Most Sensitive Indicators

It is undeniable that manganese is an essential nutrient and that occupational exposures to manganese may cause neurological toxicity. However, the threshold above which symptoms of neurological damage begin to appear is unknown. Moreover, there is considerable uncertainty about the nature of the most sensitive neurological end points.

Mergler et al. describe a continuum of dysfunction due to manganese exposure, including behavioral and emotional effects as well as the better known movement disorders.[19] Detailed neurological testing of exposed workers showed significant differences from controls in alternating and rapid movements, hand steadiness, cognitive flexibility, emotional state, and sense of smell. Yet, none of these participants exceeded the critical cutoff score of a standardized test battery that discriminates neurological patients from controls. In other words, even in the absence of obvious clinical neurological disorders, deficits are apparent with detailed testing.

Iregren reviewed four epidemiological studies of workers that used behavioral methods to look for early signs of manganese neurotoxicity after low-level exposures.[20] He concluded that there is a coherent picture of effects from low-level manganese exposure on response speed, motor functions, and memory. Alterations in mood and other subjective symptoms are inconsistently reported in these studies.

Roels et al. developed an integrative exposure assessment that recognizes the importance of cumulative exposures over time and calculated a tentative permissible dose that would produce no negative effects on hand steadiness in 95 percent of workers (micrograms manganese/cubic meter × years of exposure < 3,575 for total dust or 730 for respirable dust).[21]

The reversibility of neurological damage due to manganese is unpredictable, and better understanding awaits more complete knowledge of the mechanisms and natural history of the disorder. In most cases, overt symptoms of neurological toxicity evolve only after a prolonged period of exposure. Iregren reported that once neurotoxic effects from manganese exposure are clinically expressed, the damage to the central nervous system is essentially irreversible and, in some cases, may be progressive. One explanation for delayed development of symptoms holds that symptoms develop only with aging as the normal attrition of neurons unmasks underlying damage. This would explain a long latent period between significant manganese exposures and development of symptoms that depend on neuronal decline.

Adding MMT to the Gasoline Supply?

Some data are available to help predict environmental levels of manganese resulting from adding MMT to the gasoline supply. A study of airborne manganese levels in an urban setting in Quebec, Canada, where MMT was being used, compared results from high-density and low-density traffic areas.[22] Respirable manganese and total manganese were significantly higher in the high-density area than in the low-density area. The average exposure by inhalation was estimated to range from 0.001 to 0.030 micrograms per kilogram of body weight per day (respirable manganese) and 0.001 to 0.050 micrograms per kilogram of body weight per day (total manganese).

EPA monitoring studies in Riverside, California, where MMT was being used, estimated that about one-half of the population would be exposed to manganese levels higher than 0.05 micrograms per cubic meter, 5 to 10 percent to levels higher than 0.1 micrograms per cubic meter, and 1 percent to levels higher than 0.15 micrograms per cubic meter.[23] With daily respiratory volume of 20 cubic meters per day in adults, this implies increased pulmonary and nasal exposures combined to 1 to 3 micrograms per day for half of the population. Compared to the retained amount of dietary absorption, this represents up to a 10 percent increase. And since this additional exposure is via the pulmonary and intranasal route, it may be of greater significance than a similar dietary increase.

Blood manganese levels are of limited value as a monitor of actual exposure levels since they probably fail to reflect manganese deposited directly into the brain, and their utility for predicting adverse neurological effects is uncertain. Nevertheless, a population-based study of nonoccupationally exposed individuals in Quebec where MMT was being used in gasoline showed the zone of highest manganese blood levels was within the zone of highest water manganese levels and also highest manganese airborne levels.[24]

In their study of early neurological symptoms in workers exposed to manganese, Mergler et al. found that some members of the general population have blood manganese levels at or above the mean levels of occupationally exposed workers in whom alterations of motor function, cognitive flexibility, and emotional state were found.

Most Susceptible Populations

Not only do infants and children absorb more and excrete less dietary manganese than adults, but their immature blood–brain barrier also allows bloodborne manganese more ready access to the brain. Furthermore, children's weight-adjusted respiratory exchange is considerably larger than adults, making inhalation of ambient airborne manganese proportionately larger. Moreover, to the extent that an integrated lifetime cumulative exposure to manganese is related to the likelihood of developing neurological symptoms, children will be at larger risk than adults over a lifetime of exposure.

The brains of fetuses and newborns may be more susceptible to the toxic effects of manganese, and organisms exposed to manganese developmentally may be at risk of unique neurological effects.[25] Fetuses may be somewhat protected from elevated maternal levels by placental sequestration of manganese, though a limited amount crosses to the fetal circulation.[26]

Animals exposed to excessive manganese early in life show depressed levels of the neurotransmitters dopamine, norepinephrine, and serotonin.[27] Tagliaferro et al. have shown that gestational serotonin depletion in rodents causes much more extensive structural change in the brain of offspring than similar levels of depletion in adults.[28] This finding is entirely consistent with the well-known roles that neurotransmitters play in the developing brain, directing cellular migration, differentiation, and synapse formation.

A review of the published literature on manganese neurotoxicity in rodents identified seven studies in which test animals were exposed during the developmental period.[29] Three studies investigated behavioral outcomes, and each reported increased activity levels in offspring.

Several studies have reported a relationship between manganese hair levels in children and hyperactivity or learning disabilities.[30] As mentioned previously, infant formula contains considerably more manganese than breast milk, and soy-based formula has the highest concentrations of manganese. Collipp et al. found that the concentration of manganese in the hair of formula-fed normal newborn infants increased from 0.19 micrograms per gram of hair at birth to 0.965 micrograms per gram at six weeks, declining to 0.685 micrograms per gram at four months of age. In breast-fed infants, the hair levels increased only to 0.330 micrograms per gram at four months of age. They also found hair manganese levels of 0.434 micrograms per gram in hyperactive children and 0.268 micrograms per gram in age-matched controls from 7–10 years of age. Pihl reported hair manganese levels of 0.83 micrograms per gram in hyperactive children compared with 0.58 micrograms per gram in controls. This study also found higher lead levels in hyperactive children.

Crinella et al. recently reported the results of a pilot study that also found that children with attention-deficit hyperactivity disorder (ADHD) have significantly higher levels of manganese in head hair than matching age and demographical controls.[31] The investigators intend to study a larger group of ADHD children, half of whom are also unusually aggressive, to further evaluate this relationship.

People who are destined to develop Parkinsonism later in life may also be particularly susceptible to manganese toxicity. To the extent that manganese migrates to the brain, either directly through nasal tissues or more slowly across the blood–brain barrier, toxic effects may speed the onset or increase the severity of Parkinsonism. The degree to which manganese contributes to classical Parkinsonism remains uncertain and is a matter of debate.[32]

AREAS OF UNCERTAINTY

Uncertainty about the risks of adding MMT to the gasoline supply results from the following:

1. There is no good method for easily and accurately assessing brain exposure to manganese. The relative contributions of the multiple pathways of exposure to brain manganese levels are unknown. Blood manganese levels are likely to reflect current dietary exposures and have little utility for estimating cumulative brain exposures. Consequently, data necessary to correlate low-level airborne exposures to neurological effects are lacking.

2. Important aspects of the neurological toxicity of manganese are not well understood. Whether or under what conditions neurological damage is progressive or the result of multiple factors is uncertain. Investigators have only recently begun to look for the most sensitive measures of early neurological toxicity. There is, as yet, no general agreement on the nature of the earliest effects, how best to test for them, or the reasons for a latent period before symptoms appear.

   Some of the effects for which there are preliminary data are those about which we care deeply but find the most difficult to examine. Neurobehavioral toxicology is an evolving science with no clear consensus on test methods, end points, or the validity of extrapolating from one species to another.

3. The susceptibility of the developing brain to manganese toxicity deserves special attention. Neurological development and neuroendocrine function are dependent on appropriate neurotransmitter levels. Neurotransmitters are altered by developmental exposure to manganese. Data show that infants and children are at risk of health effects

different from those in adults. The relationship between hair manganese levels and ADHD while supported by several studies needs further investigation. It is unclear whether manganese deposited in the brains of infants and children increases their risk of neurological disease later in life. There are no published studies in which animals were exposed to manganese by inhalation, followed by examination of effects on neurobehavioral end points.

Early-life exposures to manganese may have an impact on hormone levels that may not be apparent until reproductive maturation years later. For example, prolactin levels are sometimes elevated in workers exposed to low-level manganese in industrial settings.[33] It is unknown if developmental exposures to manganese may have a similar effect.

4. Many papers, including a 1990 analysis from the Ethyl Corporation, report manganese oxides as the primary inorganic combustion product of MMT in gasoline.[34] However, more recent reports state that total particulate samples contain primarily manganese phosphate with manganese sulfate and only minor amounts of manganese oxides.[35] Varying solubilities of these different compounds are likely to influence the speed of direct uptake from the respiratory tract and may affect migration of manganese along the direct olfactory pathway.

5. Though manganese is an abundant element in the Earth's crust, levels of exchangeable manganese, the form available for uptake into plants, are orders of magnitude lower than total manganese in soils. There is a relationship between MMT use and manganese contamination in soils.[36] Plant manganese uptake depends on a number of factors including soil type, pH, and concentrations of other minerals. Human and wildlife exposure to manganese may increase from MMT use because of bioaccumulation of manganese in plants or surface water deposition.

6. The Department of Energy reports delivery of 7.73 million barrels of petroleum product to the transportation industry for gasoline daily. EPA has approved the addition of 0.03125 grams of MMT (as manganese) per gallon of gasoline. If this volume of gasoline were consumed in the United States daily, and if all gasoline were to contain MMT at this level, over 2,900 tons of manganese as phosphates, oxides, and sulfates would be released to the atmosphere annually in the United States. How this general dusting might alter the cycling of manganese through the biogeosphere is uncertain.

7. An important related issue deserves immediate attention. Since infants absorb a much larger percentage of ingested manganese than adults, excrete less of the absorbed manganese, and have a poorly developed

blood–brain barrier, it may be wise to reconsider the wisdom of feeding infants formulas that are supplemented with manganese or that contain naturally high concentrations of manganese. Concentrations of manganese in breast milk may be low for a good reason. To assume that infant formulas must be supplemented with manganese because it is an essential nutrient and because the adult diet contains larger amounts of manganese is to ignore much of what we now know about the pharmacokinetics of manganese. Widespread use of soy-based infant formulas in the first few months of life may be unwise, not only because of the high manganese content, but also because of the presence of phyto-estrogens, whose potential health effects in this age group are poorly understood.

CLEAN AIR ACT AND PRECAUTION

Widespread MMT use in gasoline will effectively expose every member of the population and all ecosystems to inorganic manganese combustion products. The neurological effects of manganese exposure for which there are well-established as well as preliminary data are not trivial. Therefore, better understanding of manganese neurotoxicity and the threshold at which it begins to occur should be critical to policy decisions.

There does not appear to be any public health benefit from adding MMT to fuel. The effect of MMT on automobile exhaust emissions has been studied.[37] There is no difference in carbon monoxide emissions from cars burning fuel with and without MMT. Nitrogen oxide emissions are slightly lower in cars using MMT fuel while total hydrocarbon emissions are slightly higher. Therefore, the advantages of MMT use appear to be limited to improved automobile performance resulting from octane enhancement and improved profitability for the Ethyl Corporation.

The Clean Air Act requires the EPA to protect the public with an ample margin of safety. Ordinarily, the agency develops a standard for a given environmental contaminant that is an estimate of exposure likely to be without adverse health effects in the general population, including those who are especially susceptible. A review of research findings demonstrates that the information necessary to develop a health-protective standard for airborne manganese is currently insufficient. The effects of up to a 10 percent increase in manganese uptake, much of this by inhalation with the potential for direct transport to the brain, are uncertain. The developing organism is particularly susceptible to the toxicity of manganese and, compared to adults, infants, and children, would be disproportionately exposed to airborne manganese because of their relatively larger respiratory volume. The impact of a population-wide increase in airborne manganese exposure on the incidence and severity of neurological disorders in the future is unknown.

Based on experimental and epidemiological evidence that currently exists, a precautionary policy would require that the safety of MMT be demonstrated prior to its use in gasoline. Adequate toxicity testing must include not only attention to the various routes of exposure but also comprehensive examination for subtle or delayed neurological and developmental effects resulting from long-term, low-dose exposures.

A precautionary approach would also require answers to the obvious questions: Is this product necessary? Are there alternatives? Is it prudent to allow a substance into the marketplace for widespread use when there is preliminary evidence that the route of exposure may enhance its low-dose toxicity—all for the sake of saving a few cents per gallon on a fuel that is already considerably cheaper than bottled drinking water?

Public and environmental health would be better protected if a full toxicological analysis were consistently a premarketing requirement for all chemicals to which the public and ecosystems are to be exposed. Then, perhaps, the true costs might be compared to expected benefits before a product is ever brought into the marketplace.

NOTES

1. Solomon, G.M., Huddle, A.M., Silbergeld, E.K., and Herman J. Manganese. Gasoline: are we repeating history? *New Solutions* 17–25, Winter 1997.
2. Ter Haar, G.L., Griffing, M.E., Brandt, M., et al. Methylcyclopentadienyl manganese tricarbonyl as an antiknock: Composition and fate of manganese exhaust products. *Journal of Air Pollution Control Association* 25:858–860, 1975.
3. Greger, J.L., Davis, C.D., Suttie, J.W., and Lyle, B.J. Intake, serum concentrations, and urinary excretion of manganese by adult males. *American Journal of Clinical Nutrition* 51:457–461, 1990.
4. Dorner K., Dziadzka, S., Hohn, A., et al. Longitudinal manganese and copper balances in young infants and preterm infants fed on breast-milk and adapted cow's milk formulas. *British Journal of Nutrition* 61(3):559–572, 1989.
5. Lonnerdal, B. Nutritional aspects of soy formula. *Acta Pediatric Suppl.* 402: 105–108, 1994.
6. Tabacova, S. Maternal exposure to environmental chemicals. *Neurotoxicol.* 7(2):421–440, 1986.
7. Cotzias, G.C., Horiuchi, K., Fuenzalida, S., and Mena, I. Chronic manganese poisoning: Clearance of tissue manganese concentrations with persistence of the neurological picture. *Neurology* 18:376–382, 1968.
8. Mena, I. The role of manganese in human disease. *Ann. Clin. Lab. Sci.* 4(6):487–491, 1974.
9. Dorner, K., Dziadzka, S., Hohn, A., et al.

10. Lynam, D.R., Pfeifer, G.D., Fort, B.F., and Gelbcke, A.A. Environmental assessment of MMT fuel additive. *Sci. Tot. Environ.* 93:107–114, 1990.
11. Andersen, M.E., and Clewell, H.J. Bioavailability and toxicokinetics of manganese. *Neurotoxicol.* 19(3):452, 1998.
12. Ibid.
13. Mena, 1974.
14. Tjalve, H., Henriksson, J., Tallkvist, J., et al. Uptake of manganese and cadmium from the nasal mucosa into the central nervous system via olfactory pathways in rats. *Pharmacol. Toxicol.* 79:347–356, 1996. Tjalve, H., Mejare, C., Borg-Neczak, K. Uptake and transport of manganese in primary and secondary olfactory neurons in pike. *Pharmacol. Toxicol.* 77:23–31, 1995.
15. Roels, H., Lauwerys, R., Buchet, J.P. et al. Epidemiological survey among workers exposed to manganese: Effects on lung, central nervous system, and some biological indices. *Am. J. Ind. Med.* 11:307–327, 1987.
16. Laskey, J.W., Rehnberg, G.L., Hein, J.F., Carter, S.D. Effects of chronic manganese (Mn3O4) exposure on selected reproductive parameters in rats. *J. Toxicol. Environ. Health* 8:677–687, 1982. Gray, L.E., and Laskey, J.W. Multivariate analysis of the effects of manganese on the reproductive physiology and behavior of the male mouse. *J. Toxicol. Environ. Health* 6:861–867, 1980.
17. Rodier, J. Manganese poisoning in Moroccan miners. *Br. J. Ind. Med.* 12:21–35, 1955.
18. Goyer, R.A. Toxic effects of metals. In *Casarett and Doull's Toxicology* 5th edition, ed. Klaassen, C.D. McGraw-Hill, New York, 1996.
19. Mergler, D., Huel, G., Bowler, R., et al. Nervous system dysfunction among workers with long-term exposure to manganese. *Environ. Res.* 64:151–180, 1994.
20. Iregren, A. Using psychological tests for the early detection of neurotoxic effects of low level manganese exposure. *Neurotoxicol.* 15(3):671–678, 1994.
21. Roels, H.A., Ghyselen, P., Buchet, J.P., et al. Assessment of the permissible exposure level to manganese in workers exposed to manganese dioxide dust. *Brit. J. Ind. Med.* 49:25–34, 1992.
22. Zayed, J., and Loranger, S. Environmental contamination and human exposure to airborne total and respirable manganese in Montreal. *Neurotoxicol.* 19(3):464, 1998.
23. Hudnell, K. Effects from environmental exposures: A review of the evidence from case studies and environmental studies. U.S. EPA. *Neurotoxicol.* 19(3):454, 1998.
24. Baldwin, M., Mergler, D., Larribe, F. et al. Bioindicator and exposure data for a population based study of manganese. Abstract. *15th Annual Neurotoxicol.* 19(3):472, 1998.
25. Chandra, S.V., and Shukla, G.S. Manganese encephalopathy in growing rats. *Environ. Res.* 15–28, 1978.
26. Fechter, L.D. Distribution of manganese during development. *Neurotoxicol.* 19(3):453, 1998.
27. Singh, J., Husain, R., Tandon, S.K., et al. Biochemical and histopathological

alterations in early manganese toxicity in rats. *Environ. Physiol. Biochem.* 4:16–23, 1974.

28. Tagliaferro, P., Ramos, A.J., Lopez, E.M., et al. Comparative neurotoxic effects of serotonin depletion in adult and neonatal rat brain. *Neurotoxicol.* 19(3):473, 1998.

29. Boyes, W.K., Miller, D.B. (U.S. EPA and CDC/NIOSH). A review of rodent models of manganese neurotoxicity. *Neurotoxicol.* 19(3):468, 1998.

30. Pihl, R.O., and Parkes, M. Hair element content in learning disabled children. *Science* 198:204–206, 1977. Collipp, P.J., Chen, S.Y., and Maitinsky, S. Manganese in infant formulas and learning disability. *Ann. Nutr. Metab.* 27:488–494, 1983.

31. Crinella, F.M., Cordova, E.J., and Ericson, J.E. Manganese, aggression, and attention-deficit hyperactivity disorder. *Neurotoxicol.* 19(3):468–469, 1998.

32. Feldman, R.G. Manganese as possible ecoetiologic factor in Parkinson's disease. *Ann. N.Y. Acad. Sci.* 648, 1992.

33. Smargiassi, A., and Mutti, A. Peripheral biomarkers and exposure to manganese. *Neurotoxicol.* 19(3):455–456, 1998.

34. Lynam, D.R., Pfeifer, G.D., Fort, B.F., and Gelbcke, A.A. Environmental assessment of MMT fuel additive. *Sci. Tot. Environ.* 93:107–114, 1990.

35. Dorman, D.C. The pharmacokinetics of manganese phosphate, a MMT combustion product, in rats and monkeys. *Neurotoxicol.* 19(3):459, 1998. Wong, J., Deutsch, S.E., Colemnares, C.A. et al. Manganese particulate from vehicles using MMT fuel. *Neurotoxicol.* 19(3):472, 1998.

36. Brault, N., Loranger, S., Courchesne, F., et al. Bioaccumulation of manganese by plants: Influence of MMT as a gasoline additive. *Sci. Tot. Environ.* 153:77–84, 1994.

Chapter 20

CLEANER PRODUCTION AND THE PRECAUTIONARY PRINCIPLE

Ken Geiser

Since the early 1970s, the primary technological approach to managing environmental pollution has focused on pollution control technologies that could be installed at the end of pollution discharge pipes. These so-called "end-of-pipe" technologies were mandated by various government laws and regulations to bring pollution releases into compliance with government-issued pollution permits. The permits were typically based on the best available control technology at the time and the capacity of receiving environmental medium to dilute and assimilate the pollution at levels that science demonstrated to be below levels of concern to human health or ecological systems.

Beginning in the mid-1980s, a new approach surfaced. This approach sought to address pollution generation not at the end of the pipe, but rather by transforming the processes of production that generated the pollution in the first place. Following a seminal report by the U.S. Congressional Office of Technology Assessment in 1986, this new approach was called "pollution prevention in the United States."[1] In international settings, the approach had several different names until 1989 when the United Nations Environment Program (UNEP) adopted the term "cleaner production." Since then, the term cleaner production has been generally accepted in international fora.

CLEANER PRODUCTION AND THE PREVENTIVE APPROACH

In adopting the term cleaner production, UNEP offered a definition of the concept that serves as a point of reference for its advocates. For UNEP:

> Cleaner production means the continuous application of an integrated preventive environmental strategy to processes and products to reduce risks to humans and the environment.
>
> For production processes cleaner production includes conserving raw materials and energy, eliminating toxic raw materials, and reducing the quantity and toxicity of all emissions and wastes before they leave a process. For products the strategy focuses on reducing impacts along the entire life cycle of the product, from raw material extraction to ultimate disposal of the product. Cleaner production is achieved by applying know-how, by improving technologies, and by changing attitudes.[2]

The pollution prevention or cleaner production approach involves redesigning the processes of production. In order to promote this objective, advocates of cleaner production have developed an array of techniques (referred to as a "toolkit") to reduce pollution by substituting more benign materials for hazardous materials, by optimizing production technologies, by conserving raw materials and energy, by improving operations and maintenance, by redesigning the products of production, and by closing process lines to form loops that recycle and reuse materials.[3] To promote these techniques some international programs encourage environmental audits. In the United States, pollution prevention programs have been promoted at the state level. Thirty states have passed some kind of pollution prevention law. Most of the laws are perscriptive and nonregulatory instruments offering technical assistance to firms that volunteer to participate. A handful of states enacted so-called "toxics use reduction" laws focused on reducing the use and release of toxic chemicals. Many of these states encourage or require facility-level pollution prevention or "toxics use reduction" plans. These audits and plans are instruments for identifying and analyzing options that can guide a continuous search for the cleanest and safest means of producing products or services.

Cleaner production programs involve a reconsideration of products as well as production processes. New techniques for considering the environmental impacts of products over their lifetime have emerged in the form of "life cycle assessment." Life cycle assessments conducted by firms on their products have proven useful in identifying priority points of environmental impact and in comparing the environmental attributes of various products or product components. The use of life cycle assessments has spurred an increase in the firm adopting "extended producer responsibility" programs that commit the firm to assist in the management of its products throughout their useful life and at the point of final disposal.

When cleaner production and pollution prevention techniques are implemented at the industrial facility, they have often produced significant results that reduce both environmental burdens and operating costs. The cost reductions have resulted from reduced expenditures for raw materials or energy, reduced costs for government compliance and the purchase of pollution control technologies, avoided costs for waste treatment services, reduced liability costs, and improved efficiencies in operating systems.

Cleaner production has been promoted by governments through various "demonstration programs" that began in Sweden and The Netherlands in the early part of the 1990s. These cleaner production demonstration programs bring managers of groups of firms in a region together to learn about cleaner production, conduct audits in search of opportunities, implement specific projects, and assess and document results. The U.S. state programs have also been aggressive in working closely with firms, but the focus has been less geographic and more on certain industrial sectors or specific production processes. Cleaner production demonstration programs have been launched all over Europe and are now common in countries as distant as India, Thailand, China, Mexico, Brazil, and Mozambique.

Throughout the 1990s, there has been a remarkable array of published articles and reports that document the environmental and financial savings that have resulted from the implementation of pollution prevention and cleaner production projects within industrial facilities. The professional trade literature is full of case studies of firms that have succeeded with pollution prevention or cleaner production programs, and there are now journals specifically focused on pollution prevention, cleaner production, and total environmental quality management. The United Nations Environment Program has published a compendium of case studies from around the world demonstrating environmental and cost benefits, including examples of:

- Recovery and sale of protein from potato starch effluent in a Dutch firm through the installation of a reverse osmosis process;
- Reduction of overspray wastes in the painting processes of a Polish metal steel plant through conversion to a pressure atomized electrostatic spray;
- Elimination of sulphide in the effluent of an Indian textile dying operation through the substitution of hydrol from the maize starch industry; and
- Reduction of volatile air emissions from a British firm's adhesive products by converting to water-based adhesive chemistries.[4]

Those who have been leaders in the promotion of pollution prevention and cleaner production have contrasted this new approach to the old pollution control approach. They argue that the new approach is best seen as a "paradigm shift" in which a new professional thesis has replaced an older conceptual approach. The old approach relied on "command and control" regulations to enforce compliance with permits that assured that pollution would not exceed the "assimilative capacity" of the environment. This is viewed as a "permissive approach," because it permits a certain amount of pollution to continue as long as it does not exceed certain regulation-defined discharge limits. In contrast, the new "preventive approach" relies on the continuous improvement of production processes so as to reduce the generation of pollution. This new approach is not compliance driven but instead relies on the will and ingenuity of industrial plant managers who seek to lower operating costs by improving material and energy efficiencies and by reducing the production of pollutants that must be managed. Such a approach could be considered as precautious.

The Concept of Precaution

Discussion of a Precautionary Principle began among environmental policy makers during the late 1980s in Europe. This principle was derived from the concept of *Vorsorgeprinzip* written into German water pollution law during the 1980s and promoted in the Ministerial Declaration of the 1987 Second International Conference on the Protection of the North Sea, which stated:

> . . . [I]n order to protect the North Sea from possibly damaging effects of the most dangerous substances, a precautionary approach is necessary which may require action to control inputs of such substances even before a causal link has been established by absolutely clear scientific evidence. . . .[5]

The principle has become an important element in environmental advocacy because it appears to require two commitments: first, to act cautiously even in the face of uncertain scientific knowledge; and second, to shift the burden for scientific justification from those opposed to an action onto those who promote the action.

Arguments about the usefulness of the Precautionary Principle have been particularly lively in regard to the discharge of pollutants into the marine environment.[6] Those who promote the concept argue that the scientific knowledge about the assimilative capacity of the ocean is at best uncertain, scientific knowledge about the health and environmental effects of many pollutants is inadequate, and the effects of pollutants in marine environments is still incomplete. Therefore, generators of pollutants should err on the side of caution and discontinue releasing pollutants that will eventually enter the sea. If the generators of pollutants persist in their desire to release these pollutants, then it is their burden to prove that the release of the pollutants will not cause harm. Such an argument is particularly vexing to industrial managers, because it would appear quite difficult to prove that the release of most pollutants would not cause at least some harm, and, therefore, they see the consequences of the Precautionary Principle as requiring the curtailment of any release of those pollutants.

THE PRECAUTIONARY CONCEPT IN THE PREVENTIVE APPROACH

The debate about the Precautionary Principle is meaningful within the context of the conventional permissive paradigm associated with pollution control. Within this approach, the focus is centered on the point of release of a pollutant. The substantive question is how much pollution can be released without damage to public health or the environment. But, the prevention paradigm implied by pollution prevention and cleaner production would shift and recast this debate. The focus of attention in this newer approach is not on the end of the pipe, but rather on the point of production and the design of the product. Here the question is not about the quantity of pollution to emit, but rather on how the process or product can be modified so as to reduce or eliminate the pollution. The search is not about finding an acceptable level of risk, but rather, about how much risk is preventable. Instead of a heroic struggle to predict future consequences of a hazardous act, the cleaner production approach is more humble in its ambition: It promotes a sequence of actions that would step-by-step reduce the hazard. The focus is on how much contamination can be avoided by considering an array of options that could produce the desired product or activity. Cleaner production does not force industrial managers into defensive corners where they feel their only option is to cease production; instead, it challenges managers to continue production and to use their ingenuity to find techniques and technologies that promote efficiencies and reduce environmental impacts.

Cleaner production is preventive by definition and by implementation. The UNEP definition notes the preventive element, and the projects that appear in the demonstration programs typically prevent pollution. But, are they precautious? First, do cleaner production projects promote caution in the face of uncertainty, and, second, do they shift the burden of justification for health and environmental effects?

The second question is the easiest to answer. Cleaner production is solely the responsibility of the industrial firm that otherwise would generate pollution. In many U.S. states and international demonstration programs, government agents or government-sponsored consultants may provide assistance to firms attempting to implement cleaner production programs, but the responsibility is clearly located with the management of the firm.

The answer to the first question is less clear. Uncertainty is present in the cleaner production paradigm. Industrial managers typically have inadequate knowledge about the effects of the pollutants that they focus on. But that uncertainty is not used to deter the search among alternative options. Instead, typical cleaner production programs urge managers to identify alternative options for environmental improvement and set priorities among them based on self-defined hazard, technology, and financial criteria. Because managers are urged to consider technical and financial considerations as well as hazard criteria in setting priorities, they could be seen as less cautious than if they focused on hazardousness alone. Indeed, even in Massachusetts where there is a tough toxics use reduction law, facility managers who are required to identify and review cleaner production options are not required to implement them. But, it is argued that in promoting a fair and thorough search of alternatives, there is a stronger likelihood of adoption than were the process more mandatory. A recent evaluation of the Massachusetts program bears this out: Eighty-one percent of the 420 firms surveyed reported that they would implement at least some of the options identified.[7]

Cleaner production programs are only as good as the scope of their goals. For instance, inappropriate boundaries can result in undesirable externalities. A cleaner production project that improves one aspect of production can leave opportunities for other unintended consequences. A project that reduces emissions at the plant by converting from solvent-based cleaning to water-based cleaning that requires a drying oven might inadvertently increase emissions from a distant power-generating station. A project that reduces pollution by increasing manual operations might inadvertently increase ergonomic risks for employees. The UNEP definition and the toolkit of many active government programs promote life cycle assessment and substitution analyses to account for these potentially unconsidered aspects, but there remains work to be done on these new methodologies.[8]

Recognizing these limits does not negate the preventive approach; rather, it heightens the role of the precautionary concept in cleaner production projects. At the core of a cleaner production assessment is a decision-making procedure that sets appropriate boundaries, identifies important criteria, encourages innovation and creativity in identifying options, and establishes a means for measuring performance and encouraging continuous improvements. The search among alternative options for the sequence of steps that will improve the efficiency of production and reduce the environmental impact is goal oriented and improvement driven and it admits to immediate imperfections. Cleaner production is, therefore, best seen as a continuous process, or, as some advocates say, as a "highway," and the journey is developed in an exploratory and site-specific manner. Well-founded decision making is critical to maintaining a steady course. The Precautionary Principle could be one of the pragmatic tests that is employed to steer cleaner production programs. Translating the principle into a clear protocol with specific steps would permit it to be taught to industrial managers and used as a credible decision-making algorithm.[9]

Over the years of implementation, cleaner production advocates have learned many lessons that are often written about. These include:

- Importance of upper management commitment to the process;
- Value of collective (team) thinking that crosses management divisions within the facility and includes "shop floor" employees;
- Value of appropriate, valid, and current information on new technologies and practices;
- Usefulness of internal leadership (so-called "champions");
- Value of starting with small, low-cost projects (so-called "low-hanging fruit") and building up on a base of successes;
- Importance of full-cost accounting that reveals the true costs of production including the somewhat "hidden" costs of environmental and health factors;
- Importance of an open and creative search for alternative options; and
- Value of metrics for measuring performance and evaluating success.

Whether facility managers initiate projects to replace solvent-based cleaning technologies with aqueous chemistries that reduce the emission of volatile organic compounds, or install better process controls to reduce the generation of off-specification paints, the action is taken by the generator of a potentially hazardous industrial process who stands to benefit by the process. The intention is to reduce the hazardousness of the process, even when there is no scientific certainty about the degree of hazard involved. There is little in these efforts to assess human exposures or to conduct lengthy risk assessments. The focus is on hazard reduction in the expectation that overall risk will also be reduced.

A tough scientific posture might find these initiatives to be somewhat crude because the scientific justification is not solid. Yet, from the practical perspective of a facility manager and, often, from an economic perspective, as well, these projects appear justified. Such pragmatics suggests that one improves what one can, particularly if it makes sense, and one leaves the causal proofs until later.

PROCESS IMPROVEMENTS AND
TECHNOLOGICAL DEVELOPMENT

Cleaner production need not be bound by current technologies. The pollution control regulations established over the past twenty years have been significant catalysts for the development of new waste management technologies. Similarly, cleaner production programs can promote innovation in management practices and in technological development.[10] A brief review of facility case studies and demonstration projects reveals many clever and innovative solutions. In Massachusetts, a diaper laundry service worked with its customers to promote an alternative to zinc oxide used as a diaper rash inhibitor in order to reduce the zinc oxide contaminants in the laundry rinse waters discharged into the Boston Harbor. The Toxics Use Reduction Institute worked with an entrepreneur in Massachusetts to assist in opening the first drop-off garment cleaning service that fully avoids the use of perchloroethylene in favor of a water-based technology. Indeed, the drive to reduce the use of chlorinated solvents in industrial parts cleaning and degreasing has permitted the expansion of a whole new industry promoting new aqueous and semiaqueous cleaning chemistries.

Cleaner production projects have promoted the development and use of many new technologies. For instance, closed-loop processes in electroplating have led to countercurrent washing techniques that permit the reuse of rinse water for progressively more contaminated operations. Fluxless soldering and low- and no-lead solders have been used to replace harsh soldering materials in the electronics industry. Ozone and peroxide pulp bleaching and oxygen delignification have reduced some of the most extensive environmental loading from pulp and paper plants. Metal cutting and stamping operations have been converted to vegetable oil cooling and cutting fluids. Hazardous paints and coatings have been replaced by electrostatic powder and radiation-cured coatings. Newspaper printing has converted to soy-based and low metal inks. Water-soluble and biodegradable polymers are replacing ethylene-based plastics in packaging that is likely to be disposed in marine environments. Many of these technologies would not have achieved acceptance were there not an increasing value placed on cleaner and safer production and products.

If cleaner production means a continuous process for improving efficiencies and reducing wastes, then it could be guided by a goal of full optimization and zero emissions. This is the direction that Gunter Pauli at the United Nations University's Zero Emissions Research Initiative is now promoting.[11] Zero emissions is in keeping with other industry efficiency objectives such as zero defects, zero accidents, and zero inventory.[12] After all, the most precautious approach to pollution would be to produce none at all. There is evidence for this vision in some writings on industrial ecology. Industrial ecology is a concept developed over the past decade originating in the engineering disciplines.[13] It has both a broad and narrow definition. Broadly, it implies a restructuring of the industrial production system so that its materials and energy flows fit nondestructively into the material and energy flows of the planet. More narrowly, industrial ecology is demonstrated in the siting of industrial facilities so that the wastes of one facility can be recycled into raw material inputs for another.

This "cascading" management of wastes prolongs the useful life of materials and delays release of those wastes as pollutants. The goal is to maximize the continued use of materials. This management of material flows to a point of optimal use would suggest that waste is reduced to zero by continuous recycling and reuse. (In reality, there can be no true zero waste, because thermodynamic principles dictate that there is always some loss from any material or energy transformation. Yet, for speculative purposes, zero waste could be considered as a goal.) The prospect of zero emission industries is therefore made possible not at the facility level, but at the industry or, at least, industrial park level. Indeed, industrial ecology seeks to eliminate the focus on waste and replace it with a focus on the reuse of potentially valuable materials. Unfortunately, industrial ecology offers little guidance on how to reduce the toxicity of the materials being recycled. While a zero emissions industrial system may function so as to contain toxic materials, this may do little to reduce harm to those who work inside those contained systems.[14]

In terms of precaution, the preventive approach, therefore, shifts the industrial activity from the technologies of waste management to the development and use of technologies that improve production. The focus of environmental policy is shifted from risk assessment to technology options assessment, but there remains much more to do. The broad concept of industrial ecology provides a vision, but the technologies of production remain largely uninformed by a precautious commitment.

PRODUCT IMPROVEMENT AND PRECAUTIOUS CONSUMPTION

Much of the early effort in promoting cleaner production focused almost exclusively on industrial production processes. In the United States, there has been an effort to broaden the focus of the early pollution prevention pro-

grams by extending the idea to agriculture, transportation, and energy production. In Europe, there has been significant developments in addressing
the products of production. The serious interest in life cycle assessment and
"product chain management" has allowed cleaner production to merge with
developing programs in ecological product design ("eco-design") and
"extended producer responsibility," whereby product producers are urged to
provide management for their products throughout the life of the product.
In Germany, this has led to the principle of "product take back" under which
firms must make arrangements for the return of the product once the user
has finished with it. German law now requires that product producers must
arrange for the return of product packaging, and soon there will be programs
for the take back of electronic components and automobiles.

This focus on products is clearly within the parameters of the UNEP definition of cleaner production, and the UNEP program has been aggressive in
its promotion of life cycle assessment, the sponsorship of conferences on
extended producer responsibility, and the establishment of a new center on
sustainable product design. The more advanced cleaner production projects
carried out at the facility level include consideration of product as well as
process design. Environment and health factors are integrated into customer
preferences, marketing objectives, and design specifications. Because this
requires an environmentally sensitive customer and supply chain, vendors,
suppliers, contractors, and customers are brought into a collective process by
which cleaner production projects are developed.

The precautious concept extends the preventive approach here as well.
Product users need to be cautious in the disposal of products about which
they have little information. At issue is not only the safest and cleanest
means of manufacturing a product, but also the most environmentally
friendly way of using and disposing of the product. Product recycling plays a
role here, as does product material composition and energy requirements.
Products intended for rapid disposal need to degrade easily. Products purchased for a long-use life need to be high quality and durable.

The Precautionary Principle suggests that products themselves need to
be considered skeptically. Just as a possible goal of precaution in production is optimal efficiency and zero emissions, a possible goal in consumption is extended use and zero disposal. The useful life of a product can be
extended by attention to its durability, adaptability, and repairability.
Attention needs to be given to product reuse and secondary-product markets and to building an economy based on servicing products rather than
making and wasting them. Indeed, Walter Stahel at the Product Life Institute in Geneva has been writing and advising on the restructuring of an

economy based on services, rather than products, for some time.[15] In this vision, firms would seek to fill customers needs, rather than simply sell them products. This more "functionalist" approach would reduce the need to move and transform materials and consume energy, in favor of performing functions.

While this vision may seem too utopian, some firms such as Xerox, Sony, and Allied Signal have adopted interesting new programs for exploring the area. All are experimenting with leasing equipment and materials and providing services to customers that once only purchased products. Xerox has set out goals that lead toward the zero-waste factory and the zero-waste office with programs that promote "asset management" for its leased products and image transfer for its customer service. As pace setting as these experiments are, they reveal the vast gulf between where they are and where the vast majority of commodity producers currently are. In most of the world, a precautionary approach to products and consumption is barely more than a misty vision.

(RE)AFFIRMING THE PRECAUTIONARY CONCEPT IN CLEANER PRODUCTION

Much of this more visionary thinking is promoted as good for the environment and good for business. There is such an attractive duality to this approach that some promoters of cleaner production glibly speak of its "win-win" character. Cleaner production is often promoted inside firms as a cost saving approach, and there are plenty of calculations that demonstrate cost avoidance in operations and capital investments with promising internal rates of return and satisfyingly short payback periods. As promoters struggle to integrate cleaner production into conventional business culture, one of the most rapidly expanding areas of development has been in the conventional business areas of accountancy and commercial banking. The World Business Council for Sustainable Development has sponsored a series of conferences and workshops on "eco-efficiency." A new text on eco-efficiency promotes environmental protection practices almost singularly on the basis of economic efficiency.[16]

There is much to be praised in these efforts to integrate cleaner production into conventional business management culture. Yet, it is reasonable to be cautious about a severe reductionism that too narrowly reshapes cleaner production into simply one more business planning factor. The duality of economic and environmental benefits allows for the optimizing of two important values that are not necessarily complementary. The challenge of cleaner production is to find those solutions that advance both values simultaneously, not to reduce one value into the other. It is one thing to find those clever business or technology options that improve both environmental and economic performance; it is quite another to assume that there is no differ- ence between these two objectives.

In the face of such reductionism, it is useful to overtly declare that pre- caution is a central tenet of cleaner production. Affirming precaution is a way of assuring that even the most well-intended economic investments in cleaner production will not veer off toward a one-dimensional focus on profit maximization. Affirming precaution in economic development poli- cies can be used to blunt the worst environmental consequences of market forces alone shaping production programs. This is particularly true where cleaner production programs may have negative impacts on other parts of the economy. Achieving higher levels of energy efficiency, reducing the use of toxic chemicals (chlorinated solvents, for instance), or converting from products to services will have undesired effects on economic sectors that generate energy, manufacture toxic chemicals (chlorine, for instance), or sell products. In such cases, the Precautionary Principle will provide an impor- tant foundation for environmental advocacy.

In the future, precaution must be more soundly embraced as a foundation principle of cleaner production. For some advocates of the preventive approach, the concept of precaution has always been assumed. Yet, a search of much of the literature on pollution prevention or cleaner production reveals little evidence of the precautionary concept in writings or references. The Precautionary Principle needs to be affirmed or reaffirmed by those who promote cleaner production and the preventive approach. Specific actions could be initiated that would move the precautionary idea forward. These include:

- The Precautionary Principle needs to be more aggressively promoted by the nongovernmental environmental and health advocacy organizations;
- Case studies of government and industrial experiences in using the pre- cautionary concept need to be written and published;
- The precautionary concept needs to be developed as a practical decision- making tool for use by industrial managers in purchasing, process design, and work organization decisions;
- Cleaner production and pollution prevention training and guidance man- uals need to include and integrate the concept of precaution;
- Professional and academic education programs in engineering, business, and law needs to integrate the precautionary concept into curricula; and
- Precautionary language needs to be written into statutes and international agreements promoting pollution prevention and cleaner production.

Cleaner production should be the manifestation of the Precautionary Principle in industrial production. The close parallel between cleaner production and the precautionary concept has led to the inclusion of cleaner production policies in international agreements that promote precaution, such as the London Dumping Convention, the International Declaration on the Protection of the North Sea, and the Bamako Convention on Transboundary Transportation of Hazardous Wastes. But, it is not enough to assume that prevention implies precaution. Too many cleaner production programs remain limited in scope and vision. A precautionary perspective extends prevention toward zero emissions and zero disposal. While these absolute goals may be functionally unobtainable, striving for such objectives is likely to bring current activities more in line with the carrying capacity of the planet. Precaution embedded within the practices of cleaner production must be one of the fundamental principles of a sustainable future.

NOTES

1. U.S. Congress, Office of Technology Assessment, *Serious Reduction of Hazardous Waste*, Washington, D.C.: U.S. Government Printing Office, 1986.
2. United Nations Environment Program, Industry and Environment Program, *Cleaner Production Program*, unpublished brochure, Paris, 1992.
3. See H. Freeman, ed., *Industrial Pollution Prevention Handbook*, New York: McGraw-Hill, 1995; and T. Jackson, ed., *Clean Production Strategies: Developing Preventive Environmental Management in the Industrial Economy*, London: Lewis Publishers, 1993.
4. United Nations Environment Program, Industry and Environment Activity Center, *Cleaner Production Worldwide*, Paris, 1993.
5. Second International Conference on the Protection of the North Sea, *Ministerial Declaration*, London, November 1987, article VII.
6. See, for instance, V. Dethlefsen, "Marine Pollution Mismanagement: Towards a Precautionary Concept," *Marine Pollution Bulletin*, 17:2, 1986, pp. 54–57, and A. Stebbing, "Environmental Capacity and the Precautionary Principle," *Marine Pollution Bulletin*, 24:6, 1992, pp. 287–295.
7. Massachusetts Toxics Use Reduction Program, *Evaluating Progress: A Report on the Findings of the Massachusetts Toxics Use Reduction Program Evaluation*, unpublished, Toxics Use Reduction Institute, March, 1997.
8. A recent study of cleaner production case studies found many to have neglected consequences for worker health and safety and a few that might have aggravated such conditions. See N. Ashford, I. Banoutas, K. Chistiansen, B. Hummelmose, and D. Stratikopollos, *Evaluation of the Relevance for Worker Health and Safety of Existing Environmental Technology Data-Bases for Cleaner and Inherently Safer Technology*, unpublished report, Center for Business Technology and Policy, Massachusetts Institute of Technology, Cambridge, Massachusetts, 1996.

9. James Hickey and Vern Walker propose four characteristics that could assist in formalizing the Precautionary Principle for decision making. These include clarity of goals, boundary of effects, identification of "covered" activities, and identification of evaluative indicators. See "Refining the Precautionary Principle in International Environmental Law," *Virginia Law Review*, 14, 1995, pp. 423–454.

10. See N. Ashford, "Understanding Technological Responses of Firms to Environmental Problems: Implications for Government Policy." In: K. Fischer and J. Schot, ed., *Environmental Strategies for Industry*, Washington, D.C.: Island Press, 1993.

11. See G. Pauli, "Zero Emissions: The Ultimate Goal of Cleaner Production," *Journal of Cleaner Production*, 5:1–2, 1997, pp. 109–113.

12. R. Pojasek, "Focusing Your P2 Program on Zero Waste," *Pollution Prevention Review*, 8:3, 1998, pp. 97–105.

13. See B. Allenby and T. Graedel, *Industrial Ecology*, Englewood Cliffs, NJ: Prentice-Hall, 1995.

14. See K. Oldenberg and K. Geiser, "Pollution Prevention and . . . or Industrial Ecology?" *Journal of Cleaner Production*, 5:1–2, 1997, pp. 103–108.

15. See O. Giarini and W. Stahel, *The Limits to Certainty*, Dordrecht, The Netherlands: Kluwer Academic Publishers, 1989.

16. See C. Fussler, *Driving Eco-Innovation*, London: Pitman Publishing, 1997.

Chapter 21

~❧~

THE PRECAUTIONARY PRINCIPLE: APPLICATION TO POLICIES REGARDING ENDOCRINE-DISRUPTING CHEMICALS

Peter L. deFur[1]

> Knowledge of what is does not open the door directly to what should be.
>
> —*Albert Einstein*

In 1992, Colborn and Clement edited a little-publicized volume summarizing and reviewing data on the effects of chemicals on reproduction and development in wildlife.[2] The basic tenet was that some chemicals alter reproductive function and developmental outcomes in a variety of animals, demonstrably birds and fish from the Great Lakes. A subsequent paper in the scientific literature[3] summarized the original volume and proposed that some chemicals are able to disrupt the hormonal system in startling and fundamental ways. Birds with crossed beaks and fish with both male and female reproductive organs are some of the problems that have come to light and are currently the subject of public interest and scientific investigation.

But the hypothesis went further than simply warning that anthropogenic chemicals were harming wildlife. On the basis of a wealth of comparative biological data, Colborn and others suggested that all animals, including humans, are susceptible to these effects and that environmental programs and regulations for chemicals need to be re-tooled to address these concerns. Indeed, data on diethylstilbestrol (DES) exposure of pregnant women during the 1950s and 1960s[4] demonstrated the harmful effects on human reproductive systems from a chemical known to be estrogenic. Now, some authors suggest that these phenomena may explain declines in sperm count and quality and male reproductive tract problems.[5] Many of the observed effects in wildlife and humans are caused by the alteration of normal hormonal control of reproductive function, including development, hence the chemicals known or suspected to exert these effects have been termed "endocrine disruptors."[6]

The hypothesis put forward in Colborn and Clement's original volume, and supported by a group of world-renowned research scientists, is that a class of compounds—endocrine-disrupting chemicals (EDCs)—can disrupt normal hormonal systems by mimicking hormones, blocking hormones, or triggering actions out of the normal hormonal sequence. Since then, a wealth of literature has emerged in the peer-reviewed journals and in book form. Recent volumes on wildlife[7] document that some effects in vertebrate wildlife can indeed be traced to chemical perturbations of reproduction and/or development. Data on humans have not been as extensively published, owing in no small part to the time and difficulty associated with human health and epidemiological studies. The National Research Council is completing (in 1999) a study that reviews the scientific data on the endocrine disruptor hypothesis, and the U.S. Environmental Protection Agency has a number of efforts underway to further investigate specific aspects of the issue and to determine the ramifications for various regulations. One of these efforts was the Endocrine Disruptor Screening and Testing Advisory Committee (EDSTAC),[8] an official federal advisory committee. EDSTAC was charged with advising on the creation of a program that the EPA could use to protect human health and the environment from the effects of EDCs, especially under existing laws that cover pesticides and chemicals in commerce.

Amidst the efforts to investigate the pure and applied scientific questions raised by EDCs, serious policy implications challenge the efficacy of the programs to regulate chemicals and protect the environment and human health. Chemicals once thought to pose little or no threat to human health or the environment are now considered possible EDCs, raising the specter of re-registration of thousands of chemicals in the United States. Because hormonal systems occur throughout the animal kingdom, and control every aspect of animal life from reproduction to migrations to growth, policy makers and regulators are challenged to identify how to test chemicals for the myriad effects. Traditional testing regimes seem impractical because of the time, expense, and complexity of testing one chemical at a time in individual species of animals. Such approaches are unable to address the very real issues of complex mixtures, low-dose exposures, timing of exposure, and the sensitivity of millions of species of animals not tested.[9] This chapter addresses policy approaches to regulating chemicals and suggests five principles that emerge from the lessons taken from EDCs.

REGULATING TO PROTECT THE ENVIRONMENT

Most authors and students of environmentalism agree that the modern environmental movement began in the late 1960s or early 1970s. At this time, it became hard to ignore deteriorating environmental conditions that included the decline of osprey populations on Long Island, the extensive pollution of Lake Erie, and the discovery of illegal and unregulated toxic waste dumps.[10] Lake Erie was so polluted that fish could not survive; dense mats of blue-green algae covered the surface. A slick of solvent on the Cuyahoga River in Cleveland caught fire in 1969; the pictures of the river afire linger still. Chemical wastes were indiscriminately dumped, as when Kepone contaminated 30 miles of the James River in Virginia. PCBs leaked into the Hudson River, contaminating striped bass, with subsequent closure of the fishery. National environmental policy was set in this climate of environmental disaster to restore and protect water and air quality, endangered species, and habitats. Congress enacted a series of laws to protect air, water, and soil from further pollution and to regulate a great variety of sources of pollution.

Environmental protection in this era sought to control pollution released from or caused by a facility or an activity. The Clean Water Act regulated the amount of chemical pollutants that entered waters from discharge pipes, and the Clean Air Act similarly put limits on how much pollutant could legally be emitted from the smokestack. This approach is termed the "end-of-pipe" strategy and was sufficient to deal only with the grossest level of pollution such as large particulate matter in air emissions and water discharges and the unauthorized disposal of toxic chemicals and wastes. There was a limit to how much could be filtered out of the end of the pipe, so to speak, and it was clear that newer strategies were needed. The end-of-the-pipe era was followed by the "pollution prevention" era, based on the simple premise that it is easier and cheaper to not create pollution in the first place, rather than clean it up afterward.

The pollution prevention approach, known as "P2," worked well in new facilities or systems that were redesigning or engineering but did not really catch on in the vast majority of facilities, especially aging ones. The next phase in fundamental strategy to protect the environment was the appearance of "market-based" approaches, promoted largely by the national environmental organizations.[11] Market approaches included debt for nature swaps, air emissions trading and marketing programs, and mitigation banks for wetlands.

Federal environmental laws are based on straightforward goals to restore and protect the environment, but specific laws use quite different regulatory concepts. The Clean Air Act uses a technology-based strategy to set standards for emission controls. The Clean Water Act uses an environmental quality-based approach on the basis of meeting the conditions necessary to maintain healthy living systems. The Superfund Act required sites cleaned to background levels. The pesticide control laws require the EPA to balance the risks against the benefits from using the chemicals to control pests. Registration of chemicals used in commerce under the Toxic Substances Control Act (TSCA) requires submission of basic chemical data, but the application for use is automatically approved after a certain time unless EPA shows why the chemical should be restricted.

The federal laws employ three different strategies for controlling or regulating toxic chemicals to protect human health and the environment: (1) technology controls; (2) meeting environmental quality standards; and (3) risk-based permits or standards. The three are summarized below.

- *Technology-based standards:* Best Available Control Technologies (BACT) and Maximum Achievable Control Technologies (MACT) set the environmental performance that a facility must achieve. Permits are based on what comes out the end of the stack (or pipe) and how much pollutant a facility releases in a given time. Technology-based systems determine the best technology for a given type of facility and then require all facilities in that category of facility to meet that performance standard. Such standards need to be revisited periodically to account for engineering improvements.
- *Environmental quality or public health standards:* The ambient conditions necessary to maintain environmentally healthy ecosystems, populations, or human health must be known, described, and quantified. Discharges or other releases must not put any more into the environmental medium (air, water, or soil) than the system can handle. Permits and standards are based on maintaining a given level of environmental quality or restoring ambient conditions.
- *Risk-based decisions:* This approach seeks to balance the known or predicted risks of a chemical or activity with the intended benefits from having and/or using the chemical or activity. Regulation of pesticides uses a risk approach, as do a number of contaminated cleanup programs. Many permits are based on an assessment of risks, and most federal programs

now use some sort of risk assessment process in their decision-making procedure. In truth, risk assessment remains a controversial method in environmental and human health protection.[12]

Until the appearance of EDC contamination issues, these three strategies, though not without problems, seemed to work because the observed conditions fit the "paradigm" in each case.

The regulatory programs that employed these approaches were based on a few key assumptions. First, human health and the environment can be protected on the basis of what is presently known about chemicals and the technologies that produce them. This assumption is the basis for technology-based regulations and control strategies. A technological focus also feeds into and supports the notion or assumption that technological advances will keep ahead of the problems resulting from technology. End-of-the-pipe controls gave way to pollution prevention under the same philosophy—technology improvements in processes can fix the problems. Decision are based on what is known, rather than on what is not known.

Second, scientists have now identified most of the worst problems, or "what we don't know won't hurt us." Society used the precision of scientific measurement to assure that we were monitoring the chemicals, the effects of those chemicals, and the fate of the chemicals and that regulators acted on that information to protect the environment and human health.

Third, the solution to pollution is dilution. Of course, this also depends on an infinite dilution (or assimilative) capacity of the environment. Air emissions and ocean discharges were based on the idea that the upper atmosphere and deep oceans were too large to pollute.

Finally, human activities are basically local and not sufficient to affect the condition of the globe. The two great receiving bodies, the ocean and atmosphere, are huge in volume compared to the amount of the discharge; regulators and policy makers never believed society could contaminate anything as large as the ocean. This assumption is related to the infinite dilution capacity of the water and air noted earlier, but adds a time dimension to the dilution assumption. Regulatory policy included the assumption in this case that some pollutants would go away in time, either from natural processing or by being sequestered.

In applying these assumptions to the regulation of toxic chemicals, there are specific elements taken directly from the field of toxicology. These assumptions are integral to both the science of toxicology and the fields of risk assessment and risk management of toxic chemicals. They are discussed on the following pages.

First, and most important, the principle tenet of toxicology, "the dose makes the poison," at least implies that low-level exposure is not toxic. At higher doses, the toxic effect is greater; at lower doses, the effect is diminished or absent. A corollary to this assumption is that with no exposure, there is no effect.

Second, there is a dose below which even toxic chemicals will show no effect (the threshold). In a few cases, notably cancer, even if there is no real threshold, there will be an effect level too low to detect.

Third, scientists can (and do) measure the necessary effects and exposures and determine the impacts or harm from high to low doses. Testing animals starting with high doses is based on the notion that the effect is easier to measure and more apparent at high concentrations, so that by reducing the dose scientists can monitor both concentration and effect at progressively lower doses. A corollary here is that the background is zero for both dose and effect in either the human population or in experimental animals, sometimes in both.

Fourth, protecting humans from the most sensitive end point, cancer, is sufficient and protective for all end points in all species. This assumption has two parts: cancer is the most sensitive end point, and protecting humans from cancer also protects other species as well.

Policies and regulations governing toxic chemicals have, for the most part, relied on laboratory testing of individual chemicals under controlled conditions. The reductionist and isolationist approach most specifically identifies the chemicals, effects, and relations between the two approaches. But testing individual chemicals is logistically impractical and has been applied to only a few of the more than 75,000 chemicals in commerce in the United States.[13] Pesticides and a few other chemicals, numbering in the hundreds, have been tested in the lab; most have not been subjected to rigorous testing.[14] Yet the control measures rely on the applicability of toxicological assumptions to regulatory strategies.

Current information on endocrine-disrupting chemicals casts doubt on these assumptions on which U.S. regulatory policy and practice are based. New findings show that low-level exposure to reproductive and developmental toxicants during key windows of sensitivity can produce profound adverse effects on offspring, organ growth, and nervous systems, for example.[15] But if these assumptions can no longer be taken as true, then the resulting policies and regulations must be questioned. Some of the recent findings on EDCs that must now be considered in new environmental policies include the following:

(1) Pollution is global. Toxic chemicals have been found in the Arctic[16] in albatross from the high seas[17] and in deep lake and ocean sediments. A recent U.S.G.S. survey of carp in the United States revealed altered reproductive hormone levels, showing the pervasive nature of the effects.[18] The global distribution of EDCs especially demonstrates that society has contaminated the farthest ends of the Earth, with unanticipated and often irreversible ecological consequences.[19]

(2) Low levels of chemical contaminants exert actions that are not frank and obvious in short-term lab tests, but show up later, often in the next generation. The effects may be seen in ways that are subtle in the short term but that magnify with time and population. For example, Gray and co-workers[20] and Peterson[21] and co-workers demonstrated that single doses of the toxic chemical dioxin alters male reproductive development and function in offspring when administered to the pregnant dam during gestation. Similar intergenerational and developmental effects have been observed in wildlife.[22]

(3) The low dose and threshold effects are challenged by no-threshold, noncarcinogenic toxic effects[23] and by chemical effects exerted at specific times in the life history of an animal. If the determinant is not the amount of the chemical, but the timing of exposure, then the animal, not the chemical, is the critical step and it is the time, not the dose that is critical. The "dose makes the poison" may have to be replaced by or supplemented with "the timing makes the poison."

(4) The chemicals once considered "safe" by virtue of dose or action may be active. The discovery that estrogenic chemicals may leach out of a plastic vessel into the liquid in that vessel forever abolished the myth that plastics were truly "inert."[24]

(5) The background is no longer zero. It is not just that Arctic and Pacific populations of wildlife carry body burdens of a variety of toxic chemicals, such as PCBs, DDT, and dioxins. Some biological responses to these chemicals have been measured at current exposure levels. Human activities have not just reached every corner of the globe and every segment of the Earth, but have affected basic biological functions in animals throughout the biosphere.

(6) Living systems function in fundamentally similar ways, at cellular, animal, and ecosystem levels. Our knowledge of these events and processes in wildlife can be considered predictive or anticipatory for humans. Cellular level activities include gene responses, metabolic pathways, and molecular evolution of species. On the positive side, left alone,

some recovery should occur; on the down side, we may be engaged in an artificial genetic selection experiment of immense magnitude. Cairns[25] concluded that biological and ecosystem integrity is necessary maintain ecosystems in a life-sustaining fashion. The effects of widespread EDC exposure, in my estimation, threaten that integrity.

Scientists simply do not know enough about how animals work and cannot make such predictions with enough accuracy and confidence to support these important decisions. The information accumulating as federal agencies and scientific groups seek to understand EDCs clearly shows the lack of knowledge regarding effects of these chemicals. At EPA, the EDSTAC acknowledged that there are no adequate biological assays to test chemicals for EDC effects on fetuses and neonates, nor to protect many species of fish, birds, and almost none for invertebrate animals that constitute about 97 percent of the animals on Earth.[26] An international scientific workshop in 1997[27] concluded much the same thing for wildlife, noting information gaps in endocrinology of many wildlife animals, in understanding hormone and chemical metabolism and in knowledge of reproductive endocrinology of invertebrates. The scientific understanding of reproduction and development in animals, and how chemicals affect these processes, are simply not sufficiently adequate to predict how these chemicals will affect hormonal processes.

POLICY PRINCIPLES

The Precautionary Principle and four subprinciples that incorporate lessons from EDC experiences will guide the development of new policies for protecting the environment and human health.

The Precautionary Principle

As described in this volume, the Precautionary Principle provides that we should take anticipatory action in the face of scientific uncertainty and the possibility of harm. EDCs are among the most significant threats to the environment and public health. As demonstrated previously, they have grave potential harm and enormous scientific uncertainty. Therefore, anticipatory regulatory programs need to be based on the assumption that chemicals can and will have adverse effects and human activities will have unintended and undesirable consequences. After using some chemicals for years under the false assumption that there were no adverse effects, scientists now are finding that these compounds are active and exert adverse effects.

Actions of Society Need to Improve the Environment

Society needs to look actively for ways to clean up the problems created over many decades. Cleanup can and should be a part of every program and activity. Society can no longer just accept the status quo or assume that if harmful activities are prevented then the rest of the problems will take care of themselves. The next generation of environmental decisions needs to address mitigation for past practices. While this appears to be reactive, rather than anticipatory action, it is in fact anticipating impacts on future generations.

Every Action Causes Some Response

The laws of thermodynamics apply to environmental decisions; for every action there will be an equal and opposite reaction. Creating, using, and releasing chemicals will have some response in the ecosystem of which humans are a part. The experience with EDCs shows that even "inert" chemicals can have effects; no one looked for these effects at first. Now the regulatory bodies must anticipate the unexpected—the best approach is assuming that chemicals have effects unless there is some evidence to the contrary. This subprinciple couples scientific uncertainty and anticipatory action to prevent harm. A corollary to this point is that less contamination is better. The fewer toxic chemicals in the environment, the less likely that there will be a harmful response. With time and normal ecological processes, some of society's past errors will be covered and no longer accessible, others will be biologically processed, but slowly. Unfortunately, some, such as lead and dioxin, will remain forever.

The Timing Makes the Poison

The history of the field of toxicology is that the dose makes the poison. Now the evidence indicates that seemingly inactive chemicals can exert powerful effects when administered at a particular time. Past toxicological practice has been that the dose makes the poison. But now we find chemicals that exert an effect only when administered at precise times, and perhaps, or likely without threshold. We cannot guarantee control over timing (another element of uncertainty); therefore, we cannot make any guarantees over protection or safety. Accordingly, we must act to prevent all exposures.

Burden Shifting

As described in the Wingspread Statement on the Precautionary Principle, the applicant or proponent of an activity or process or chemical needs to demonstrate to the satisfaction of the public and the regulatory community that the environment and public health will be safe. The proof must shift to the party or entity that will benefit from the activity and that is most likely to have the information. This, in effect, reduces the uncertainty that attends the potential harm created by activities, processes, and chemicals.

EDCs raise a number of challenges to toxicology and hence to the resulting regulatory policies. As described here, the traditional assumptions are now challenged, and new policies will have to fit a new paradigm. As Kuhn[28] noted, in coining the term "paradigm shift," when a new set of observations cannot be explained by the current model or "paradigm," the time has come for a new paradigm. Society has reached that point regarding understanding of and actions toward the environment. Now, new processes and practices that extract, use, manufacture, and release fewer or no toxic chemicals are needed. These facilities need to be low impact and the process not based on the notion that more and bigger is better but on the idea that most of what has been done in industrialization has had and will continue to have long-term profound impacts, notably of the sort posed by EDCs.

NOTES

1. Acknowledgments: An early version of this chapter is in press in the journal *Biotechnology International*. This work was funded in part through a grant to Virginia Commonwealth University from the W. Alton Jones Foundation. I am grateful to the editors for their outstanding work in preparing the Wingspread Conference and all that made it successful. Thanks to Carolyn Raffensperger and Pete Myers for all the discussion on the subject and to Sharon deFur for teaching me the lessons from another discipline.

2. Colborn, T., and C. Clement, eds., *Chemically-Induced Alterations in Sexual and Functional Development: The Wildlife/Human Connection.* Princeton, NJ: Princeton Science Publishers, 1992.

3. Colborn, T., F. vom Saal, and A. Soto. "Developmental Effects of Endocrine-Disrupting Chemicals in Wildlife and Humans." *Environmental Health Perspectives* 101: 378–384, 1993.

4. McLachlan, J.A., R.R. Newbold, C.T. Teng, and K.S. Korach. "Environmental Estrogens: Orphan Receptors and Genetic Imprinting." In: T. Colborn and C. Clement, eds., *Chemically-Induced Alterations in Sexual and Functional Development: The Wildlife/Human Connection.* Princeton, NJ: Princeton Science Publishers, 1992, pp. 107–112.

5. Swan, S. H., E.P. Elkin, and L. Fenster. "Have Sperm Densities Declined? A Reanalysis of Global Trend Data." *Environmental Health Perspectives* 105(11): 1228–1232. Colburn and Clements, 1992.

6. Colburn and Clements, 1992.

7. Rolland, R. M., M. Gilbertson, and R.E. Peterson. *Chemically Induced Alterations in Functional Development and Reproduction of Fishes.* Pensacola, FL:

SETAC, 1997. See also Kendall, R.J., R.L. Dickerson, J.P. Geisy, and W.P. Suk, eds., *Principles and Processes for Evaluating Endocrine Disruption in Wildlife*. Pensacola, FL: SETAC, 1997.

8. "Endocrine Disruptor Screening and Testing Advisory Committee," report to EPA, September 1998. In Press. EPA Office Pollution Prevention and Toxic Substances, Washington, D.C.

9. Tattersfield, L., P. Mathiessen, P. Campbell, N. Grandy, and R. Lange, eds. 1997. *SETAC-Europe/OECD/EC Expert Workshop on Endocrine Modulators and Wildlife: Assessment and Testing*. Brussels. SETAC-Europe.

10. Dowie, M. *Losing Ground: American Environmentalism at the Close of the Twentieth Century*. Cambridge, MA: MIT Press, 1996. See also Marine, G. *America the Raped*. New York: Simon and Schuster, 1969.

11. See Dowie, 1996.

12. National Research Council. *Understanding Risk*. Washington, D.C.: National Academy Press, 1996.

13. Zeeman, M., and J. Guilford. "Ecological Hazard Evaluation and Risk Assessment Under EPA's Toxic Substances Control Act (TSCA): An Introduction." In: W.G. Landis, J.S. Hughes, and M.A. Lewis, eds., *Environmental Toxicology and Risk Assessment Vol. 1*. Philadelphia, PA. ASTM, 1993. ASTM STP 1179; pp. 7–21.

14. Zeeman and Guilford, 1993.

15. Colburn and Clement, 1992; and Colburn, vom Saal, and Soto, 1993.

16. Iwata, H., S. Tanabe, N. Sakai, and R. Tatsukawa. "Distribution of Persistent Organochlorines in Oceanic Air and Surface Seawater and Role of Ocean in Their Global Transport and Fate." *Environmental Science and Technology* 27: 1080–1098, 1993.

17. Auman, H.J., J.P. Ludwig, C.S. Summer, D.A. Verbrugge, K.L. Froese, T. Colborn, and J.P. Geisy, Jr. "PCBS, DDE, DDT, and TCDD-Eq in Two Species of Albatross on Sand Island, Midway Atoll, North Pacific Ocean." *Environmental Toxicology and Chemistry* 16: 498–504, 1997.

18. Goodbred, S.L., R.J. Gilliom, T.S.Gross, N.P. Denslow, W.L. Bryant, and T.R. Schoeb. "Reconnaissance of 17B-Estradiol, 11-Ketotestosterone, Vitellogenin, and Gonad Histopathology in Common Carp of the United States Streams: Potential for Contaminant-Induced Endocrine Disruption." U.S. Geological Survey Open File Report 96-627. USGS, Denver, CO 80225. 1977.

19. Schindler, D.W. "Ecosystems and Ecotoxicology: A Personal Perspective." In: M.C. Newman and C.H. Jague, eds., *Ecotoxicology, A Hierarchical Treatment*. Boca Raton, FL: CRC Press, 1996, pp. 371–398.

20. Gray, L.E., Jr. W.R. Kelce, E. Monosson, et al. "Exposure to TCDD During Development Permanently Alters Reproductive Function in Male Long Evans Rats and Hamsters: Reduced Ejaculated and Epididymal Sperm Numbers and Sex Accessory Gland Weights in Offspring with Normal Androgenic Status." *Toxicology and Applied Pharmacology* 131: 108–118, 1995. See also Peterson, R.E., R.W. Moore, T.A. Mably, D.L. Bjerke, and R.W. Goy. "Male Reproductive System Ontogeny: Effects of Perinatal Exposure to 2,3,7,8 Tetra-

chlorodibenzo-p-dioxin." In: T. Colborn and C. Clement, eds. *Chemically-Induced Alterations in Sexual and Functional Development: The Wildlife/Human Connection*. Princeton, NJ: Princeton Science Publishers, 1992, pp. 175–194.

21. Gilbertson, M., and G.A. Fox. "Pollutant-Associated Embryonic Mortality of Great Lakes Herring Gulls." *Environmental Pollution* 12: 211–216, 1977.

22. Birnbaum, L.S. "The Mechanism of Dioxin Toxicity: Relationship to Risk Assessment." *Environmental Health Perspectives* 102 (Suppl. 9) 15–167, 1994. See also, Colburn and Celment, 1992.

23. Ibid. See also Tattersfield et al., 1997 and Dowie, 1996.

24. Soto, A.M., H. Justicia, J.W. Wray, and C. Sonnenschein. "P-Nonyl-phenol: An Estrogenic Xenobiotic Released from Modified Polystyrene." *Environ. Health Perspectives* 92: 167–173, 1996. See also Steinmetz, R., N.G. Brown, D.L. Allen, R.M. Bigsby, and N. Ben-Jonathen. "The Environmental Estrogen Biphenol A Stimulates Prolactin Eelease In Vitro and In Vivo. *Endocrinology* 138: 1780–1786, 1997.

25. Cairns, J., Jr. "Determining the Balance Between Technological and Ecosystem Services." In: P.C. Schulze, ed. *Engineering Within Ecological Constraints*. Washington, D.C.: National Academy Press, 1996, pp. 13–30.

26. Rolland, Gilbertson, and Peterson, 1997.

27. Endocrine Disruptor Screening and Testing Advisory Committee, in press.

28. Kuhn, T. S. *The Structure of Scientific Revolutions*, 2nd ed. Chicago: University of Chicago Press, 1970, pp. 210.

Appendix A

LESSONS FROM WINGSPREAD

Margaret Mead once said that we shouldn't underestimate the power of a small group of people because nothing else has ever changed the world. On January 23, 1998, in the snowy woods of Wisconsin, a small group of people gathered to discuss the Precautionary Principle. We had invited this group because each person had thought deeply about the Precautionary Principle (or similar principles with different names) and was exceptionally creative. The group collectively could cast a wide-angled lens on the enormous environmental and public health problems we face in the world and possible ways of solving them. We wanted to break out of old molds and find new ways to think about thorny issues. So we brought a farmer, doctors, an artist, community and labor activists, and a fair number of lawyers and scientists to the Frank Lloyd Wright–designed Wingspread Conference Center in Racine.

We offer the following observations and thoughts of the Wingspread participants at the conference. They capture the large and elegant contours of the Precautionary Principle. We invite you to listen in and participate vicariously through these notes. We end with the Wingspread Statement.

The Meaning of Precaution

Precaution challenges the historical assumptions of limitlessness by bounding human hubris. The core of precaution can be articulated with the Greek verb "prosecho"—meaning to take care and to take notice—suggesting that humans have two roles, those of stewards and watchdogs. Precaution is about a functionally respectful relationship with nature. It is about affirmative, anticipatory action to protect public and ecosystem health. Precaution is one of many principles guiding human activities; but it incorporates parts of others such as justice, equity, respect, common sense, and prevention.

Precaution is culturally framed, so its meaning will differ depending on who uses it. For some, precaution means considering and fostering a responsibility to future generations in all decisions. For others, precaution is consistent with Rawl's notion of fairness and justice and should promote empowerment of those least empowered. Still, for others, precaution means avoiding regrets and holding back when there is some uncertainty about the possible impacts of an activity. Finally, some feel that precaution should be more than avoiding harm ("what we can get away with") but about how we restore the integrity of ecosystems and human health. However, Wingspread participants agreed that precaution is a simple concept rooted in common sense and the primacy of environmental and public health. According to the group, there are four major components of a precautionary approach: decision making in the face of uncertainty; shifting burdens of proof; a full analysis of alternatives to potentially harmful activities; and democratic decision-making structures.

Scientific Uncertainty—The Heart of Precaution

Scientific uncertainty and ignorance about the effects of anthropogenic stress on ecological systems are the underlying rationale behind a Precautionary Principle. Precaution demands that we are open and honest about uncertainty. We must define uncertainty in broader terms than what we know and do not know. There is uncertainty about exposure, but there is also uncertainty about the models used to relate exposure to disease. Finally, as each individual is unique, variability poses another form of uncertainty. Wingspread participants discussed another type of uncertainty that may have far greater implications than knowledge uncertainty: politically imposed uncertainty. Unless we begin to expose uncertainty as an unavoidable component of decisions involving environmental and public health harm, we run the risk of making truly ignorant decisions. As discussed by Wingspread participants, uncertainty becomes the reason for taking action to prevent harm and for shifting the benefit of the doubt to those beings and systems that might suffer harm.

Wingspread participants agreed that precaution challenges science in fundamental ways. The traditional model of academic or laboratory science uses a high standard for establishing conclusive knowledge. However, for problems relating to environmental hazards, the ideals of laboratory science often cannot be achieved or are only achieved at the expense of prevention-oriented actions. The traditional model of science also suppresses speculation and cross-discipline studies. Thus, some fundamental changes in science will be needed if precaution is to be embedded in research design and public policy. These changes include a change in the incentives awarded to scientists that allow them to examine problems and hypotheses outside the boundaries of "normal" science; a need to encourage scientists to make policy conclusions not only on the basis of what they know statistically and scientifically, but also based on what they believe; minimization of Type II errors in decision making; and the augmentation of other types of legitimation other than pure science.

PRECAUTION, DEMOCRACY, AND HUMAN RIGHTS

Precaution was viewed by many at the Wingspread Conference as an issue of ethics, morality, and truth. For those affected by environmental harms, taking precaution is a simple issue of right and wrong. Precaution is about protecting future generations, who have no power over the decisions made today, and protecting those who are most vulnerable or with the least power in society. For example, decisions about toxic chemicals should ask the basic question of whether exposure is safe for a six-week-old embryo; if not, then the activity should not occur. Decisions about harm to human health are public decisions and thus require the maximum feasible participation of people affected by decisions.

Precaution is also about human rights. For example, there is a disconnect between those who benefit from harmful activities and those who suffer. There is also a lack of consent among those who suffer the burden of "acceptable risks." This differs widely from medical ethics, where testing should only occur with the express permission of those involved and only when there is no other alternative. Is there a difference between the types of experiments conducted to test drugs and the experimentation that occurs everyday on humans and ecosystems from exposure to untested, synthetic chemicals?

Sandra Steingraber outlined two violations of human rights involved in the way toxic chemicals are currently used and released into the environment: (1) toxic tresspass, where toxic chemicals enter our bodies without our permission; a deliberate introduction of toxic chemicals into the environment, especially when that risk is not accepted or known to those affected can be considered a crime against those who suffer the consequences; and (2) a violation of the human right to enjoy the environment and not fear adverse consequences; for example, a father coming home from work should be able to hug his children without the fear that his children are exposed to the chemicals to which he was exposed in the workplace, or Native Americans should be able to achieve subsistence livelihoods without fear that fish will not be available or contaminated with PCBs. The human rights approach focuses on basic rights that have been removed by environmental contamination and other ecosystem harm and places the onus back on those who create hazards.

Integrating precaution into public health and environmental decision making will require large-scale changes in power structures, a reinvigorization of democracy and structures for allowing greater public participation in decisions affecting their lives. Those who are at risk of suffering from environmental degradation are much more likely to employ a commonsense, precautionary approach than the government, which must defend its decisions in the courts or those who stand to gain (either in the short term or long term) from an activity. While Wingspread participants discussed some of the methods available to promote more democratic decision making, such as community research networks, consensus conferences, and campaign finance reform, they realized that these methods will take a long period to become institutionalized.

MOVING FORWARD WITH PRECAUTION

More research and outreach will be needed to ensure that the principle is used not only to guide decisions but also in the decision-making process itself. Perhaps of greatest importance to the implementation of the Precautionary Principle is shifting of questions on which we base our environmental protection efforts. Given our uncertainty and ignorance and the vast complexity of ecological systems, we can no longer ask what level of harm is safe. We need to question basic human activities (consumption and materialism, resource exploitation) and ask questions about how we can avoid harm and live in sync with our environment, learning from millions of years of ecological self-regulation.

We will also need to set broad goals as a society and strive toward them, rather than trying to predict the consequences of our actions. Simple goals, such as "no children shall be born with persistent toxic chemicals in their bodies by the year 2005," provide a clear milestone and basis for efforts to protect human health and the environment. There is little room for debating quantitative predictions when such straightforward goals are set. Goals and targets also challenge human ingenuity and innovation and enable us to focus our efforts on restoration and prevention rather than on justifying actions that might cause harm. Humans (and the environment) do not have to accept risk in order to live a prosperous, healthy life.

Precaution must become a moral imperative, of equal or greater importance than economic growth or military security. The elevation of precaution will drive science toward solving problems for the public good. However, to achieve this elevation, our government agencies will need to shift their focus from protector and mediator of interests to public (and environmental) trustee. In the long run, growth (in a more holistic sense of the word), sustainability, and prosperity will be a reflection of the extent to which we are precautious.

At the end of the Wingspread Conference, participants drafted the following statement. We include it here in full with all the names of the signers.

Wingspread Statement on the Precautionary Principle

January 25, 1998

The release and use of toxic substances, the exploitation of resources, and physical alterations of the environment have had substantial unintended consequences affecting human health and the environment. Some of these concerns are high rates of learning deficiencies, asthma, cancer, birth defects and species extinctions; along with global climate change, stratospheric ozone depletion and worldwide contamination with toxic substances and nuclear materials.

We believe existing environmental regulations and other decisions, particularly those based on risk assessment, have failed to protect adequately human health and the environment—the larger system of which humans are but a part.

We believe there is compelling evidence that damage to humans and the worldwide environment is of such magnitude and seriousness that new principles for conducting human activities are necessary.

While we realize that human activities may involve hazards, people must proceed more carefully than has been the case in recent history. Corporations, government entities, organizations, communities, scientists, and other individuals must adopt a precautionary approach to all human endeavors.

Therefore, it is necessary to implement the Precautionary Principle: When an activity raises threats of harm to human health or the environment, precautionary measures should be taken even if some cause-and-effect relationships are not fully established scientifically.

In this context the proponent of an activity, rather than the public, should bear the burden of proof.

The process of applying the Precautionary Principle must be open, informed and democratic and must include potentially affected parties. It must also involve an examination of the full range of alternatives, including no action.

Wingspread Participants:
(Affiliations are noted for identification purposes only.)

Dr. Nicholas Ashford, Massachusetts Institute of Technology

Katherine Barrett, University of British Columbia

Anita Bernstein, Chicago-Kent College of Law

Dr. Robert Costanza, University of Maryland

Pat Costner, Greenpeace

Dr. Carl Cranor, University of California, Riverside

Dr. Peter deFur, Virginia Commonwealth University

Gordon Durnil, attorney

Dr. Kenneth Geiser, Toxics Use Reduction Institute, University of Massachusetts, Lowell

Dr. Andrew Jordan, Centre for Social and Economic Research on the Global Environment, University of East Anglia, United Kingdom

Andrew King, United Steelworkers of America, Canadian Office, Toronto, Canada

Dr. Frederick Kirschenmann, farmer

Stephen Lester, Center for Health, Environment and Justice

Sue Maret, Union Institute

Dr. Michael M'Gonigle, University of Victoria, British Columbia, Canada

Dr. Peter Montague, Environmental Research Foundation

Dr. John Peterson Myers, W. Alton Jones Foundation

Dr. Mary O'Brien, environmental consultant

Dr. David Ozonoff, Boston University

Carolyn Raffensperger, Science and Environmental Health Network

Dr. Philip Regal, University of Minnesota

Hon. Pamela Resor, Massachusetts House of Representatives

Florence Robinson, Louisiana Environmental Network

Dr. Ted Schettler, Physicians for Social Responsibility

Ted Smith, Silicon Valley Toxics Coalition

Dr. Klaus-Richard Sperling, Alfred-Wegener-Institut, Hamburg, Germany

Dr. Sandra Steingraber, author

Diane Takvorian, Environmental Health Coalition

Joel Tickner, University of Massachusetts, Lowell

Dr. Konrad von Moltke, Dartmouth College

Dr. Bo Wahlstrom, KemI (National Chemical Inspectorate), Sweden

Jackie Warledo, Indigenous Environmental Network

Appendix B

～

USES OF THE PRECAUTIONARY
PRINCIPLE IN INTERNATIONAL TREATIES AND
AGREEMENTS IN U.S. LEGISLATION

Much of the text of these agreements was taken from: Hickey, J. and V. Walter. 1995. "Refining the Precautionary Principle in International Environmental Law. *Virginia Environmental Law Journal* 14: 423–436.

OZONE LAYER PROTOCOL

Parties to this protocol . . . determined to protect the ozone layer by taking precautionary measures to control equitably total global emissions of substances that deplete it, with the ultimate objective of their elimination on the basis of developments in scientific knowledge, taking into account technical and economic considerations. (Protocol on Substances that Deplete the Ozone Layer, Sept. 16, 1987, 26 ILM 1541.)

SECOND NORTH SEA DECLARATION

In order to protect the North Sea from possibly damaging effects of the most dangerous substances . . . a precautionary approach is addressed which may require action to control inputs of such substances even before a causal link has been established by absolutely clear scientific evidence. (Ministerial Declaration Calling for Reduction of Pollution, Nov. 25, 1987, 27 ILM 835.)

UNITED NATIONS ENVIRONMENT PROGRAMME

Recommends that all governments adopt "the principle of precautionary action" as the basis of their policy with regard to the prevention and elimination of marine pollution. (Report of the Governing Council on the Work of its Fifteenth Session, United Nations Environment Programme, UN GAOR, 44th Sess. Supp No 25, 12th mtg at 153, UN DOC A44/25 (1989).)

NORDIC COUNCIL'S CONFERENCE

And taking into account . . . the need for an effective precautionary approach, with that important principle intended to safeguard the marine ecosystem by, among other things, eliminating and preventing pollution emissions where there is reason to believe that damage or harmful effects are likely to be caused, even where there is inadequate or inconclusive scientific evidence to prove a causal link between emissions and effects. (Nordic Council's International Conference on Pollution of the Seas: Final Document Agreed to Oct. 18, 1989, in Nordic Action Plan on Pollution of the Seas, 99 app. V (1990).)

PARCOM RECOMMENDATION 89/1 — 22 JUNE 1989

The Contracting Parties to the Paris Convention for the Prevention of Marine Pollution from Land-Based Sources:

Accept the principle of safeguarding the marine ecosystem of the Paris Convention area by reducing at source polluting emissions of substances that are persistent, toxic, and liable to bioaccumulate by the use of the best available technology and other appropriate measures. This applies especially when there is reason to assume that certain damage or harmful effects on the living resources of the sea are likely to be caused by such substances, even where there is no scientific evidence to prove a causal link between emissions and effects (the principle of precautionary action).

THIRD NORTH SEA CONFERENCE

The participants . . . will continue to apply the Precautionary Principle, that is to take action to avoid potentially damaging impacts of substances that are persistent, toxic, and liable to bioaccumulate even where there is no scientific evidence to prove a causal link between emissions and effects. (Final Declaration of the Third International Conference on Protection of the North Sea, Mar. 7-8, 1990. 1 YB Int'l Envtl Law 658, 662-73 (1990).)

BERGEN DECLARATION ON SUSTAINABLE DEVELOPMENT

In order to achieve sustainable development, policies must be based on the precautionary principle. Environmental measures must anticipate, prevent, and attack the causes of environmental degradation. Where there are threats of serious or irreversible damage, lack of full scientific certainty should not be used as a reason for postponing measures to prevent environmental degradation. (Bergen Ministerial Declaration on Sustainable Development in the ECE Region. UN Doc. A/CONF.151/PC/10 (1990), 1 YB Intl Envtl Law 429, 4312 (1990).)

SECOND WORLD CLIMATE CONFERENCE

In order to achieve sustainable development in all countries and to meet the needs of present and future generations, precautionary measures to meet the climate challenge must anticipate, prevent, attack, or minimize the causes of, and mitigate the adverse consequences of, environmental degradation that might result from climate change. Where there are threats of serious of irreversible damage, lack of full scientific certainty should not be used as a reason for postponing cost-effective measures to prevent such environmental degradation. The measure adopted should take into account different socioeconomic contexts. (Ministerial Declaration of the Second World Climate Conference (1990). 1 YB Intl Envtl Law 473, 475 (1990).)

BAMAKO CONVENTION ON TRANSBOUNDARY HAZARDOUS WASTE INTO AFRICA

Each party shall strive to adopt and implement the preventive, precautionary approach to pollution problems which entails, inter alia, preventing the release into the environment of substances which may cause harm to humans or the environment without waiting for scientific proof regarding such harm. The parties shall cooperate with each other in taking appropriate measures to implement the Precautionary Principle to pollution prevention through the application of clean production methods, rather than the pursuit of a permissible emissions approach based on assimilative capacity assumptions. (Bamako Convention on Hazardous Wastes within Africa, Jan. 30, 1991, art. 4, 30 ILM 773.)

OECD COUNCIL RECOMMENDATION C(90)164 ON INTEGRATED POLLUTION PREVENTION AND CONTROL— JANUARY 1991

The recommendation is accompanied by guidance which is an integral part of the recommendation. It lists some essential policy aspects including: The absence of complete information should not preclude precautionary action to mitigate the risk of significant harm to the environment.

Maastricht Treaty on the European Union

Community policy on the environment . . . shall be based on the Precautionary Principle and on the principles that preventive actions should be taken, that environmental damage should as a priority be rectified at source and that the polluter should pay. (Treaty on the European Union, Sept. 21, 1994, 31 ILM 247, 285-86.)

Helsinki Convention on the Protection and Use of Transboundary Watercourses and International Lakes

The Precautionary Principle, by virtue of which action to avoid the potential transboundary impact of the release of hazardous substances shall not be postponed on the ground that scientific research has not fully proved a causal link between those substances, on the one hand, and the potential transboundary impact, on the other hand. (Convention on the Protection and Use of Transboundary Watercourses and International Lakes, Mar. 17, 1992, 31 ILM 1312.)

The Rio Declaration on Environment and Development

In order to protect the environment, the precautionary approach shall be widely applied by states according to their capabilities. Where there are threats of serious or irreversible damage, lack of full scientific certainty shall not be used as a reason for postponing cost-effective measures to prevent environmental degradation. (Rio Declaration on Environment and Development, June 14, 1992, 31 ILM 874.)

Climate Change Conference

The parties should take precautionary measures to anticipate, prevent, or minimize the causes of climate change and mitigate its adverse effects. Where there are threats of serious or irreversible damage, lack of full scientific certainty should not be used as a reason for postponing such measures, taking into account that policies and measures to deal with climate change should be cost-effective so as to ensure global benefits at the lowest possible cost. To achieve this, such policies and measures should take into account different socioeconomic contexts, be comprehensive, cover all relevant sources, sinks, and reservoirs of greenhouse gases and adaptation, and comprise all economic sectors. Efforts to address climate change may be carried out cooperatively by interested parties. (Framework Convention on Climate Change, May 9, 1992, 31 ILM 849.)

UNCED Text on Ocean Protection

A precautionary and anticipatory rather than a reactive approach is necessary to prevent the degradation of the marine environment. This requires inter alia, the adoption of precautionary measures, environment impact assessments, clean production techniques, recycling, waste audits and minimization, construction and/or improvement of sewage treatment facilities, quality management criteria for the proper handling of hazardous substances, and a comprehensive approach to damaging impacts from air, land, and water. Any management framework must include the improvement of coastal human settlements and the integrated management and development of coastal areas. (UNCED Text on Protection of Oceans. UN GAOR, 4th Sess., UN Doct A/CONF.151/PC/100 Add. 21 (1991).)

Energy Charter Treaty

In pursuit of sustainable development and taking into account its obligations under those international agreements concerning the environment to which it is a party, each contracting party shall strive to minimize in an economically efficient manner harmful environmental impact occurring either within or outside its area from all operations within the energy cycle within its area, taking proper account of safety. In doing so each contracting party shall act in a cost-effective manner. In its policies and actions each contracting party shall strive to take precautionary measures to prevent or minimize environmental degradation. The contracting parties agree that the polluter in the areas of contracting parties should, in principle, bear the cost of pollution, including transboundary pollution, with due regard to the public interest and without distorting investment in the energy cycle or international trade. (The Draft European Energy Charter Treaty Annex I, Sept 14, 1994, 27/94 CONF/104.)

U.S. President's Council on Sustainable Development

There are certain beliefs that we as council members share that underlie all of our agreements. We believe: (number 12) even in the face of scientific uncertainty, society should take reasonable actions to avert risks where the potential harm to human health or the environment is thought to be serious or irreparable. (President's Council on Sustainable Development. Sustainable America: A New Consensus, 1996.)

Commonwealth of Massachusetts House Bill No. 3140, 1997

An act to establish the Principle of Precautionary Action as the guideline for developing environmental policy and quality standards for the commonwealth.

Be it enacted by the Senate and House of Representatives in General Court assemblies, and by the authority of the same, as follows:

The Precautionary Principle shall be applied to all policy and regulatory decisions of the administration in order to prevent threats of serious or irreversible damage to the environment.

The Precautionary Principle shall be applied when there are reasonable grounds for concern that a procedure or development may contribute to the degradation of the air, land and water of the Commonwealth.

Lack of full scientific certainty shall not be used as a reason for postponing cost-effective measures to prevent costly environmental degradation.

The Precautionary Principle, by virtue of which preventive measures are to be taken when there are reasonable grounds for concern that substances or energy introduced, directly or indirectly, into the environment may bring about hazards to human health, harm living resources and ecosystems, damage amenities or interfere with other legitimate uses even when there is no conclusive evidence of a causal relationship between the inputs and the effects.

All state entities and contracting parties shall take all necessary steps to ensure the effective implementation of the Precautionary Principle to environmental protection and to this end they shall:

(a) encourage prevention of pollution at source, by the application of clean production methods, including raw materials selection, product substitution and clean product technologies and processes, and waste minimalization throughout society;

(b) evaluate the environmental and economic consequences of alternative methods, including long-term consequences; and

(c) encourage and use as fully as possible scientific and socioeconomic research in order to achieve an improved understanding on which to base long-term policy options.

Afterword

~~

WHY THE PRECAUTIONARY PRINCIPLE?
A MEDITATION ON POLYVINYL CHLORIDE (PVC)
AND THE BREASTS OF MOTHERS*

Sandra Steingraber

Those of you who know me know that when I talk on these topics I usually speak out of two identities: biologist and cancer activist. My diagnosis with bladder cancer at age 20 makes more urgent my scientific research. Conversely, my Ph.D. in ecology informs my understanding of how and why I became a cancer patient in the first place: bladder cancer is considered a quintessential environmental disease. Links between environment and public health became the topic of my third book, *Living Downstream*, but since I have been given the task of speaking about the effect of toxic materials on future generations, I'm going to speak out of another one of my identities—that of a mother.

I'm a very new mother. I gave birth in September 1998 to my daughter and first child. So, I'm going to speak very intimately and in the present tense. You know it's a very powerful thing for a person with a cancer history to have a child. It's a very long commitment for those of us unaccustomed to looking far into the future. My daughter's name is Faith.

---

* Remarks delivered at the Lowell Center for Sustainable Production's workshop, Building Materials into the Coming Millenium, Boston, November 1998.

I'm also learning what all parents must learn, which is a new kind of love. It's a love that's more than an emotion or a feeling. It's a deep physical craving like hunger or thirst. It's the realization that you would lay down your life for this eight-pound person without a second thought. You would pick up arms for them. You would empty your bank account. It's love without boundaries and were this kind of love directed at another adult, it would be considered totally inappropriate. A kind of fatal attraction. Maybe, when directed at babies, we should call this "natal attraction."

I say this to remind us all what is at stake. If we would die or kill for our children, wouldn't we do anything within our power to keep toxics out of their food supply? Especially if we knew, in fact, there were alternatives to these toxics?

Of all human food, breast milk is now the most contaminated. Because it is one rung up on the food chain higher than the foods we adults eat, the trace amounts of toxic residues carried into mothers' bodies become even more concentrated in the milk their breasts produce. To be specific, it's about 10 to 100 times more contaminated with dioxins than the next highest level of stuff on the human food chain, which are animal-derived fats in dairy, meat, eggs, and fish. This is why a breast-fed infant receives its so-called "safe" lifetime limit of dioxin in the first six months of drinking breast milk. Study after study also shows that the concentration of carcinogens in human breast milk declines steadily as nursing continues. Thus the protective effect of breast feeding on the mother appears to be a direct result of downloading a lifelong burden of carcinogens from her breasts into the tiny body of her infant.

When it comes to the production, use, and disposal of PVC, the breasts of breast-feeding mothers are the tailpipe. Representatives from the vinyl industry emphasize how common a material PVC is, and they are correct. It is found in medical products, toys, food packaging, and vinyl siding. What they don't say is that sooner or later all of these products are tossed into the trash, and here in New England, we tend to shovel our trash into incinerators. Incinerators are de facto laboratories for dioxin manufacturer, and PVC is the main ingredient in this process. The dioxin created by the burning of PVC drifts from the stacks of these incinerators, attaches to dust particles in the atmosphere, and eventually sifts down to Earth as either dry deposition or in rain drops. This deposition then coats crops and other plants, which are eaten by cows, chickens and hogs. Or, alternatively, it's rained into rivers and lakes and insinuates itself into the flesh of fish. As a breast-feeding mother, I take these molecules into my body and distill them in my breast tissue. This is done through a process through which fat globules from throughout my whole body are mobilized and carried into the breast lobes, where, under the direction of a pituitary hormone called prolactin, they are made into human milk. Then, under the direction of another pituitary hormone called oxytocin, this milk springs from the grape-like lobes and flows down long tubules into the nipple, which is a kind of sieve, and into the back of the throat of the breast-feeding infant. My daughter.

So, this, then, is the connection. This milk, my milk, contains dioxins from old vinyl siding, discarded window blinds, junked toys, and used I.V. bags. Plastic parts of buildings that were burned down accidentally are also housed in my breasts. These are indisputable facts. They are facts that we scientists are not arguing about. What we do spend a lot of time debating is what exactly are the health effects on the generation of children that my daughter belongs to. We don't know with certainty because these kids have not reached the age at which a lot of diseases possibly linked to dioxin exposure would manifest themselves. Unlike mice and rats, we have long generational times. We do know with certainty that childhood cancers are on the rise, and indeed they are rising faster than adult cancers. We don't have any official explanation for that yet.

Let me tell you something else I've learned about breast feeding. It's an ecstatic experience. The same hormone (oxytocin) that allows milk to flow from the back of the chest wall into the nipple also controls female orgasm. This so-called let-down reflex makes the breast feel very warm and full and fizzy, as if it were a shaken-up Coke bottle. That's not unpleasant. Moreover, the mouths of infants—their gums, tongues, and palates—are perfectly designed to receive this milk. A newborn's mouth and a woman's nipple are like partners in a tango. The most expensive breast pump—and I have a $500 one—can only extract about half of the volume that a newborn baby can because such machines cannot possibly imitate the intimate and exquisite tonguing, sucking, and gumming motion that infants use to extract milk from the nipple, which is not unpleasant either.

Through this ecstatic dance, the breast-fed infant receives not just calories, but antibodies. Indeed the immune system is developed through the process of breast feeding, which is why breast-fed infants have fewer bouts of infectious diseases than bottle-fed babies. In fact, the milk produced in the first few days after birth is almost all immunological in function. This early milk is not white at all but clear and sticky and is called colostrum. Then, from colostrum you move to what's called transitional milk, which is very fatty and looks like liquid butter. Presumably then, transitional milk is even more contaminated than mature milk, which comes in at about two weeks post-partum. Interestingly, breast milk is so completely digested that the feces of breast-fed babies doesn't even smell bad. It has the odor of warm yogurt and the color of French mustard. By contrast, the excretions of babies fed on formula are notoriously unpleasant.

What is the price for the many benefits of breast milk? We don't yet know. However, one recent Dutch study found that schoolchildren who were breast fed as babies had three times the level of PCBs in their blood as compared to children who had been exclusively formula fed. PCBs are probably carcinogens. Why should there be any price for breast feeding? It should be a zero-risk activity.

If there was ever a need to invoke the Precautionary Principle—the idea that we must protect human life from possible toxic danger well in advance of scientific proof about that danger—it is here, deep inside the chest walls of nursing mothers where capillaries carry fat globules into the milk-producing lobes of the mammary gland. Not only do we know little about the long-term health effects of dioxin and PCB exposure in newborns, we haven't even identified all the thousands of constituent elements in breast milk that these contaminants might act on. For example, in 1997 researchers described 130 different sugars unique to human milk. Called oligosaccharides, these sugars are not digested but function instead to protect the infant from infection by binding tightly to intestinal pathogens. Additionally, they appear to serve as a source of sialic acid, which is essential to brain development.

So, this is my conclusion. Breast feeding is a sacred act. It is a holy thing. To talk about breast feeding versus bottle feeding, to weigh the known risks of infectious diseases against the possible risks of childhood or adult cancers is an obscene argument. Those of us who are advocates for women and children and those of us who are parents of any kind need to become advocates for uncontaminated breast milk. A woman's body is the first environment. If there are toxic materials from PVC in the breasts of women, then it becomes our moral imperative to solve the problem. If alternatives to PVC exist, then it becomes morally imperative that we embrace the alternatives and make them a reality.

About the Contributors

Nicholas A. Ashford is professor of technology and policy at the Massachusetts Institute of Technology, where he teaches courses in law, technology, and the environment. He was a public member and chairman of the National Advisory Committee on Occupational Safety & Health and served as chairman of the Committee on Technology Innovation & Economics of the EPA National Advisory Council for Environmental Policy and Technology.

Katherine Barrett is a doctoral student at the University of British Columbia, Vancouver, in the Department of Botany. She is currently studying ethical and scientific issues surrounding genetically engineered crops.

Anita Bernstein is professor of law at Chicago-Kent College of law. The first legal academic to win a Fulbright research grant in European Community Affairs, Bernstein spent the 1992–1993 academic year as a Jean Monet Fellow at the European University Institute, resident in Florence, where she studied European laws pertaining to products liability and sexual harassment.

Laura Cornwell has worked on policy tools for environmentally sustainable development for over ten years, including as a policy analyst in the regulatory innovations branch at the U.S. Environmental Protection Agency and as an advisor on natural resource and land-use planning issues in Canada, Central America, and the Caribbean. Currently, Dr. Cornwell is a research associate for the Sustainable Communities Initiative at the University of Victoria, British Columbia.

Robert Costanza is director of the University of Maryland Institute for Ecological Economics and is a professor in the Center for Environmental Science, at Solomons, and in the Zoology Department at College Park. He is co-founder and president of the International Society for Ecological Economics (ISEE) and has served on numerous government committees, including the U.S. EPA National Advisory Council for Environmental Policy and Technology (NACEPT), the National Research Council Board on Sustainable Development, and the Committee on Global Change Research.

Carl F. Cranor is professor of philosophy and associate dean of Humanities, Arts, and Sciences at the University of California, Riverside. He has served on the State of California's Proposition 65 Science Advisory Panel (1989–1992), the National Academy of Sciences' Law and Science Planning Committee (1997), and several NSF peer review panels.

Peter L. deFur is an affiliate associate professor in the Center for Environmental Studies at Virginia Commonwealth University in Richmond, where he directs a project on environmental health. Dr. deFur has held positions as a senior scientist with the Environmental Defense Fund, in Washington, D.C., and faculty positions with Southeastern Louisiana University and George Mason University, in Virginia.

Gordon K. Durnil is a Republican leader who is concerned about adverse environmental health effects on our children. He served eight years as the Indiana Republican State Chairman, was on the Executive Committee of the Republican National Committee throughout the Reagan years, and was appointed by President George Bush as chairman of the U.S.–Canada International Joint Commission.

Ken Geiser serves as associate professor in the Department of Work Environment at the University of Massachusetts, Lowell, where he also serves as director of the Toxics Use Reduction Institute (TURI) and director of the Center for the Study of Public Policy, director of the Center for Environmentally Appropriate Materials, and director of the Lowell Center for Sustainable Production. Dr. Geiser serves on several national and international committees focused on promotion and implementation of clean production.

Wes Jackson is a geneticist and founder and president of the Land Institute in Salina, Kansas. His books include *Altars of Unhewn Stone, New Roots for Agriculture,* and *Becoming Native to This Place.*

Andrew Jordan is a senior research associate at the Centre for Social and Economic Research on the Global Environment (CSERGE), which is jointly located at the University of East Anglia, Norwich, and University College London. His specialization includes British and E.U. environmental policy and "new" institutional political theory.

Frederick Kirschenmann is an organic farmer and philosopher who runs a 3,100-acre grain and livestock operation in North Dakota. He was a founding member of the Northern Plains Sustainable Agriculture Society, serves on the Board of the World Sustainable Agriculture Society, and is president of the Board of the Henry A. Wallace Institute.

Sanford Lewis is an environmental attorney whose clients include environmental, community, and labor organizations, including advice on the negotiation of good neighbor agreements and other strategies for ensuring corporate accountability. He directs the Good Neighbor Project and chairs the national Network Against Corporate Secrecy. He was assisted on his chapter by Good Neighbor Project interns Brian Milder and Keri Funderburg.

R. Michael M'Gonigle holds the Eco-Research Chair in Environmental Law and Policy in the Faculty of Law at the University of Victoria, British Columbia, where he is cross-appointed to the Department of Environmental Studies. He has written widely on international law, law of the sea, and environmental issues and is a cofounder of Greenpeace International and the Sierra Club Legal Defense Fund.

Peter Montague is director of the Environmental Research Foundation in Annapolis, Maryland. He edits *Rachel's Environment & Health Weekly,* which provides grassroots environmental activists with technical information (often about human health and chemical contamination) in an understandable form.

Mary O'Brien has worked for the past seventeen years on issues of alternatives to toxics and risk assessment, public interest science, and ecosystem policy as a staff scientist with several grassroots organizations: Northwest Coalition for Alternatives to Pesticides, Environmental Law Alliance Worldwide, Environmental Research Foundation, and Hells Canyon Preservation Council. She currently chairs the Eugene (Oregon) Toxics Board implementing the first local hazardous substance materials accounting law in the nation.

Timothy O'Riordan is professor of environmental sciences at the University of East Anglia and is associate director of the Centre for Social and Economic Research on the Global Environment. He has published extensively on environmental policy and politics, including an edited book on *Interpreting the Precautionary Principle* (Earthscan, London, 1994).

David Ozonoff is the chair of the Department of Environmental Health in the Boston University School of Public Health. His research focuses on health effects to communities of various kinds of toxic exposures, especially from hazardous waste sites; new approaches to understanding the results of small case-control studies; and the effects of exposure misclassification in environmental epidemiology.

Carolyn Raffensperger is executive director of the Science and Environmental Health Network (SEHN), a national consortium of about fifty environmental groups dedicated to the use of science to protect the environment and public health. She came to SEHN from a background of working for the environmental community, including nine years as the Field Representative for the Sierra Club in Chicago, Illinois.

David Santillo, Paul Johnston, and *Ruth Stringer* are scientists at the Greenpeace Research Laboratories, based at the University of Exeter, United Kingdom. The laboratories provide analytical and informational support on a range of scientific issues to the organization worldwide. The Precautionary Principle is one of several key issues on which the group has worked and published over a number of years.

Madaleine L. Scammel coordinates the Loka Institute's effort to establish a nationwide and international Community Research Network.

Ted Schettler is science director for the Science and Environmental Health Network and is cochair of the Human Health and Environment Project of Greater Boston Physicians for Social Responsibility, addressing the health effects of environmental contamination and toxic exposures. He is a member of the Environmental Committee of Physicians for Social Responsibility and was on the U.S. EPA Endocrine Disruptor Screening and Testing Advisory Committee.

Dr. Richard E. Sclove, author of the award-winning book *Democracy and Technology* (New York: Guilford Press, 1995), is the founder and president of the Loka Institute, a nonprofit organization concerned with making science and technology responsive to democratically decided social and environmental concerns. In 1996–1997, he initiated the first pilot consensus conference for citizen-based technology assessment in the United States.

Scottish Natural Heritage is a government agency that was formed in 1992 to secure the conservation and enhancement of, and foster understanding and facilitate the enjoyment of, the natural heritage of Scotland, as well as to have regard for the desirability of securing that anything done, whether by SNH or any other person, in relation to the natural heritage of Scotland is undertaken in a sustainable manner.

Sandra Steingraber, poet, biologist, and cancer survivor, is author of *Living Downstream: A Scientists Personal Investigation of Cancer and the Environment* (New York: Vintage, 1997). She was named a Ms. magazine woman of the year in 1997 and in 1998 received the Will Solimene Award for Excellence in Medical Communications from the New England chapter of the American Medical Writers Association.

Joel A. Tickner is a doctoral candidate in the Work Environment Program at the University of Massachusetts, Lowell, where his research focuses on the development of a decision-making framework for operationalizing the Precautionary Principle. He is also a research associate in the Lowell Center for Sustainable Production. An EPA STAR Fellow, Joel has been active for several years on pollution prevention, risk assessment, and toxic chemicals policy issues, having served as an advisor to numerous nonprofit environmental groups and trade unions both in the United States and abroad.

Bo Wahlström is the international advisor for the Swedish Chemical Inspectorate, where he is presently working on persistent organic pollutants. He has held numerous other government and academic positions and is known for advancing the concept of chemical sunseting internationally.

Index